The Risks of Nuclear Energy Technology

Science Policy Reports

The series Science Policy Reports presents the endorsed results of important studies in basic and applied areas of science and technology. They include, to give just a few examples: panel reports exploring the practical and economic feasibility of a new technology; R & D studies of development opportunities for particular materials, devices or other inventions; reports by responsible bodies on technology standardization in developing branches of industry.

Sponsored typically by large organizations – government agencies, watchdogs, funding bodies, standards institutes, international consortia – the studies selected for Science Policy Reports will disseminate carefully compiled information, detailed data and in-depth analysis to a wide audience. They will bring out implications of scientific discoveries and technologies in societal, cultural, environmental, political and/or commercial contexts and will enable interested parties to take advantage of new opportunities and exploit on-going development processes to the full.

For other titles published in this series, go to
http://www.springer.com/series/8882

Günter Kessler • Anke Veser •
Franz-Hermann Schlüter • Wolfgang Raskob •
Claudia Landman • Jürgen Päsler-Sauer

The Risks of Nuclear Energy Technology

Safety Concepts of Light Water Reactors

Springer

Günter Kessler
Franz-Hermann Schlüter
Stutensee
Germany

Anke Veser
Eggenstein
Germany

Wolfgang Raskob
Claudia Landman
Jürgen Päsler-Sauer
Institut für Kern- und Energietechnik (IKET)
Karlsruher Institut für Technologie (KIT)
Eggenstein-Leopoldshafen
Germany

ISSN 2213-1965 ISSN 2213-1973 (electronic)
ISBN 978-3-662-51251-7 ISBN 978-3-642-55116-1 (eBook)
DOI 10.1007/978-3-642-55116-1
Springer Heidelberg New York Dordrecht London

Printed on acid-free paper

Springer is part of Springer Science+Business Media (www.springer.com)

Preface

The work on the present book **Safety of Light Water Reactors** was begun upon suggestion of Dr. L. Ascheron, Scientific Editor, Springer Verlag after the Fukushima reactor accident. The main task of this book is, in our opinion, to describe the scientific results of the past decades and the comparably high safety standard of the current/modern international reactor safety engineering.

This includes scientific results and technical developments that minimize the consequences of accidents to the population.

In the first part an overview of the nuclear power capacities, as well as the capacities for enrichment and reprocessing installed worldwide is provided. After a short presentation of the fundamentals of reactor physics, the radiological threshold values needed for understanding the Light Water Reactors still in operation in Germany are described as an example. In the case of Pressurized Water Reactors this includes the so-called Konvoi Series, as Boiling Water Reactors the SWR-72 of Kraftwerkunion (Siemens). As further European examples the new European Pressurized Water Reactor (EPR) and the new European Boiling Water Reactor SWR-1000 (KERENA) that were developed since 1995 by German and French reactor engineers are introduced. For the USA and Japan exemplarily the Pressurized Water Reactors AP1000 of Westinghouse and the US-APWR of Mitsubishi as well as the Boiling Water Reactors ABWR and ABWR-II (General Electric, Toshiba and Hitachi) are presented.

Broad room is then dedicated to the results of the safety research programs on core melt accidents performed at the former Kernforschungszentrum Karlsruhe (Nuclear Research Center Karlsruhe, now KIT) during the past two decades. Via the German and French reactor safety commissions these results became part of the new safety concept of the EPR and the SWR-1000.

The reactor accidents of Three-Mile-Island (USA), Chernobyl (Ukraine) and Fukushima (Japan) are described in detail. The safety concept of the German Light Water Reactors still in operation, including the plant internal emergency measures that were introduced after the Chernobyl-accident, as well as the new safety concept of EPR and SWR-1000 are then thoroughly compared and discussed with the conclusions of the severe reactor accidents that occurred so far, especially the Fukushima accident.

Since the September 11 attacks on the American World Trade Center in 2001 it is intensely discussed in public how nuclear power plants are designed against a postulated airplane impact and which hazards for the population result from such an event. For this reason this topic is covered by a special section.

Despite the high safety standards of Light Water Reactors the plant internal emergency measures are an integral part of nuclear safety culture. By means of the decision support system RODOS (Real-time On-line DecisiOn System) potential protective and countermeasures are presented that are available for the decision maker to minimize the consequences of an accident to the population. The book further describes which scientific methods and models are used to analyse the radiological situation and initiate the appropriate measures. Thereby it is not restricted to the so-called early emergency management measures but also describes model approaches that can be used for predictions of long-term prevention measures. The Fukushima accident is used exemplarily as an application of the RODOS system.

The described further development of computer-assisted decision support systems is mainly based on European research approaches. Therefore this chapter ends with a short outlook on the development of scientific and institutional aspects of the nuclear emergency managements.

May 2014

Günter Kessler
Anke Veser
Franz-Hermann Schlüter
Wolfgang Raskob
Claudia Landman
Jürgen Päsler-Sauer

Contents

Part I
The Physical and Technical Safety Concept of Light Water Reactors

Günter Kessler, Anke Veser

Chapter 1
Introduction

Abstract This chapter lists the capacity of commercial nuclear power plants built and operated in different countries of the world in 2013. About 80 % of all operating nuclear power plants are Light Water Reactors (LWRs), predominantly Pressurized Water Reactors (PWRs) and Boiling Water Reactors (BWRs). An additional 11 % are Heavy Water Reactors (HWRs) and 4 % are advanced gas cooled, graphite moderated nuclear power reactors (AGRs). Only about 3.4 % are Russian retrofitted RBMK1000 reactors still operating in Russia. One prototype Fast Breeder Reactor (FBR) was operating in Russia, one became operational in India and one experimental FBR was operated in Japan.

The resources of natural uranium were assessed in 2007 by IAEA and OECD/NEA to be 5.47 million tons (reasonably assured and inferred). An additional 7.77 million tons of speculative and about 4.2 million tons in the Chattanooga Shales in the USA are listed.

These uranium resources are then contrasted with the uranium consumption of each nuclear power reactor which is 171 tons per GW(e) and year for LWRs. If plutonium recycling in a closed fuel cycle is applied this uranium consumption is reduced by a factor of 1.55. FBRs would consume only 1.7 tons of U-238 per GW(e) and year which would extend the time period for nuclear energy application (uranium and thorium resources) to thousands of years.

For LWRs and other commercial nuclear reactors the natural uranium must be enriched. This is done predominantly by the gaseous diffusion and the gas centrifuge process. The laser enrichment process (SILEX) is still under deployment in the USA. Commercial spent fuel reprocessing facilities were built and are operated in France, Great Britain, Russia and Japan. This reprocessing capacity in the world can reprocess the spent fuel of about half of the presently operating LWR capacities. The majority of nuclear power plants built and operated in the world today is used for electricity generation. Such nuclear power reactors are built in unit sizes of about 1 and 1.6 GW(e) and operated for economical reasons mainly in the so-called base load regime.

G. Kessler et al., *The Risks of Nuclear Energy Technology*, Science Policy Reports, DOI 10.1007/978-3-642-55116-1_1,

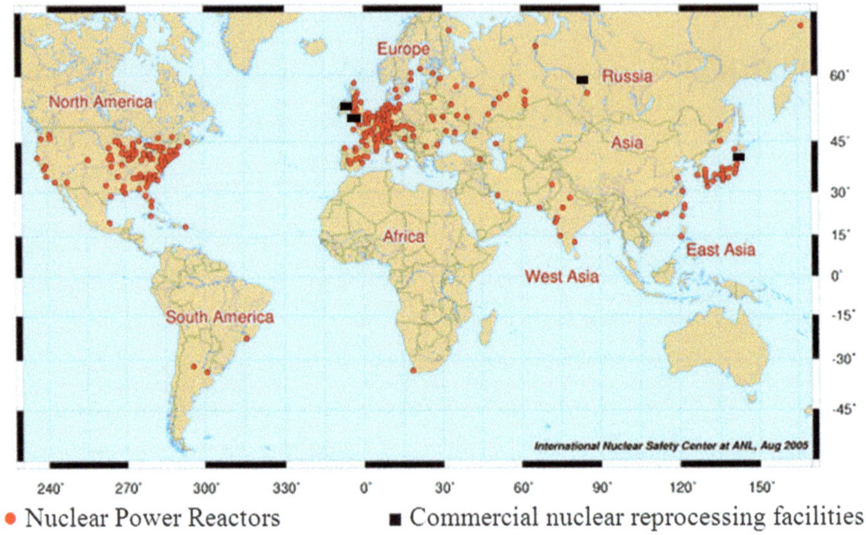

Fig. 1.1 Map of nuclear power reactors and commercial nuclear reprocessing facilities operating in the world by 2012 [2] adapted. *Red circle*—Nuclear Power Reactors, *Black square*—Commercial nuclear reprocessing facilities

In April 2013 there were 433 nuclear power reactors with a total power capacity of about 370 GW(e) operating in the world (Fig. 1.1). These nuclear power reactors produced about 16 % of the world's electrical energy consumption. Those countries having the highest number of nuclear power reactors installed and operating by 2013 are listed in Table 1.1. However, there were also many countries in Central- and South-America, in Africa, Asia, Australia and Europe which had not decided yet to rely on electricity generation by nuclear power reactors. In Western Europe, e.g. such countries are Portugal, Denmark, Norway, Italy, Austria etc. for different reasons [1].

About 80 % of all operating nuclear power reactors are Light Water Reactors (LWRs); predominantly Pressurized Water Reactors (PWRs) and Boiling Water Reactors (BWRs). An additional 11 % are Heavy Water Reactors (HWRs) and 4 % of all nuclear power reactors are advanced gas cooled, graphite moderated nuclear power reactors (AGRs). Only 11 RBMK-1000 reactors (Chernobyl-type, graphite moderated light water cooled), i.e. 3.4 % are still operating near St. Petersburg, Smolensk, and Kursk (Russia). However, this type of nuclear power reactor will be taken out of operation in the near future [1].

One prototype power reactor of the future Fast Breeder Reactor (FBR) type was also operating in Russia and one experimental Fast Breeder operated in Japan.

In addition to these 433 nuclear power reactors presently in operation there are 103 nuclear power reactors with a power capacity of 103 GW(e) under construction in the USA (10), France (1), Belarus (2), Slovakia (2), Finland (1), Russia (11), Ukraine (3), Romania (2), India (7), China (42), Taiwan (2), Pakistan (2),

Table 1.1 Nuclear power reactor capacity installed in the world by 2013

Country	Number of nuclear power reactors	Nuclear power reactor capacity GW(e)	Share of nuclear energy in total electrical energy (%)
USA	103	103.198	20
France	58	63.130	76
Japan[a]	50	44.104	23
Russia	33	23.642	18
Canada	19	13.473	14
South-Korea	23	20.697	38
Great Britain	16	9.213	12
Ukraine	15	13.107	47
China	15	11.658	3
Sweden	10	9.303	42
Germany	9	12.058	16
Spain	7	7.066	16
Rest of the World	75	40.931	-
Sum	**433**	**371.580**	-

[a]All reactors in Japan, except for two PWRs, were under safety review before restart in early 2013

South-Korea (5), Japan (3), Argentina (1), Brazil (1), United Arab. Emirates (4) and Turkey (4). Again 88 % of these nuclear power reactors are LWRs, predominantly of the PWR type. Modern LWRs have a yearly power availability factor of about 85–93 %. They are predominantly operated in the base load regime, but can also be operated in partial load. Especially in Russia they were and are also used for heat generation for district heating and desalination (BN-350) [1].

During the past decades nuclear power reactors were designed for an operation time of 35–40 years. Modern LWRs, however, are designed for an operation time of 60 years.

1.1 Uranium Resources

Natural uranium is found in uranium ores in concentrations from around fractions of a percent to several percent. Natural uranium can be bought on the world market from uranium resources and uranium mines in Australia, Canada, Kazakhstan, Niger, Namibia, Russia, Uzbekistan, USA and other countries. Natural uranium contains 0.7204 % of the isotope U-235, 99.2742 % of the isotope U-238 and 0.0054 % of the isotope U-234. For LWRs this natural uranium must be enriched in the isotope U-235 up to a concentration of 4–5 %.

The available uranium resources are assessed on a yearly basis by IAEA and OECD/NEA and listed in different categories. The uranium resources were assessed in 2007 by IAEA and OECD/NEA to be 5.47 million tons (reasonably assured and inferred). "Reasonably assured" means that these uranium resources can be mined, "inferred" means that additional investigations are required until the uranium ores can be mined. At the same time IAEA and OECD/NEA prognosticated additional speculative 7.77 million tons of uranium ores and further 4.2 million tons in the Chattanooga Shales in USA [3, 4].

1.2 Uranium Consumption

A present LWR with a power capacity of 1 GW(e) consumes about 171 tons of natural uranium (availability factor of 93 %) per year. This means, that e.g. about 370 GW(e) presently in operation (assumed all nuclear power reactors would be LWRs) will consume over 80 years about 5 million tons of natural uranium. Correspondingly a future 480 GW(e) nuclear power capacity (assumed all nuclear power reactors would be LWRs) would consume in 180 years about 15 million tons of natural uranium. Heavy Water Reactors or Light Water Reactors with plutonium recycling would have by a factor of 1.55 lesser natural uranium consumption and would extend the above time period correspondingly [5].

The fission neutrons originating from the fission process are moderated or slowed down by the collision with atoms of a moderator or coolant, e.g. light or heavy water, in the cores of LWRs and HWRs to so-called thermal energy of 0.025 eV. This corresponds to neutron velocities of 2,200 m/s. In liquid metal cooled Breeder Reactors the fission neutrons originating from the fission process are slowed down only to 0.2 MeV as the moderator or the coolant (sodium, lead or lead-bismuth) is of medium or high atomic mass. In this range of neutron energies of 0.2 MeV and higher the nuclear reactions for breeding of Pu-239 from U-238 are favorable. This newly generated Pu-239 can be utilized as artificial fissionable nuclear fuel in e.g. LWRs or FBRs.

Fast Breeder Reactors are started initially with uranium/plutonium fuel in their core and uranium fuel in their blankets. They consume per GW(e) and year only 1.7 tons of U-238 (either natural uranium or depleted uranium from uranium enrichment plants). The technical feasibility of sodium cooled Fast Breeder Reactors has been proven already in the USA, Russia, UK, France, India and Japan during the past decades. Fast Breeder Reactors require a closed fuel cycle with spent fuel reprocessing and fuel refabrication [5, 6].

This by a factor of about 100 lower fuel consumption (1.7 tons per year and GW (e) for Fast Breeder Reactors compared to 171 tons per year and GW(e) for Light Water Reactors) of fast Breeders would extend the above given time periods accordingly. As the nuclear breeding is also possible for the Th-232/U-233 nuclear fuel cycle, the available resources of natural uranium and thorium and the application of Fast Breeder Reactors would prolong the above discussed 180 years to many thousand years [5].

1.3 Uranium Enrichment

For present Light Water reactors the initial fuel of the core must be enriched from the 0.72 % of U-235 of natural uranium to an enrichment of 4–5 % U-235 (depending on the fuel burnup) in U-235/U-238 uranium dioxide (UO_2) fuel. This is achieved first by chemical conversion of the uranium ores U_3O_8 into uranium hexafluoride, UF_6, which is gaseous above a temperature of 55 °C. This gaseous UF_6 is enriched presently in essentially three different commercial enrichment processes:

- gaseous diffusion process
- gas centrifuge process
- LASER enrichment process

The LASER enrichment process SILEX is deployed in a first commercial enrichment plant in the USA. The gas centrifuge process is already used in large scale plants in Russia, Europe, Japan, China and the USA. The earliest deployed large scale gaseous diffusion enrichment plants are still in operation in the USA, France and China. They will be replaced in the future by the more economical gas centrifuge enrichment plants and probably LASER enrichment plants.

The production capacity of enrichment plants is measured in kg or tons separation work units (SWU). An LWR of 1 GW(e) power requires a reload of 25 tons enriched uranium fuel with an enrichment of 4.4 % U-235. This requires 175 tons SWU [7, 8] (Table 1.2).

After enrichment in U-235 the UF_6 will be treated chemically to become UO_2. In fuel fabrication facilities UO_2 powder will be pressed and sintered to UO_2 pellets of about 1 cm diameter and 1 cm height. These pellets are filled into about 4 m long Zircaloy (zirconium-aluminum alloy) tubes (fuel rods). The tubes are then filled with helium and welded gastight on both ends. On the upper end of these fuel rods an empty space of about 10 cm length remains where the fission gases can collect during reactor operation. Fission gas pressure can rise then up to several MPa.

A number of countries with nuclear power plants operate also UO_2 fuel cycle plants. In total there were 37 uranium mines, 22 uranium conversion plants, 13 uranium enrichment plants, 40 uranium fuel fabrication plants and 5 spent fuel reprocessing plants commercially operating in the world in 2008 [5].

1.4 Spent Fuel Reprocessing

Commercial spent fuel reprocessing was deferred in the USA in 1982 for fear of proliferation of plutonium. Later this decision was revised by the US government but no commercial reprocessing industry developed in the USA up till now. Only intermediate storage and direct spent fuel disposal were pursued. Germany and Sweden did follow this example of the USA. Other countries like France, Great

Table 1.2 Worldwide installed capacity of gaseous diffusion-, gas centrifuge- and LASER-enrichment plants [8]

	Enrichment-capacity in million kg SWU/a		
Enrichment-method	In operation	Under construction	Under licensing or planned
Gaseous diffusion			
USA	11.3		
France	10.8		
China	0.2		
Gas centrifuge			
Russia	20.0	3.0	1.0
Great Britain	3.7		2.7
Netherlands	3.5	0.13	6.8
Germany	1.8		7.5
USA			1.2
China			0.25
Japan	1.0		
France	0.3		0.5
Iran	0.02		
Brazil	0.01		
LASER (SILEX)			
USA			3.5–6.0
SUM	52.63	3.13	23.45–25.45

Table 1.3 Worldwide spent fuel reprocessing capacity in tones per year [5]

Country	Facility	Fuel type	Reprocessing capacity in tons per year
France	Cap de la Hague	LWR	1,700
Great Britain	Sellafield	LWR	1,200
	Windscale	AGR	1,500
Japan	Tokai-mura	LWR	90
	Rokkasho-mura	LWR	800
Russia	Mayak	LWR	500
India	Tarapur	CANDU	100
	Kalpakkam	FBR	100
China	Lanzhou	LWR	50

LWR Light Water Reactor
AGR Advanced Gas cooled Reactor
CANDU Canadian Deuterium Uranium pressurized Heavy Water Reactor
FBR Fast Breeder Reactor

Britain, Russia or Japan did not follow this strategy but do reprocess spent fuel elements. This led to the situation that spent fuel reprocessing facilities were almost entirely built and operated in nuclear weapon states with Japan being the sole exception.

Table 1.3 shows the reprocessing capacities for spent fuel elements available in different countries of the world in 2012. The reprocessing capacities for LWR spent fuel add up to a total of 4,340 tons per year (see also Fig. 1.1). In addition reprocessing capacities of 1,200 tons per year for metallic spent fuel of AGRs, 100 tons per year capacities for CANDU spent fuel and 100 tons per year capacities for spent FBR fuel are in operation. As about 25 tons of spent uranium fuel are unloaded per GW(e) and year from a Light Water Reactor this world wide reprocessing capacity would be sufficient for spent fuel of e.g. 174 GW(e) LWR. This is about half of the presently available LWR capacities in the world.

References

1. American Nuclear Society (2013) World list of nuclear power plants. Nuclear News March 2013. American Nuclear Society, La Grange Park, IL
2. International Nuclear Safety Center at ANL-Aug 2005 (2000) http://www.ne.anl.gov/research/ierc/intnlcoop.html
3. OECD (2008) Nuclear Energy Outlook 2008. OECD-NEA No. 6348. OECD, Paris
4. OECD (2008) Uranium 2007 – resources, production and demand. OECD-NEA-IAEA 6345. OECD, Paris
5. Kessler G (2012) Sustainable and safe nuclear fission energy. Springer, Heidelberg
6. Carré F et al (2009) Overview on the French nuclear fuel cycle strategy and transition scenario studies. In: Proceedings of Global 2009, Paris, Paper No. 9439
7. Villani S (ed) (1979) Uranium enrichment: Topics in applied physics, Vol 35. Springer, Berlin
8. Laughter M (2007) Profile of world uranium enrichment programs – 2007 ORNL/TM-2007/193. Oak Ridge National Laboratory, Oak Ridge, TN

Chapter 2
Some Facts About Neutron and Reactor Physics

Abstract Chapter 2 describes some facts about neutron and reactor physics needed for the understanding of Chaps. 3–10. It starts with the radioactive decay and the definitions of the decay constant and the half-life. It continues with the explanation of the fission process for fissile nuclear isotopes, e.g. U-233, U-235, or Pu-239 and the fission energy release by creation of fission fragments (products), prompt fission neutrons and delayed neutrons and radiation (β-particles, γ-rays and antineutrinos). This is followed by the definition of reaction rates of neutrons with other atomic nuclei, the presentation of measured microscopic cross sections for absorption, capture and fission as well as the definition of the macroscopic cross section and the neutron flux.

In LWR cores the fuel is arranged heterogeneously in lattice cells together with a moderator (water) in order to slow down the fission neutrons with high kinetic energy to kinetic energies in the range of 0.025 eV (thermal energy). This is most effective if the enriched uranium fuel is put in cylindrical rods which are arranged in e.g. a square grid. The optimization of the geometrical distance between the fuel rods leads to important safety characteristics of LWR cores: the negative fuel Doppler coefficient and the negative coolant (moderator) coefficient.

The definition of the criticality factor or effective multiplication factor, k_{eff}, allows a characterization whether the reactor core is operated in steady state condition or whether it is subcritical or even supercritical. The criticality or effective multiplication factor, k_{eff}, can be changed by moving or by insertion or withdrawing of absorber material (boron, cadmium, gadolinium, indium, silver, hafnium, erbium) in the core. This allows control of the reactor. The reactor core is controlled always in a k_{eff} range where the delayed neutrons are dominating. The delayed neutrons are therefore of highest importance for the control of the reactor.

During reactor operation over months and years the initially loaded U-235 in the low enriched uranium fuel will be consumed, neutron absorbing fission products will build up or other heavy nuclei with masses above U-235 and Pu-239 will be created. This decreases the criticality of the effective multiplication factor k_{eff}. This burnup effect on the criticality factor k_{eff} is accounted for by the design of the reactor core. The enrichment of the initially loaded fuel is increased such that k_{eff}

G. Kessler et al., *The Risks of Nuclear Energy Technology*, Science Policy Reports, DOI 10.1007/978-3-642-55116-1_2,

becomes slightly >1. This is balanced by absorber materials (moveable absorber rods, burnable neutron poisons, e.g. gadolinium or boric acid) which keep the reactor core always at $k_{eff} \geq 1$.

After shutdown of the reactor the gradually decaying fission products and the radioactive decay of higher actinides creates afterheat in the reactor core. This afterheat (decay heat) must be transferred by the coolant water to outside coolant towers or to river or sea water.

Prior to the description of Light Water Reactor designs some basic characteristics of reactor physics and reactor safety will be presented. For a deeper understanding of these characteristics the literature given in the reference is recommended [1–8].

2.1 Radioactive Decay, Decay Constant and Half-Life

Radioactive decay changes the number of unstable (radioactive) isotopes, N(t), existing per cm^3 as a function of time, t. This change can be described by the exponential law of

$$N(t) = N_0 \bullet \exp(-\lambda t)$$

where λ is the decay constant and N_0 the number of atomic nuclei per cm^3 at the time $t = 0$. Instead of the decay constant, λ, one can also use the half-life, $T_{1/2} = (\ln 2)/\lambda$, which is the time by which half of the nuclei existing at $t = 0$ have decayed. The decay rate, $\lambda \cdot N(t)$, is called the activity of a specimen of radioactive material. This activity is measured in units of Curie or Becquerel [1, 2].

One Becquerel, denoted Bq, is defined as one disintegration per second. One Curie, denoted Ci, is defined as 3.7×10^{10} disintegrations per second, which is approximately the activity of 1 g of radium. Low activities are also measured in $mCi = 10^{-3}$ Ci or $\mu Ci = 10^{-6}$ Ci [1, 2].

2.2 Fission Process

If a neutron of a certain velocity (kinetic energy) is absorbed by a fissile heavy nucleus, e.g. U-233, U-235 or Pu-239, the resulting compound nucleus can become unstable and split (fission) into two or even three fragments (Fig. 2.1). The fission fragments are created essentially according to a double humped yield distribution function with mass numbers between about 70 and 165. The mass yield distribution functions are similar for heavy nuclei fissioned by neutrons with kinetic energies of 0.0253 eV (thermal spectrum reactors) up to about 0.2 MeV for Fast Breeder

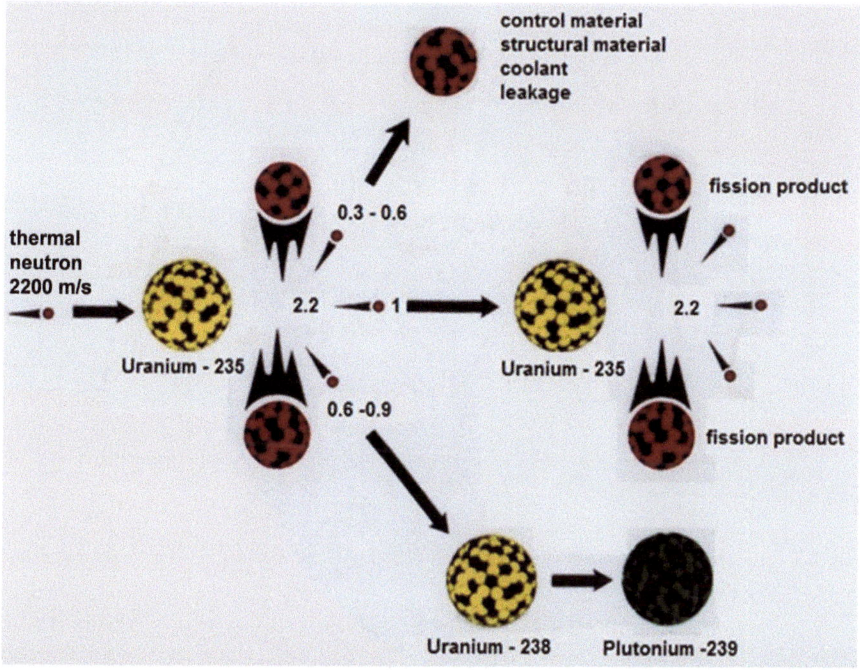

Fig. 2.1 Fission of U-235 nucleus by a thermal neutron

Reactors[1] (Fig. 2.2). They depend slightly on the kinetic energy of the incident neutrons causing fission and on the type of heavy nuclei (U-233, U-235, Pu-239).

In addition to the fission products (fragments), 2–3 prompt neutrons are emitted during the fission process. These prompt fission neutrons appear within some 10^{-14} s. They are created with different kinetic energies following a certain distribution curve around an average neutron energy of about 2 MeV. In some heavy nuclei with even mass numbers, e.g. Th-232 and U-238, nuclear fission can only be initiated by incident neutrons with a certain, relatively high, threshold kinetic energy (Table 2.1), whereas the uneven heavy nuclei, e.g. U-233, U-235, Pu-239 etc. can be fissioned by neutrons with all kinetic energies >0 eV. However, the even-uneven rule is not a rigorous one, e.g. Am-242m can also be fissioned by thermal neutrons.

The fission products can either be solid, volatile or gaseous. Many of the fission products decay further emitting so-called delayed neutrons, β-particles, γ-rays and antineutrinos. The delayed neutrons resulting from the decay of particular fission products—called precursors—represent less than 1 % of all released neutrons.

[1] 1 eV = 1.602×10^{-19} J is the kinetic energy acquired by an electron passing through a potential gradient of 1 V. 1 keV is equal to 10^3 eV and 1 MeV is equal to 10^6 eV. The energy of 0.0253 eV corresponds to a neutron velocity of 2,200 m/s.

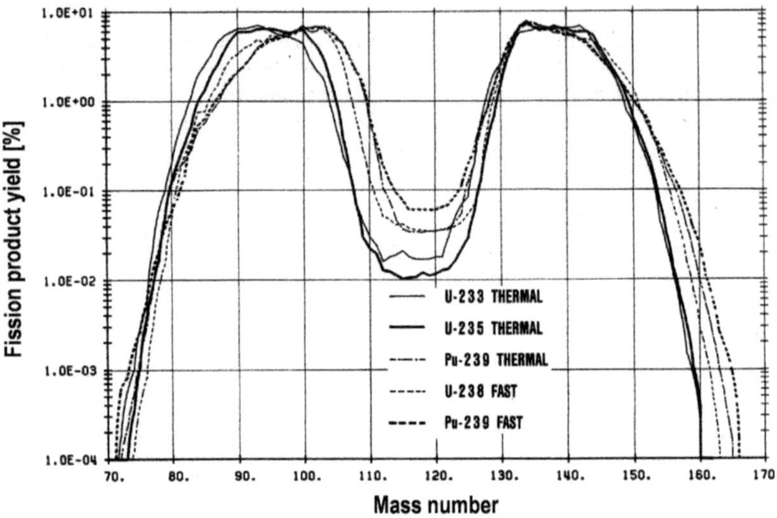

Fig. 2.2 Fission product yield (%) for fission reaction of different isotopes by thermal and fast (E > 0.2 MeV) neutrons [9]

Table 2.1 Threshold kinetic energy for incident neutrons causing substantial fission in different heavy nuclei [10]

Heavy nucleus	Th-232	U-233	U-234	U-235	U-238	Pu-239
Incident neutron kinetic energy [MeV]	>1.3	>0	>0.4	>0	>1.1	>0

The fraction β of delayed neutrons originating from fissioning by thermal neutrons (0.0253 eV) of U-235 is $\beta = 0.67$ %, and $\beta = 0.22$ % from fissioning of Pu-239. They appear following decay constants of 0.01–3 s^{-1} for U-235 and 0.01–2.6 s^{-1} for Pu-239. These delayed neutrons are of absolute necessity for the safe control and operation of nuclear fission reactors [6, 7, 11].

The total energy release per fission, Q_{tot}, appears as kinetic energy of the fission products, E_f, of the prompt fission neutrons, E_n, as β^--radiation, E_β, as γ-radiation, E_γ, or as neutrino radiation, E_ν, (Table 2.2). The neutrino radiation does not produce heat in the reactor core due to the small interaction probability of neutrinos with matter. Table 2.2 also shows the total energy, Q_{tot}, and the thermal energy, Q_{th}, released during fission of a nucleus. Some of β^--radiation and γ-radiation of the fission products is not released instantaneously, but delayed according to the decay of the different fission products.

On the average, about 194 MeV or 3.11×10^{-11} J are released per fission of one U-235 atom. Most of the fission energy is released instantaneously.

Since 1 g of U-235 meal contains 2.56×10^{21} atoms, the complete fission of 1 g of U-235 results in:

Table 2.2 Different components of energy release per fission of some heavy nuclei in MeV by incident neutrons of different kinetic energy (in the eV or MeV range) [10]

Heavy nucleus	Incident neutron energy	E_f	E_n	E_β	E_γ	E_ν	Q_{tot}	Q_{th}
U-235	0.025 eV	169.75	4.79	6.41	13.19	8.62	202.76	194.14
	0.5 MeV	169.85	4.8	6.38	13.17	8.58	202.28	193.7
U-238	3.10 MeV	170.29	5.51	8.21	14.29	11.04	206.24	195.2
Pu-239	0.025 eV	176.07	5.9	5.27	12.91	7.09	207.24	200.15
	0.5 MeV	176.09	5.9	5.24	12.88	7.05	206.66	199.61
Pu-240	2.39 MeV	175.98	6.18	5.74	12.09	7.72	206.68	198.96

$$7.96 \times 10^{10}\text{J or } 2.21 \times 10^{4}\text{kWh} \quad \text{or} \quad 0.92\ \text{MWd}_{th}\ \text{thermal energy}$$

For other fissile materials like U-233 or Pu-239 the energy release per fission is similar. Also fission by neutrons with thermal energies (0.025 eV) or by energies of 0.5 MeV leads to almost equal energy releases.

Therefore, it is usually assumed that the fission of the mass of 1 g of fissile material, e.g. U-235 or Pu-239 produces roughly 1 MWd_{th} and the measure of burnup in MWd_{th} per tonne of fuel also corresponds roughly to the number of grams of, e.g. U-235 fissioned in 1 ton of spent fuel [12].

2.3 Neutron Reactions

Neutrons produced in nuclear fission have a certain velocity or kinetic energy and direction of flight. In a fission reactor core, e.g. with U-235/U-238 fuel they may be scattered elastically or inelastically or absorbed by different atomic nuclei. In some cases the absorption of neutrons may induce nuclear fissions in heavy nuclei (U-235 etc.) so that successive generations of fission neutrons are produced and a fission chain reaction is established.

2.3.1 Reaction Rates

If n $(\vec{r}, v, \vec{\Omega})$ is the number of neutrons at point $\vec{r}$, with velocity v and the direction of flight $\vec{\Omega}$, then these neutrons can react within a volume element dV with $N \cdot dV$ atomic nuclei (N being the number of atomic nuclei per cm^3 of reactor volume). The number of reactions per second e.g. scattering or absorption, is then proportional to

$$\mathrm{v} \cdot \mathrm{n}\left(\vec{r}, \mathrm{v}, \vec{\Omega}\right) \quad \text{and to } \mathrm{N} \cdot \mathrm{dV}$$

The proportionality factor σ(v) is a measure for the probability of the nuclear reactions and is called microscopic cross section of the nucleus for a specific type of reaction. The microscopic cross section σ(v) is measured in $10^{-24}\ \mathrm{cm}^2 \triangleq 1$ barn. It is a function of the velocity, v, or kinetic energy, E, of the neutron and of the type of reaction and differs for every type of atomic nucleus. As for an absorption reaction the neutron can either remain captured or lead to fission of a heavy nucleus the relation

$$\sigma_a(\mathrm{v}) = \sigma_c(\mathrm{v}) + \sigma_f(\mathrm{v})$$

Is valid with
$\sigma_a(\mathrm{v})$ microscopic absorption cross section
$\sigma_c(\mathrm{v})$ microscopic capture cross section
$\sigma_f(\mathrm{v})$ microscopic fission cross section
The reaction rate can be written

$$\mathrm{R}\left(\vec{r}, \mathrm{v}\right) = \sigma(\mathrm{v}) \cdot \mathrm{N}\left(\vec{r}\right) \cdot \mathrm{v} \cdot \mathrm{n}\left(\vec{r}, \mathrm{v}\right) = \Sigma\left(\vec{r}, \mathrm{v}\right) \cdot \phi\left(\vec{r}, \mathrm{v}\right)$$

The quantity $\Sigma\left(\vec{r}, \mathrm{v}\right) = \mathrm{N} \cdot \sigma(\mathrm{v})$ is called macroscopic cross section.

The quantity $\phi\left(\vec{r}, \mathrm{v}\right) = \mathrm{v} \cdot \mathrm{n}\left(\vec{r}, \mathrm{v}\right)$ is called the neutron flux.

Figure 2.3 shows the microscopic fission cross section as a function of the neutron kinetic energy for the heavy nuclei U-235, U-238 and Pu-239. The fission cross sections for U-235 and Pu-239 increase with decreasing kinetic energies. In the energy region of about $0.1\text{–}10^3$ eV this behavior is superposed by resonance cross sections [12–14].

The capture cross section for U-238 is shown in Fig. 2.4.

Neutron capture in U-238 leads to U-239 and after β^--decay to Np-239 which again decays to Pu-239.

$$^{238}_{92}\mathrm{U} \xrightarrow{\mathrm{n},\gamma} {}^{239}_{92}\mathrm{U} \xrightarrow[23.5\mathrm{min}]{\beta^-} {}^{239}_{93}\mathrm{Np} \xrightarrow[2.35\ \mathrm{d}]{\beta^-} {}^{239}_{94}\mathrm{Pu}$$

Such microscopic cross sections (measured in barn) are steadily compiled, evaluated, supplemented and revised in nuclear data libraries [15–17].

The capture cross section of U-238 (Fig. 2.4) shows distinct narrow resonance peaks above about 5 eV. At medium neutron energies (keV range) the resonance peaks become smaller and above about 10 keV—in the so-called unresolved resonance energy range—they cannot be resolved any more by experiments because of resonance overlapping. These resolved and unresolved resonance peaks broaden if the temperature of the U-238 fuel increases. This phenomenon

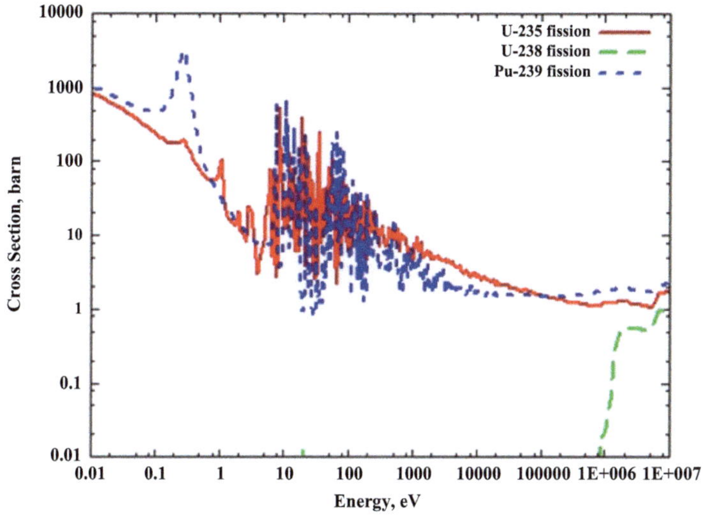

Fig. 2.3 Microscopic fission cross sections (measured in barn) for U-235, U-238 and Pu-239 [13]

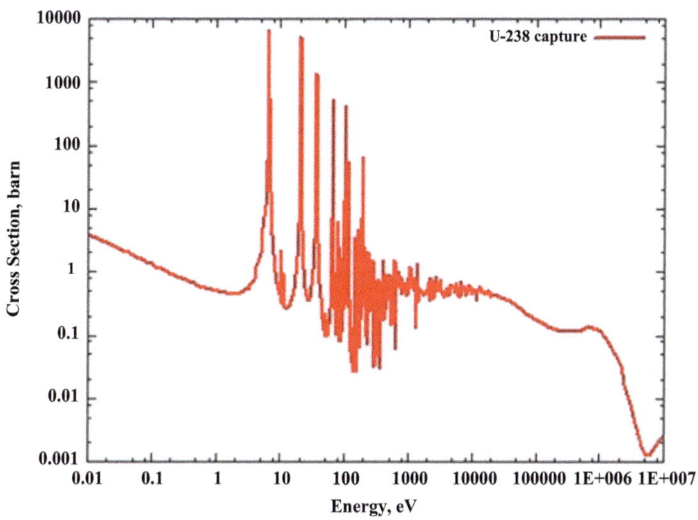

Fig. 2.4 Microscopic capture cross section for U-238 [13]

is very important for the fuel-Doppler-temperature coefficient, which determines—among other temperature coefficients—the safety characteristics of nuclear power reactors [12, 18].

The microscopic fission cross sections of U-235 or Pu-239 (Fig. 2.3) become higher—with the exception of the energy range where the resonances occur—at low neutron energies (about <0.1 eV).

Therefore, the U-235/U-238 fuel in Light Water Reactor cores is mixed with materials of low mass number (moderator) in order to slow down the fission neutrons having high kinetic energy by a number of elastic and inelastic collisions to kinetic energies in the range <1 eV (Fig. 2.3). This is most effective if the enriched uranium fuel is put in cylindrical rods which are arranged in e.g. a square grid. This lattice of fuel rods arranged with certain distances must be cooled by a flowing coolant which can be identical with the moderator as in case of water in Light Water Reactors. The fission neutrons after a few collisions with the fuel atoms then fly with high velocity into the surrounding water/moderator. They are slowed down by collisions within a short distance to so-called thermal energies of 0.025 eV (Fig. 2.5). The neutrons are then in thermal equilibrium with the relatively low kinetic energies of the water molecules. Another advantage of the deceleration of the neutrons in the surrounding water is given by the fact that the probability is lower that the neutrons can be captured in the resonance region of U-238 inside the fuel rods (Fig. 2.4) [1–5].

After the neutrons are thermalized within the moderator region they migrate back by diffusion processes into the fuel rods. As they have lower kinetic energies now also the microscopic fission cross sections are much higher (in the 0.025 eV energy range) than those for fission neutrons. The consequences are more fission reactions. Also the ratio between fission and capture reactions becomes more favorable in the fuel.

An optimum volume ratio between moderator and fuel for the grid of fuel rods is found around 2 for light water (H_2O). This optimal volume ratio can be achieved by adaption of the distance between the fuel rods. For heavy water (D_2O) as a moderator this optimal volume ratio is about 20 and for graphite as a moderator it is found to be around 54 [1–5].

Light water (H_2O) has a higher microscopic capture cross section than heavy water or pure graphite. Consequently it is possible to build and operate nuclear power reactors with natural uranium (0.72 % U-235 enrichment) if heavy water or graphite are used as moderator. In fact the first reactor used natural uranium as fuel and graphite as moderator. With light water as moderator in Light Water Reactors the uranium fuel must be enriched to 3–5 % in U-235 (depending on the burnup of the uranium fuel). Structural materials which must be used for the design of the reactor core should also have low microscopic capture cross sections. Light Water Reactors, therefore, use an alloy of zirconium and aluminum (Zircaloy) for the cladding of the fuel rods and grid spacers of the fuel elements [1–4].

Uranium dioxide (UO_2) with its high melting point (2,865 °C) and its good irradiation properties in the neutron field of the reactor core is used as fuel in LWR cores.

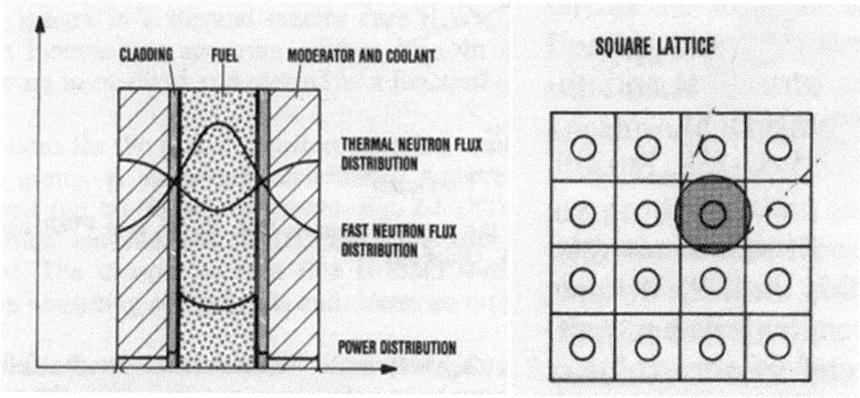

Fig. 2.5 Square lattice cell with fuel rod, cladding and moderator of a LWR-fuel element

2.4 Criticality or Effective Multiplication Factor k_{eff}

The ratio between the number of newly generated neutrons by fission and the number of neutrons absorbed in the reactor core or escaping from the reactor is called the criticality factor or effective multiplication factor, k_{eff}.

When $k_{eff} = 1$, the reactor core is critical and can be operated in steady state. At $k_{eff} < 1$ the reactor core is subcritical, e.g. with control or absorber rods fully inserted in the core.

Boron, Cadmium or Gadolinium etc. can be used as absorber materials, either as metallic alloys in control rods or as burnable poisons in ceramic form in fuel rods and special poison rods or as a fluid, e.g. boric acid in the coolant of an LWR [1–4].

For a $k_{eff} > 1$ the reactor core is supercritical. More neutrons are produced than are absorbed in the reactor core or do escape from the core. The neutron chain reaction is ascending (reaction rates and the number of neutrons and, thus, reactor power increase as a function of time).

The criticality or effective multiplication factor k_{eff} of a reactor core is determined by the proper choice of its geometrical dimensions (diameter and spacing of the fuel rods, diameter and height of the reactor core), by the choice of the moderator and coolant as well as by the choice of the fuel and structural materials. The choice of the U-235 enrichment of the fuel is of decisive importance for LWRs.

2.5 Neutron Density and Power Distribution

Figure 2.6 displays the spatial distribution of the neutron density in the range of thermal energies for a Pressurized Water Reactor (PWR). The distribution of fission reaction rates or of the power generated by fissions is essentially proportional to the

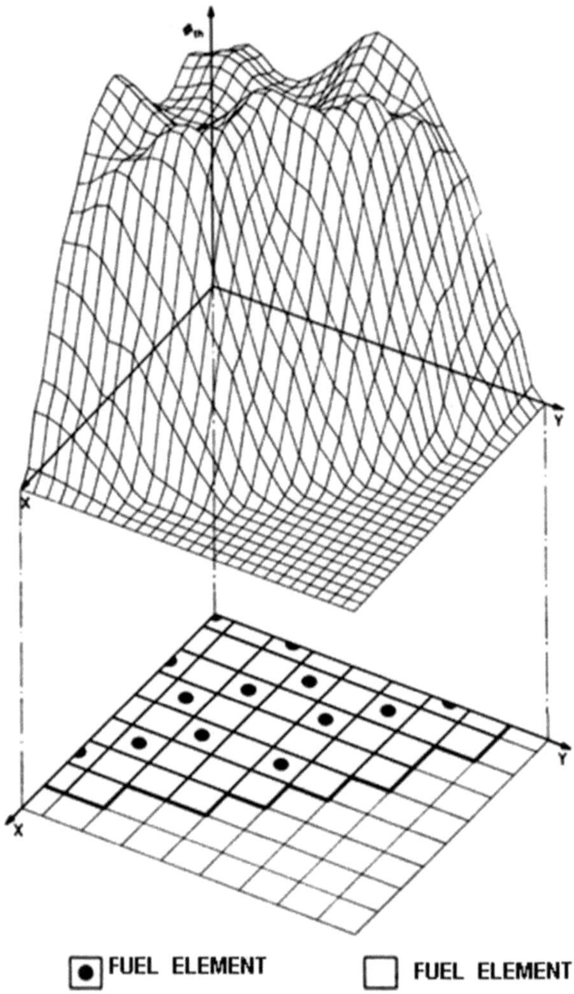

Fig. 2.6 Spatial distribution of the thermal neutron density in a Pressurized Water Reactor core with partially inserted control rods [11]

neutron density distribution. The absorber- or control-rods are partially inserted in axial direction in the PWR core. The control rods absorb neutrons and are responsible for the spatial distortions of the thermal neutron flux. They influence the criticality level and the spatial power distribution.

The spatial distribution of the neutrons with a certain velocity or kinetic energy and flight direction can be described by the Boltzmann neutron transport equation or by Monte Carlo methods [1–4]. For both cases, numerical methods in one-, two- or three-dimensional geometry were developed. Computer program packages (deterministic codes for the solution of the Boltzmann transport equation and Monte Carlo codes using stochastic solution methods) are available for various applications [1, 3–5, 18, 38].

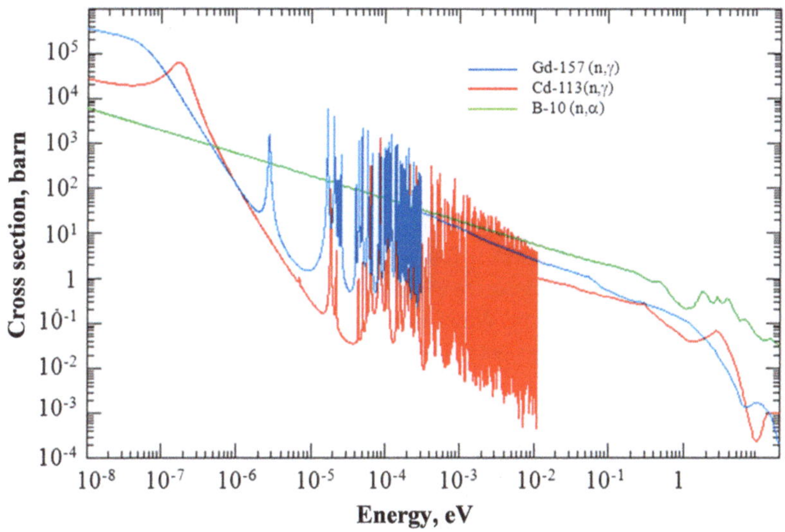

Fig. 2.7 Microscopic capture cross sections of Cadmium and Gadolinium isotopes and B-10 (n,α) cross sections [13]

For many practical applications it is sufficient to solve the neutron diffusion equation which is an approximation to the Boltzmann neutron transport equation.

The microscopic cross sections shown in Figs. 2.3, 2.4 and 2.7 are collected in a special format in cross section libraries, e.g. JEFF [15], ENDF/B [16], JENDL [17]. Their continuous energy range can be approximated and divided into a number of energy groups with specifically defined microscopic group cross sections applying codes, e.g. NJOY [19] or MC^2-3 [20, 21]. The heterogeneous cell geometry of the reactor core (Fig. 2.5) can be accounted for by codes, e.g. WIMS [22] or MC^2-3 [20, 21]. These computational methods are summarized in [23, 24].

Whole-core calculations can be done in diffusion theory by codes, e.g. DIF3D [25] or SIMULATE-4 [26]. Whole-core (Fig. 2.6) codes applying Boltzmann neutron transport theory were developed for two- and three-dimensional geometries. Examples for such computer codes are, e.g. DANTSYS [27] or PARTISN [28]. Monte Carlo Codes are available for both lattice (Fig. 2.5) and whole-core (Fig. 2.6) geometries. Such codes include, e.g. MCNP5 [29] or VIM [30].

The number of neutrons in the reactor core can be controlled by moving or adding, e.g. absorber materials (neutron poisons). This is done in a k_{eff}-range, where the delayed neutrons are dominating the transient behavior of the neutron flux. The delayed neutrons come into being in a time range of seconds. Therefore, the number of neutrons or the power in reactor cores can also be controlled safely by moving absorber materials in the time range of seconds [1, 2, 6, 11].

2.6 Neutron Poisons for the Control of the Reactor Power

Neutron poisons are used to control the number of neutrons and the power in the reactor core. Such neutron poisons have high microscopic absorption cross sections for neutrons. Neutron poisons are, e.g. Boron, Cadmium, Gadolinium etc. Figure 2.7 displays microscopic absorption cross sections for Boron, Cadmium and Gadolinium isotopes.

The neutron poisons are inserted into the reactor core as, e.g. axially moveable cylindrical control rods (Pressurized Water Reactors) or axially moveable cruciform control elements (Boiling Water Reactors). Another possibility is to add boric acid (H_2BO_3) to the cooling water, or to extract it from its solution in the cooling water.

Withdrawing the, e.g. Cadmium or the Boron carbide control elements from the reactor core changes the effective multiplication factor from $k_{eff} < 1$ to $k_{eff} > 1$. Inserting the control elements changes k_{eff} from 1.0 to $k_{eff} < 1$. This action changes the number of fission reactions and the power of the reactor, correspondingly. Similarly, the reactor can be controlled by the variation of the concentration of the boric acid in the cooling water. However, all variations of the effective multiplication factor k_{eff} are limited by design such that they remain below about half of the fraction of the delayed neutrons (see Sect. 2.10).

2.7 Fuel Burnup and Transmutation During Reactor Operation

During reactor operation over months and years the initially loaded U-235 in the low enriched uranium fuel will be consumed due to neutron fission and capture processes. As a consequence also the initial criticality or effective multiplication factor k_{eff} decreases. Neutron capture in U-235 leads to U-236. Subsequent neutron capture in U-236 leads to Np-237. Neutron capture in the fertile isotope U-238 leads to U-239 and after decay to Np-239 and further decay to the buildup of the new fissile isotope Pu-239. Subsequent neutron captures in Pu-239 lead to the higher Pu-Isotopes Pu-240, Pu-241, Pu-242. After β^--decay of Pu-241 americium is created. Neutron capture in americium leads to curium. This increases somewhat the criticality or effective multiplication factor k_{eff}. Fission products originating from the fission of fissile isotopes decrease the criticality or effective multiplication factor k_{eff} due to their absorption cross sections. The combination of these three effects results in a time dependent change—usually a decrease—of the criticality factor, k_{eff}, during reactor operation.

This burnup effect on k_{eff} is accounted for by design of the reactor core. The enrichment of the initially loaded fuel is increased such that k_{eff} becomes slightly >1. As the k_{eff} shall be equal 1 during the whole reactor operation cycle, this is balanced by absorber materials in the core (moveable absorber rods or special rods with burnable absorber material or burnable absorber materials dissolved in the coolant or mixed with the fuel). The accumulating fission products and the decreasing k_{eff} are counteracted, e.g. by moving absorber rods slowly out of the core during reactor operation. At the end of the operation cycle the absorber rods are almost withdrawn out of the core and spent fuel must be unloaded and replaced by new fuel elements.

2.7.1 Prediction of the Burnup Effects

The calculation of the change in concentration of all isotopes, actinides and fission products requires besides the power history or the equivalent time-dependent neutron flux distribution, the knowledge of the microscopic cross sections and decay constants of all isotopes as well as the yields of fission products during the reactor operation [18, 23, 31–34, 38]. The solution of a coupled system of ordinary differential equations with these data as coefficients and given initial concentrations at $t = 0$ results in the concentration of each isotope at the time t during reactor operation. Figure 2.8 shows the masses of the most important isotopes for 1 ton of initial reactor fuel after a reactor operation time of 6 years and a fuel burnup of 60,000 MWd/t.

2.8 Reactor Control and Temperature Effects

A start up of the reactor by withdrawing control elements leads to a rise in reactor power and temperatures. Temperature changes provoke changes in material densities and microscopic cross sections by the Doppler broadening of resonances (see Sect. 2.2). Also the neutron energy spectrum can be shifted by moderation of the neutrons [11, 23]. All these effects together result in changes of the criticality or effective multiplication factor, k_{eff}.

The design parameters of the reactor core are selected such that increases of power and temperatures always lead to a smaller criticality or effective multiplication factor k_{eff}. Therefore, a power increase is only possible by withdrawing absorber control elements or decreasing the concentration of boric acid in the cooling water. In Boiling Water Reactors the power can also be increased by increasing the coolant flow, which leads to smaller concentrations of steam bubbles in the reactor core.

The most important safety design requirements for LWRs are, therefore

- a negative fuel-Doppler-temperature coefficient
- a negative coolant/moderator-temperature coefficient

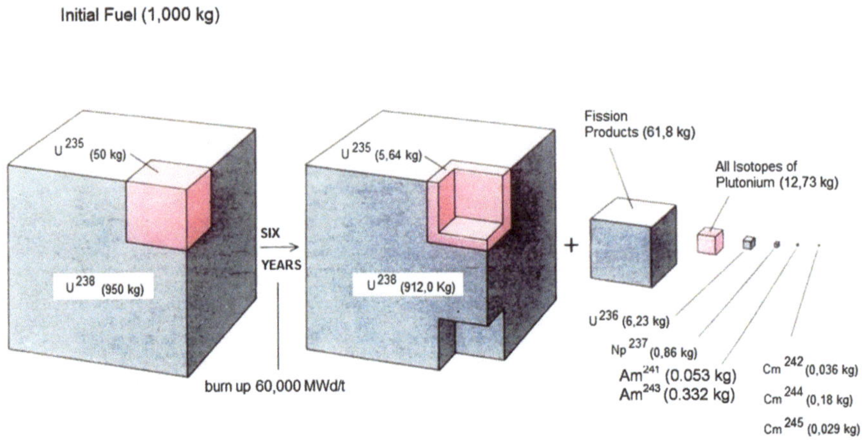

Fig. 2.8 Masses for 1 ton of initial reactor fuel after a burnup of 60,000 MW_{th}/t

2.9 Afterheat of the Fuel Elements After Reactor Shut Down

For reactor shut down the absorber/control elements are inserted into the reactor core and the coolant flow is drastically reduced. After reactor shut down the fuel elements—having reached their maximum burnup—can be unloaded. However, although the power is shut down, the gradually decaying fission products generate heat in the reactor core, even if the neutron fission chain reaction has been interrupted (after shut down of the reactor core). This afterheat, or decay heat, is composed of the contributions by the decay chains of the fission products and of contributions of radioactive decay by U-239, Np-239, and the higher actinides, which are unstable. It is a function of the power history of the reactor core before shutdown and is thus strongly influenced by the burnup of the fuel. Figure 2.9 shows the relationship between the power of the fuel elements in the reactor core of a PWR after shut down, P(t), and the power during operation, P_0. The afterheat, P(t), drops very sharply as a function of time. Shortly after shut down it is about 6 %, after 6 h it is still about 1 %, after 1 week 0.3 %, after 3 months about 0.1 %, and after 1 year it is 0.04 % of the nominal reactor power, P_0, during operation.

After reactor shut down this afterheat must be transferred to the cooling towers or a river by the normal cooling system. After unloading from the reactor core the spent fuel elements are cooled in intermediate spent fuel storage pools [11].

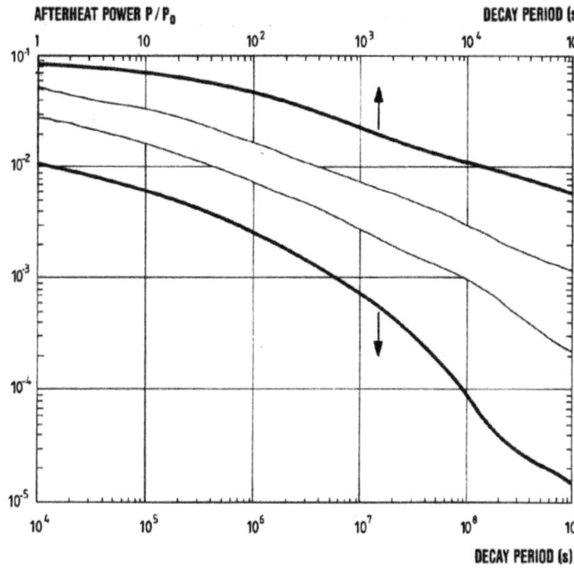

Fig. 2.9 Post-shut down afterheat of a PWR core as a function of time (initial enrichment 3.2 % U-235, burnup 32,000 MWd(th)/t) [11]

2.10 Non-steady State Power Conditions and Negative Temperature Feedback Effects

Power reactors are generally operated at constant criticality or steady power. Exemptions are: startup conditions and power rise, transition from partial to full load power, reactor shut down and accidental conditions. Accidental conditions must be analyzed and presented to licensing authorities prior to begin of reactor operation. Such accidental conditions are, e.g. inadvertent or faulty withdrawal of absorber/control elements (increase of the criticality or effective multiplication factor, k_{eff}, above 1.0) or coolant loss as a consequence of pipe rupture or a faulty opening of valves followed by primary coolant pressure loss. An increase of the criticality or effective multiplication factor, k_{eff}, provokes an increase of fission reactions and a rise of power as a function of time. The increase of power results in an increase of fuel temperature and by thermal conduction—with a certain time delay—also to an increase of the coolant temperature. As already mentioned this results in important negative feedback effects which counteract the power increase. These negative feedback effects and the delayed neutrons allow the safe control of nuclear reactors. These important negative feedback effects—already mentioned above—will be explained now in more detail.

2.10.1 The Fuel-Doppler-Temperature Coefficient

The fuel-Doppler-temperature coefficient is due to the fact that the microscopic resonance cross sections depend on the temperature of the fuel and the relative velocities, respectively, of neutrons and atomic nuclei [8, 35].

The resonance cross sections for U-238, U-235, Pu-239, etc. show very pronounced peaks at certain neutron kinetic energies (Figs. 2.3 and 2.4). An increase in fuel temperature broadens this shape of the resonance curve and lowers its peak which, in turn, results in a change in the fine structure of the neutron energy spectrum. The neutron reaction rates for capture and fission are changed as a consequence. Above all, the resonance absorption for U-238 increases as a result of rising fuel temperatures, while the effect of a temperature change in the resonance cross sections of the fissile materials, U-235 and Pu-239, is so small that it can generally be neglected if the fuel enrichment is not extremely high. For these reasons, temperature increases in the fuel result in a negative temperature feedback effect (Doppler effect) brought about by the increase in neutron absorption in U-238. The Doppler effect is somewhat less pronounced at very high fuel temperatures because adjacent resonances will overlap more and more. The resonance structure then is no longer as pronounced as at low temperatures, which leads to a reduction of the negative Doppler effect.

As a consequence of the negative fuel-Doppler-temperature coefficient the criticality or effective multiplication factor, k_{eff}, decreases at higher fuel temperatures. Typical values for the fuel-Doppler-temperature coefficient are

-for PWRs -2.5×10^{-5} change in k_{eff} per degree of fuel temperature increase
-for BWRs -2×10^{-5} change in k_{eff} per degree of fuel temperature increase

As the fuel-Doppler-temperature coefficient is coupled to the fuel temperature it is acting practically instantaneously.

2.10.2 The Moderator/Coolant-Temperature Coefficient of LWRs

The main contribution to the coefficients of moderator or coolant temperatures stem from changes in the densities of the moderator or coolant and from resultant shifts in the neutron energy spectrum. Temperature rises decrease the density of the coolant and accordingly reduce the moderation of neutrons. The neutron spectrum is shifted towards higher energies. As a result of the lower moderator density and the correspondingly higher transparency to neutrons of the core it is also possible that appreciably more neutrons will escape from the reactor core and neutron losses due to leakage rate will increase [8, 34, 36].

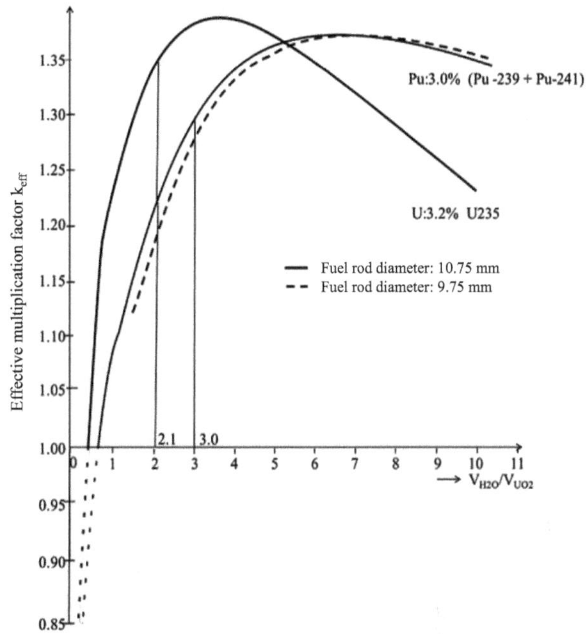

Fig. 2.10 Criticality or effective multiplication factor, k_{eff}, as a function of the volume ratio V_{H2O}/V_{UO2} in a lattice cell for two examples (3.2 % U-235 enriched U-235/U-238 fuel and 3.0 % Pu-239/Pu-241 enriched plutonium/uranium dioxide fuel [37])

For the present line of PWRs, the sum total of the individual contributions to changes in various energy ranges finally leads to a negative coefficient of the moderator temperature which, however, also depends on the concentration of boric acid dissolved in the coolant and the burnup condition of the reactor core.

Figure 2.10 shows the criticality or effective multiplication factor, k_{eff}, as a function of the volume ratio of the moderator (Light Water Reactor) to fuel in a lattice cell (Fig. 2.5) in a LWR core [37]. This volume ratio of moderator to fuel varies if the moderator density or the distance, p, between the fuel rods with diameter, d, of the lattice cell changes. Left of the maximum of the curves in Fig. 2.10 the lattice cell is called undermoderated with lower pitch to fuel rod diameter, p/d, ratio. It is called overmoderated in the range right of the maximum of the curves (higher p/d ratio) [23, 36, 37].

The three curves of Fig. 2.10 are valid for fuel with 3.2 % U-235 enriched uranium (U-235/U-238) dioxide fuel and for 3.0 % (Pu-239/Pu-241) enriched so-called mixed plutonium/uranium dioxide fuel as well as for two fuel rod diameters in case of the plutonium/uranium fuel.

LWRs are always designed with an undermoderated lattice cell in the fuel element (left side of the k_{eff} curves in Fig. 2.10). In this case a temperature increase followed by a decrease of the moderator (H_2O) density and of the effective water volume V_{H2O} as well as of the ratio V_{H2O}/V_{UO2} results in a decrease of the criticality or effective multiplication factor k_{eff} (shift to the left). For LWRs the moderator/coolant-temperature coefficient is:

for PWRs -2×10^{-4} change in k_{eff} per degree of temperature increase of the water
for BWRs -1.3×10^{-3} change in k_{eff} per% increase of steam volume (void coefficient)

The negative moderator temperature- or void-coefficient determine the safety behavior of LWRs during coolant loss accidents [11, 36]. Contrary to the fuel-Doppler-temperature coefficient which acts in case of power increases instantaneously, the coolant/moderator-temperature coefficient can react only with a certain time delay during power transients. This is due to the fact that in the case of a fuel temperature increase the moderator temperature or vapor production increase only after a certain time delay (thermal conductivity in the fuel and cladding). However, in case of moderator/coolant depressurization (pipe break or faulty opening of a valve in the primary coolant system) the void coefficient also reacts instantaneously.

Figure 2.10 shows the criticality or effective multiplication factor for low enriched uranium fuel and for low enriched plutonium/uranium fuel. The curves for these two fuel types are shifted against each other. Usually, for uranium fuel a design value of $V_{H2O}/V_{UO2} = 2.1$ is selected for PWRs to obtain a sufficiently negative moderator/coolant-temperature coefficient. For plutonium/uranium fuel a ratio $V_{H2O}/V_{UO2} = 3.0$ is chosen.

Figure 2.10 **shows another important result for the case of a molten core as a result of severe core melt accidents. In case of a molten core, the water in LWRs is evaporated, the lattice structure (Fig.** 2.5**) is destroyed and the fuel is arranged in form of a core melt. In this case $V_{H2O}/V_{fuel} \rightarrow 0$ and $k_{eff} < 0.9$, i.e. the molten rearranged core material is subcritical (see Chap. 9).**

2.11 Behavior of the Reactor in Non-steady State Conditions

As has been explained above, $k_{eff} = 1$ corresponds to the steady state condition of the reactor core, in which case the production of fission neutrons is in a state of equilibrium with the number of neutron absorbed and the number of neutrons escaping from the reactor core.

For $k_{eff} \neq 1$, either the production or the loss term become dominant, i.e., the number of neutrons n(t) and $k_{eff}(t)$ vary as a function of time.

Axial movements of the absorber rods in the core change the loss term of neutrons and influence $k_{eff}(t)$. The relative change as a function of time $k_{eff}(t)$ is called reactivity $\rho(t)$.

$$\text{Reactivity} \quad \rho(t) = \frac{k_{eff}(t) - 1}{k_{eff}(0)} = \frac{\Delta k_{eff}(t)}{k_{eff}(0)}$$

$\rho(t)$ is measured in units of the fraction of the delayed neutrons $\beta \approx 0.0064$ for U-235. Historically it was defined that $\beta = 1$ dollar (1 \$) = 100 cents.

The behavior of the reactor power, temperatures or other changes of the steady state conditions, e.g. variations of the system pressure or coolant velocity can be described by a system of differential equations [8, 11, 18, 35, 36]. These are:

- The differential equations for the instationary neutron kinetics (space- and time-dependent prompt neutron flux distribution and concentrations of the delayed neutrons and their precursor atoms)
- The differential equations for the space and time dependent temperature fields in the fuel, the cladding and coolant of the reactor core (including the material properties of the different materials, e.g. thermal conductivity, heat capacity, etc.)
- The equations for the feedback effects affecting the effective multiplication factor, k_{eff}, as, e.g. the fuel-Doppler-temperature coefficient and the moderator/coolant-temperature coefficient
- The equations for the time dependent temperatures and pressures at the inlet of the reactor core caused by perturbations on the secondary side of the steam exchangers.

Not in all cases all parts of these coupled systems of differential equations must be solved together. In case of relatively fast variations of the physical core characteristics (time range of seconds or less), e.g. the core inlet coolant temperature can be considered to remain constant, as the steam generators parameters change only slowly.

In many cases the instationary neutron kinetics can be approximated by a system of coupled ordinary differential equations with initial conditions. In this case the prompt and delayed neutrons are represented by one ordinary differential equation and six differential equations for the precursor atoms which emit the delayed neutrons by radioactive decay. This leads to seven ordinary differential equations. The solution of these systems of coupled ordinary differential equations shows that three ranges of k_{eff} are of importance:

For supercriticality $k_{eff} > 1$ two ranges of k_{eff} must be distinguished.

- The range between $k_{eff} = 1$ and $k_{eff} < 1 + \beta$ in which the multiplying chain reaction is determined by the **delayed neutrons**. In this range of keff the relatively slowly originating delayed neutrons (from the radioactive decay of precursor atoms (see Sect. 2.1)) allow the control of the nuclear reactor in a time range of seconds to minutes. Control procedures by moving control rods or changes of the concentration of boric acid in the coolant water are performed in this range as displayed by Fig. 2.11.
- The range of $k_{eff} > 1 + \beta$ in which the multiplying chain reaction is determined by the **prompt neutrons** originating promptly from the fission process

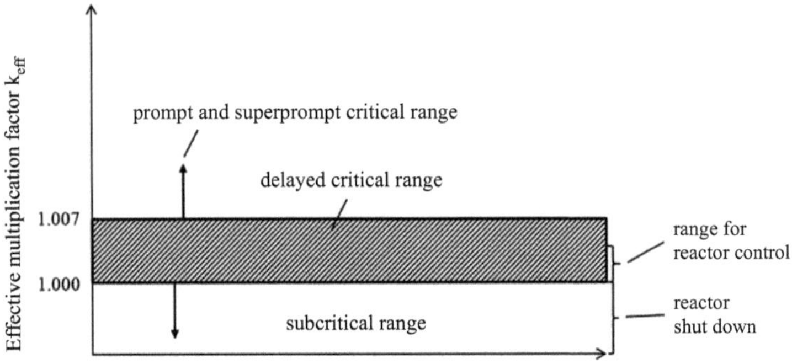

Fig. 2.11 Different ranges of the effective multiplication (criticality) factor k_{eff} for a typical LWR

(see Sect. 2.1). In this range the time difference of successive neutron generations is given by the lifetime of the prompt neutrons which is about 2.5×10^{-5} s for LWRs with U-235 enriched uranium fuel. This means that the chain reaction multiplies very fast. The very rapidly increasing number of fissions and the reactor power as well as the fuel temperatures are, however, reduced by the counteracting **negative** fuel-Doppler-temperature coefficient. This reduces the power and after having attained a certain peak level the power drops again. LWRs are designed such that the k_{eff} is limited such that the energy released by the power peak is small and limited.

– The range of $k_{eff} < 1$ if control/absorber elements are inserted into the reactor core or the concentration of boric acid is increased in the cooling water. In this case the power drops rapidly within seconds to the afterheat level.

Figure 2.11 explains the three ranges of k_{eff} which are important for the description of the non-steady or instationary behavior of nuclear reactors. All control procedures, e.g. withdrawal of control/absorber elements or increase of the boric acid concentration in the coolant water are performed in the delayed prompt critical range of k_{eff}. The design of an LWR core must guarantee that the super prompt critical range is never attained during normal operation. In case that the super prompt critical range should be attained during accidental condition, then the fuel-Doppler-temperature coefficient will limit the energy released in a power peak. In addition rapid automatic reactor shut down by control/absorber rods will limit the damage to the reactor core.

Similarly, the negative moderator/coolant-temperature coefficient or void coefficient will drastically reduce the reactor power in case of depressurization of the primary coolant system by large scale pipe breaks or faulty opening of valves in the primary system.

References

1. Lamarsh JR (1983) Introduction to nuclear reactor theory, 2nd edn. Addison-Wesley, Reading, MA
2. Glasstone S, Edlund MC (1952) Nuclear reactor theory. D. Van Nostrand, Princeton, NJ
3. Ott K et al (1983) Introductory nuclear reactor statics. American Nuclear Society, LaGrange Park, IL
4. Weinberg AM, Wigner EP (1958) The physical theory of neutron chain reactors. University of Chicago Press, Chicago, IL
5. Bell GI, Glasstone S (1970) Nuclear reactor theory. Van Nostrand Reinhold, New York, NY
6. Ash M (1965) Nuclear reactor kinetics. McGraw-Hill, New York, NY
7. Keepin GR (1965) Physics of nuclear kinetics. Addison-Wesley, Reading, MA
8. Ott KO et al (1985) Nuclear reactor dynamics. American Nuclear Society, LaGrange Park, IL
9. Broeders CHM (2010) Personal communication. KIT Karlsruhe
10. Michaudon A (1981) Nuclear fission and neutron induced fission cross sections. Pergamon, Oxford
11. Kessler G (2012) Sustainable and safe nuclear fission energy. Springer, Heidelberg
12. Dresner L (1960) Resonance absorption in nuclear reactors. Pergamon, New York, NY
13. Nuclear Data Files and Their References, Nuclear Data Center, Japan Atomic Energy Agency. http://wwwndc.jaea.go.jp/nucldata/evlrefs.html
14. Oblozinsky P et al (2010) Evaluated nuclear data. In: Cacuci DG (ed) Nuclear engineering fundamentals, vol 1, Handbook of nuclear engineering. Springer, New York, NY
15. Koning A et al (2006) The JEFF-3.1 nuclear data library. JEFF Report 21, NEA No. 6190. OECD/NEA, Paris
16. Roussin RW et al (1994) Current status of ENDF/B-VI. In: Proceedings of international conference nuclear data for science and technology, Gatlinburg, TN, vol 2, p 692
17. Kikuchi Y (1994) JENDL-3, revision 2: JENDL 3-2. In: Proceedings of international conference nuclear data for science and technology, Gatlinburg, TN, vol 2, p 685
18. Stacey W (2007) Nuclear reactor physics. Wiley, New York, NY
19. MacFarlane RE et al (1994) The NJOY nuclear data processing system version 91. LA-12740-M, Los Alamos National Laboratory
20. Yang WS et al (2010) Neutronics modeling and simulation of SHARP for fast reactor analysis. Nucl Eng Technol 42:520
21. Lee CH et al (2011) MC^2-3: multigroup cross section generation code for fast reactor analysis. ANL-NE-11-41, Argonne National Laboratory
22. Askew J et al (1966) A general description of the lattice code WIMS. J Br Nucl Energy Soc 5:564
23. Knott D et al (2010) Lattice physics computations. In: Cacuci DG (ed) Nuclear engineering fundamentals, vol 1, Handbook of nuclear engineering. Springer, New York, NY
24. Yang WS (2012) Fast reactor physics and computational methods. Nucl Eng Technol 44(2):177–198
25. Lawrence RD (1983) The DIF3D nodal neutronics option for two- and three-dimensional diffusion theory calculations in hexagonal geometry. ANL-83-1, Argonne National Laboratory
26. Bahadir T et al (2005) SIMULATE-4 multigroup nodal code with microscopic depletion model. In: Mathematics and computation, supercomputing, reactor physics and nuclear and biological applications, Avignon, France, 12–15 September
27. Alcouffe RE et al (1995) DANTSYS: a diffusion accelerated neutral particle transport code system. LA-12969-M
28. Alcouffe RE et al (2000) PARTISN abstract. In: PHYSOR 2000 ANS international topical meeting on advances in reactor physics and mathematics and computation into the next millennium, Pittsburgh, PA, 7–11 May
29. Brown FB et al (2009) MCNP5-1.51 release notes. LA- UR-09-00384, Los Alamos National Laboratory

30. Blomquist RN (1995) VIM continuous energy Monte Carlo transport code. In: Proceedings of international conference on mathematics, computations, reactor physics and environmental analysis, Portland, OR, 30 April–4 May
31. Haeck W et al (2007) An optimum approach to Monte Carlo burnup. Nucl Sci Eng 156:180–196
32. Fission Product Nuclear Data (FPND) – 1977 (1978) Proceedings of second advisory group meeting on fission product nuclear data, Energy Centrum Netherlands, Petten, 5–9 September 1977. International Atomic Energy Agency, IAEA-213, Vienna
33. ORNL (2005) SCALE A modular code system for performing standardized computer analyses for licensing evaluations. ORNL/TM-2005/39, version 5, vols I–III
34. Poston DI et al (1999) Development of a fully-automated Monte Carlo burnup code MONTEBURNS. LA-UR-99-42
35. Nicholson R et al (1968) The Doppler Effect in fast reactors. In: Advances in nuclear science and technology, vol 4. Academic, New York, NY, p 109
36. Emendörfer D et al (1993) Theorie der Kernreaktoren, Band 2: Der instationäre Reaktor, B.I.-Wissenschaftsverlag, Mannheim; Leipzig; Wien; Zürich
37. Märkl M (1976) Core engineering and performance of pressurized water reactors. Kraftwerk Union AG, Erlangen
38. Goluoglu S (2011) Monte Carlo criticality methods and analysis capabilities in SCALE. Nucl Technol 174:214

Chapter 3
The Design of Light Water Reactors

Abstract This chapter describes the designs and safety concepts of presently operating and more recently developed (future) LWRs. The chapter concentrates on LWR plants (PWRs and BWRs) with 1,300–1,700 MW(e) power output manufactured in Europe, the USA and Japan. As a presently operating PWR the standard PWR of 1,300 MW(e) of KWU (Germany) is chosen. As more recent (future) PWR designs the Advanced Pressurized Water Reactor AP1000 and the US-APWR (1,700 MW(e)) developed by Westinghouse (USA) and Mitsubishi (Japan) are described. In addition the European Pressurized Water Reactor (EPR) with 1,600 MW(e) power output of AREVA (France) is presented too.

The fuel elements, the control elements, the core design, the pressure vessel, the design characteristics of the primary system, the steam generators and steam conditions for the turbine-generator system are all very similar for all these PWRs. Most of the PWRs with 1,300 MW(e) and more power output have four redundant coolant circuits. Their emergency core and afterheat cooling systems are also fourfold redundant. An exception is AP1000 with only two redundant coolant circuits and two redundant emergency core and afterheat cooling systems. All PWRs provide emergency core cooling at three different pressure levels (high pressure (core make up tanks), medium pressure (accumulator) and low pressure). Whereas in the older KWU PWR-1,300 design the primary system can be depressurized manually by the operator, this is realized automatically by the automatic depressurization system (ADS) in the more recent designs of AP1000, US-APWR and EPR. Emergency cooling water is taken from the building sump in case of the KWU PWR-1,300, whereas the more recent designs AP1000, US-APWR and EPR are equipped with a large water volume in-containment refueling water storage tank (IRWST). The afterheat can be removed in case of a severe core melt accident either passively through the inner steel containment wall (AP1000) or by using sprinkler systems in connection with containment heat removal systems (EPR and US-APWR). In case of a core melt accident water of the IRWST can be drained into the reactor cavity (AP1000). This water cools the reactor pressure vessel from the outside and prevents the core melt from penetrating through the pressure vessel wall. In case of EPR a molten core spreading and

G. Kessler et al., *The Risks of Nuclear Energy Technology*, Science Policy Reports,
DOI 10.1007/978-3-642-55116-1_3, © Springer-Verlag Berlin Heidelberg 2014

cooling device (core catcher) provides long term cooling of the molten core. The outer containment of PWRs protects the inner components in the inner containment against external events (earthquakes, hurricanes etc.).

As presently operating standard BWR the BWR-1,300 of KWU (Germany) and the ABWR built by General Electric (USA), TOSHIBA and HITACHI (Japan) are described. In addition the more recently designed (future) SWR-1,000 (KERENA) of AREVA (France) and the ABWR-II of General Electric (USA), TOSHIBA and HITACHI (Japan) will be presented. The fuel elements, the control elements, the core design etc. of the BWR-1,300, ABWR and SWR-1,000 (KERENA) designs are very similar. The pressure suppression system with suppression pool, drywell and the containment system of the BWR-1,300 and the ABWR are again very similar. The threefold emergency core cooling and afterheat removal systems of BWR-1,300 and ABWR show only little differences. In case of the SWR-1,000 (KERENA) with fourfold cooling and emergency cooling systems more passive safety systems are installed. At depressurization the steam is discharged into four interconnected core flooding pools. These core flooding pools serve as a passive heat sink. The heat is transferred through four containment emergency condensers to an upper shielding/storage pool. Additional passive heat transfer systems are the emergency condensers within the four core flooding pools. The drywell with reactor pressure vessel is flooded passively in case of drastic losses of coolant and danger of core melt.

The ABWR-II design of General Electric (USA), TOSHIBA and HITACHI (Japan) evolved from the ABWR. It has a larger fuel element, modified emergency core cooling systems and passive core cooling as well as passive containment cooling systems. In case of core melt through the bottom of the pressure vessel the core melt can be cooled on a steel plated cavity underneath the pressure vessel.

3.1 Light Water Reactors

It was already mentioned in Chap. 1 that nuclear power generation is currently mainly based (about 80 %) on Light Water Reactors (LWRs) designed as Pressurized Water Reactors (PWRs) or Boiling Water Reactors (BWRs). LWRs use low enriched uranium fuel, which makes for greater flexibility in the choice of reactor core materials, especially allowing normal (light) water to be used as a coolant and moderator. PWRs deliver the heat generated in their reactor core to water circulating under high pressure in primary coolant circuits. From here the heat is transferred to a secondary coolant system via steam generators to produce steam driving a turbo-generator system. In BWRs the steam for the turbo-generator system is generated right in the reactor core and sent directly to the turbo-generator. PWRs and BWRs have been advanced to a high level of technical maturity. They are built now in unit sizes up to 1,200 and 1,700 MW(e) [1].

3.2 Pressurized Water Reactors

In the 1950s PWRs were developed in the USA particularly as power plants for nuclear submarines ("N.S. Nautilus" 1954). The successful application of the PWR concept then resulted in the construction of the first non-military experimental nuclear power plants in the USA and Russia.

Present PWRs have been or are built basically according to the same technical principles by a number of manufacturers in various countries (USA, Russia, France, Germany, Japan, South-Korea, China). In this chapter, the modern PWRs of manufacturers in the USA, Japan and Europe are presented. These are the standard PWR of 1,300 MW(e), which will be described as built in the 1990s by Kraftwerk Union in Germany (KWU-PWR) [1]. In addition, the Advanced Pressurized Water Reactors AP1000 and the US-APWR, developed by Westinghouse (USA) and Mitsubishi (Japan), are presented [2, 3]. Finally the 1,600 MW(e) European Pressurized Water Reactor (EPR) will be described as built by AREVA (France) [4]. PWR designs by other manufacturers have small technical differences relative to the above design concepts but these are not relevant for a general understanding of the basic principles of safety design.

Figure 3.1 shows the main design principles of a PWR. The heat generated by nuclear fission in the reactor core is transferred from the fuel elements to the coolant in the primary system. The highly pressurized water (15.5 or 15.8 MPa for AP1000, US-APWR, KWU-PWR and EPR) is circulated by coolant pumps and heated in the reactor core from inlet temperatures of 281 °C (AP1000), 288 °C (US-APWR), 292 °C (KWU-PWR) or 296 °C (EPR) to outlet temperatures of 321 °C (AP1000), 328 °C (US-APWR), 326 °C (KWU-PWR) or 328 °C (EPR) (Tables 3.1 and 3.2). It flows to two or four steam generators, where it transfers its heat to the secondary steam system. In the secondary system, steam of 5.5 (AP1000), 6.7 (US-APWR), 6.8 (KWU-PWR) or 7.8 MPa (EPR) and 273 °C (AP1000), 284 °C (US-APWR), 285 °C (KWU-PWR) or 293 °C (EPR) is generated. The steam drives the turbine and the generator. The steam exhausted by the turbine is precipitated in the condenser, and the condensate water is pumped back into the steam generators. Waste heat delivered to the condenser is discharged into the environment either to river, lake or sea water or through a cooling tower.

3.2.1 Core

The core initially contains fuel elements with three or four different levels of U-235 enrichment. In case of EPR, e.g. some fuel elements also contain gadolinium as burnable poison. The higher enriched fuel elements are arranged at the core periphery, the less enriched fuel elements are distributed throughout the interior of the core (Fig. 3.2). This provides for a relatively flat power distribution over the core and adapts to later so-called equilibrium core loadings. In later core reloadings,

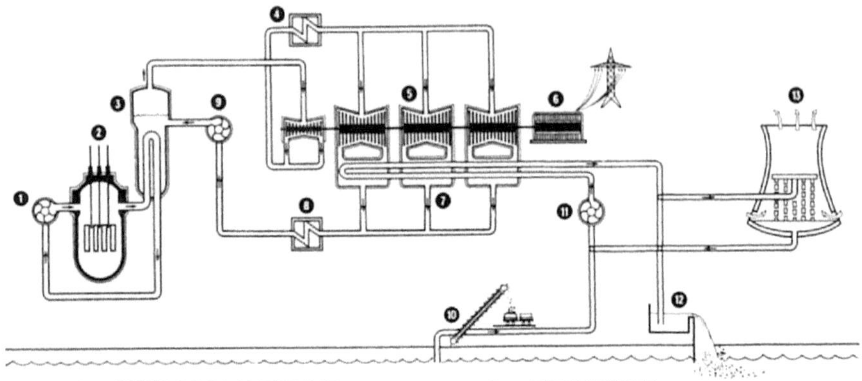

1	MAIN COOLANT PUMP	8	PREHEATER
2	PRESSURIZED WATER REACTOR	9	FEEDWATER PUMP
3	STEAM GENERATOR	10	COOLING WATER PURIFICATION SYSTEM
4	WATER SEPARATOR		
5	TURBINE	11	COOLING WATER PUMP
6	GENERATOR	12	OVERFLOW BASIN
7	CONDENSER	13	COOLING TOWER

Fig. 3.1 Functional design diagram of a pressurized water reactor power plant (Kraftwerk Union) [1]. 1. Main coolant pump, 2. Pressurized water reactor, 3. Steam generator, 4. Water separator, 5. Turbine, 6. Generator, 7. Condensor, 8. Preheater, 9. Feedwater pump, 10. Cooling water purification system, 11. Cooling water pump, 12. Overflow basin, 13. Cooling tower

fuel elements with different burnups and different enrichments are also arranged in similar patterns. The specific power density in the core is about 95 (EPR) to 110 (AP1000) kW(th)/l. Around 2008 the average maximum burnup of the fuel was about 55,000 MWd/ton over an irradiation period in the core of about 4 years. Around 2010–2020, a maximum burnup to 70,000 MWd/ton is strived for. A 1 year reloading cycle can be reached by unloading either one fourth or one fifth of the fuel elements at maximum burn up.

Uranium dioxide (UO_2) is used as a core fuel. The UO_2 powder is pressed and sintered into pellets of about 10 mm height and diameter with an average density of 10.4 g/cm^3. The pellets along with pressurized helium are placed into tubes (cladding) made of Zircaloy with an active core length of 3.6–4.2 m. This cladding material is chosen for its low neutron absorption as well as good mechanical and corrosion properties. The tubes are welded and assembled into fuel elements (Fig. 3.3).

A fuel element of a 1,300 MW(e) or 1,600 MW(e) reactor contains 236 or 265 fuel rods, respectively. The core has 157–257 fuel elements (Tables 3.1 and 3.2), with a total uranium mass of 84 tons (AP1000), 125 tons (KWU-PWR), 153 tons (EPR) and 138.5 tons (US-APWR). Some of the fuel elements contain control elements with 20–69 axially moveable absorber rods (Fig. 3.4) [6, 7]. These absorber rods are filled with boron carbide or silver-indium-cadmium as neutron absorbing materials.

Table 3.1 Design characteristics of the large PWR power plants of Kraftwerk Union and AREVA [1, 4]

		Kraftwerk Union (KWU-PWR)	AREVA (EPR)
		PWR-1,300	EPR-1,600
Reactor power			
Thermal	MW(th)	3,780	4,500
Electrical	MW(e)	1,300	1,600
Plant efficiency	%	32.8	35.6
Number of coolant loops		4	4
Reactor core			
Equivalent core diameter	m	3.6	3.77
Active core height	m	3.9	4.2
Specific core power	kW(th)/l	95	96
Density	kW(th)/kg U	36.5	28.4
Number of fuel elements		193	241
Total amount of fuel (uranium mass)	kg U	125,000	153,000
Fuel element and control element			
Fuel		UO_2	UO_2
U-235 fuel enrichment – reloading	wt%	3.5–4.5	4–5
Cladding material		Zircaloy-4	Zircaloy-M5
Cladding outer diameter	mm	10.75	9.50
Cladding, thickness	mm	0.725	0.57
Fuel rod spacing	mm	14.3	12.6
Av. specific fuel rod power	W/cm	208	156
Fuel assembly array		16 × 16	17 × 17
Control/absorber rods			
Absorber material		20 rods inserted in fuel element Ag, In, Cd and B_4C	24 rods inserted in fuel element Ag, In, Cd and B_4C
Number of fuel rods per fuel element		236	265
Burnable poison		Boric acid	Gd and boric acid
Number of control elements		61	89
Heat transfer system			
Primary system			
Total coolant flow			
Core flow rate	t/s	18.8	21.4
Coolant pressure	MPa	15.8	15.5
Coolant inlet temp.	°C	292	295.6
Coolant outlet temp.	°C	326	328.2
Steam supply system			
Steam pressure	MPa	6.8	7.8
Steam temperature	°C	285	293

(continued)

Table 3.1 (continued)

		Kraftwerk Union (KWU-PWR)	AREVA (EPR)
		PWR-1,300	EPR-1,600
Fuel cycle			
Average fuel burn up	MWd/ton	55,000	70,000
Refueling sequence		1/3 to 1/4 per year	1/5 per year
Average fissile fraction in spent fuel			
U-235	wt%	0.8	0.8
Plutonium	wt%	0.7	0.7

3.2.2 *Reactor Pressure Vessel*

The fuel elements, control elements and core monitoring instruments are contained in a large reactor pressure vessel (RPV) designed to withstand the operating pressures at operating temperatures (Fig. 3.5). The vessel has an inner diameter of about 5 m, a thickness of the cylindrical wall of about 200–250 mm, and a height of up to approx. 13 m. The cover head of the vessel, which holds all the control rods and control rod drive mechanisms (CRDM), can be removed for refueling. The water flowing at a rate of 18.8 t/s (KWU-PWR) and 21.4 t/s (EPR), respectively, enters the reactor vessel through two (AP1000) or four (KWU-PWR, US-APWR, EPR) inlet nozzles close to the top and flows downward through the annulus between the vessel and the thermal shield and neutron reflector to the core inlet near the bottom. It then returns upward through the core being heated up there and leaving the two or four outlet nozzles of the pressure vessel. The reactor vessel is made of, e.g. 22NiMoCr37 steel, in case of Kraftwerk Union PWRs and of 16 MND5 for EPR. Its inner surface is plated with austenitic steel.

3.2.3 *Coolant System*

The water leaving the outlet nozzles of the pressure vessel transports the heat generated in the reactor core through two (AP1000) or four (KWU-PWR, US-APWR, EPR) identical primary coolant circuits to two or four steam generators and is then recirculated to the pressure vessel. The inner diameter of the primary system pipes is, e.g. 750 mm for the KWU-PWR or 780 mm for EPR. Each primary coolant pump has a pressure head, e.g. of 0.8 (KWU-PWR) or 1.0 MPa (EPR) and consumes 5.4 or 9 MW(e) of electrical power, respectively. The whole primary system is also plated with austenitic steel. The pressure of 15.5 MPa (KWU-PWR) or 15.8 MPa (EPR) in the primary coolant system is maintained by a pressurizer filled partly with water and partly with steam. It has heaters for boiling the water

Table 3.2 Design characteristics of the large PWR power plants of Westinghouse (AP1000) and Mitsubishi (US-APWR) [2, 3, 5]

		Westinghouse	Mitsubishi
		AP1000	US-APWR
Reactor power			
Thermal	MW(th)	3,415	4,451
Electrical	MW(e)	1,115	1,700
Plant efficiency	%	33	38
Number of coolant loops		2	4
Reactor core			
Equivalent core diameter	m	3.04	4.22
Active core height	m	4.27	3.65
Specific core power	kW(th)/l	110	87
Density	kW(th)/kg U	40	32
Number of fuel elements		157	257
Total amount of fuel (uranium mass)	kg U	84,000	138,500
Fuel element and control element			
Fuel		UO_2	UO_2
U-235 fuel enrichment -reloading	wt%	2.4–4.5	4–5
Cladding material		ZIRLO™	Zircaloy-M5
Cladding outer diameter	mm	9.50	9.50
Cladding, thickness	mm	0.57	0.57
Fuel rod spacing	mm	12.6	12.6
Av. specific fuel rod power	W/cm	187	153
Fuel assembly array		17 × 17	17 × 17
Control/absorber rods			
Absorber material		69 rods inserted in fuel element Ag, In, Cd, stainless steel	69 rods inserted in fuel element Ag, In, Cd
Number of fuel rods per fuel element		264	264
Burnable poison		boric acid	Gd and boric acid
Number of control elements		24	89
Heat transfer system			
Primary system			
Total coolant flow			
Core flow rate	t/s	18.9	19.3
Coolant pressure	MPa	15.5	15.8
Coolant inlet temp.	°C	281	288
Coolant outlet temp.	°C	321	328.2
Steam supply system			
Steam pressure	MPa	5.5	6.7
Steam temperature	°C	271	284
Fuel cycle			
Average fuel burn up	MWd/ton	60,000	60,000

(continued)

Table 3.2 (continued)

		Westinghouse	Mitsubishi
		AP1000	US-APWR
Operating cycle length	Months	18	18
Design life	Years	60	60
Average fissile fraction in spent fuel			
U-235	wt%	0.8	0.8
Plutonium	wt%	0.7	0.7

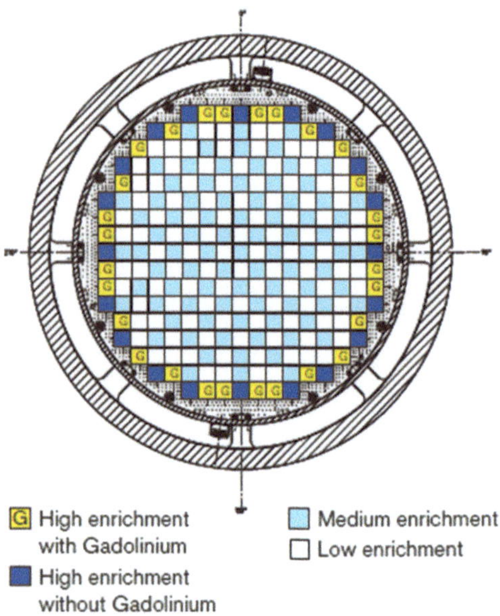

Fig. 3.2 EPR core with typical initial fuel element enrichments (AREVA) [4]

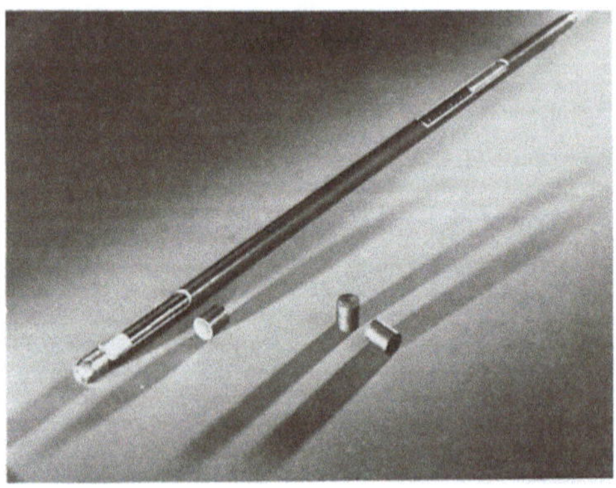

Fig. 3.3 PWR fuel rod and fuel pellets (Kraftwerk Union) [1]

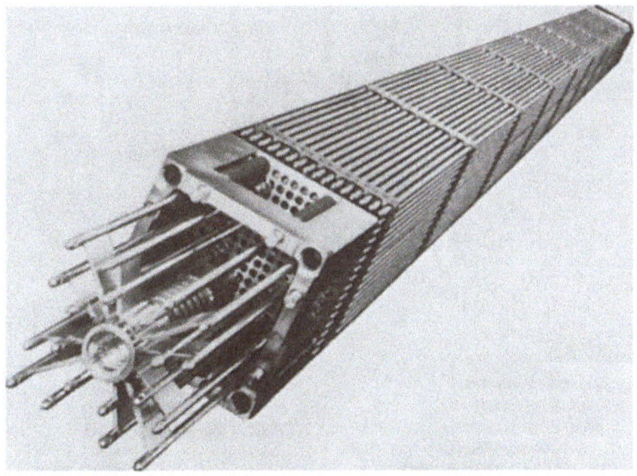

Fig. 3.4 Fuel element and control element of a 1,300 MW(e) PWR (Kraftwerk Union) [1]

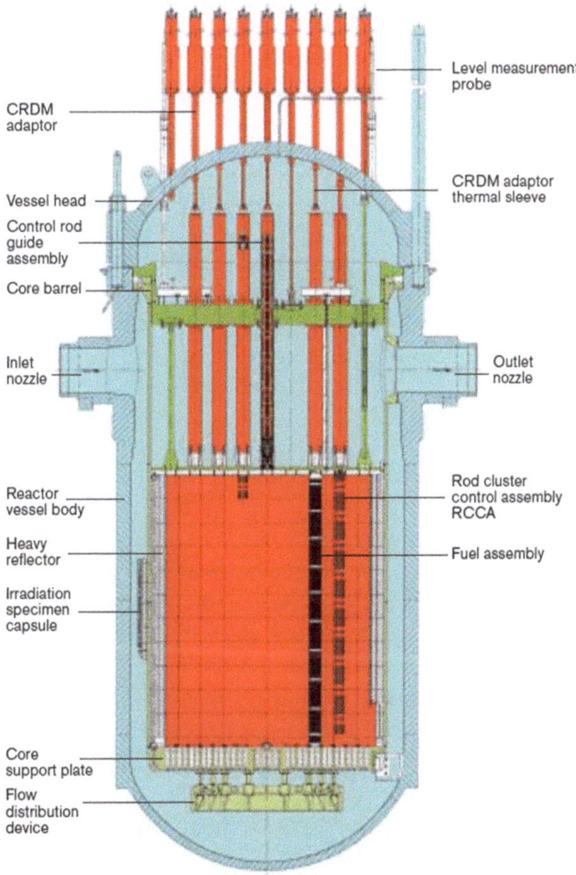

Fig. 3.5 Reactor pressure vessel with core and internal components of EPR (AREVA) [4]

and sprayers for condensing the steam to keep the pressure within specified operating limits (Fig. 3.6).

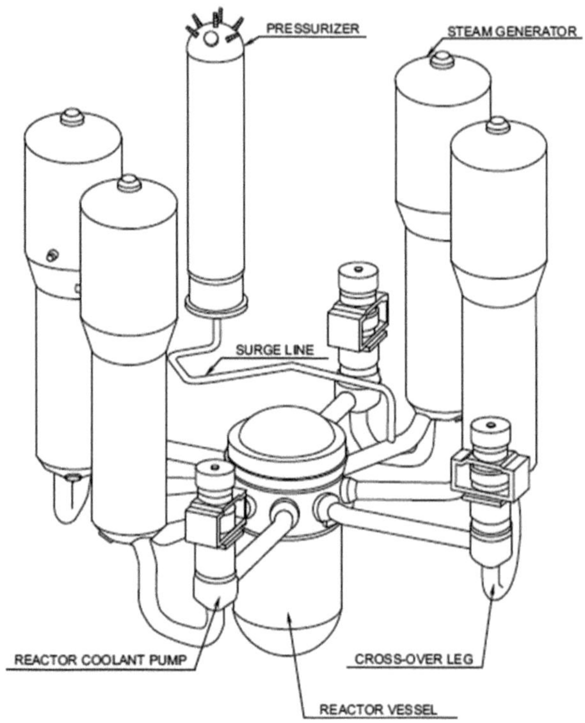

Fig. 3.6 Isometric view of the reactor coolant system US-APWR [5]

In the steam generators heat is transferred through a large number of tubes (Fig. 3.7). Feed water at a pressure of about 7 MPa is evaporated and steam is passed through separators to remove water droplets and attain a moisture content of less than about 0.015 %. In the steam generator, the primary coolant system is separated from the secondary coolant system by the tubes. The EPR steam generator (Fig. 3.7) with economizer makes it possible to reach a saturation pressure of 7.8 MPa by special guidance of the feed water flow. This leads to a thermal efficiency of up to about 36 % (EPR) [4].

The steam flows through the turbine valves (Fig. 3.1) into the high pressure section of the turbine and, after reheating, into the low pressure section of the turbine. The expanded steam is precipitated in the condenser and pumped back as feed water by the main condensate pumps into the feed water tank. The main feed water pumps move the pressurized water from the feed water tank through four main feed water pipes into the four steam generators.

In case the turbine would have to be suddenly disconnected from the grid as a result of some fault condition, steam can be directly passed into the condenser by means of bypass valves. If the condenser should not be available due to some failure, the steam can be blown off to the atmosphere by means of blow down valves and safety valves.

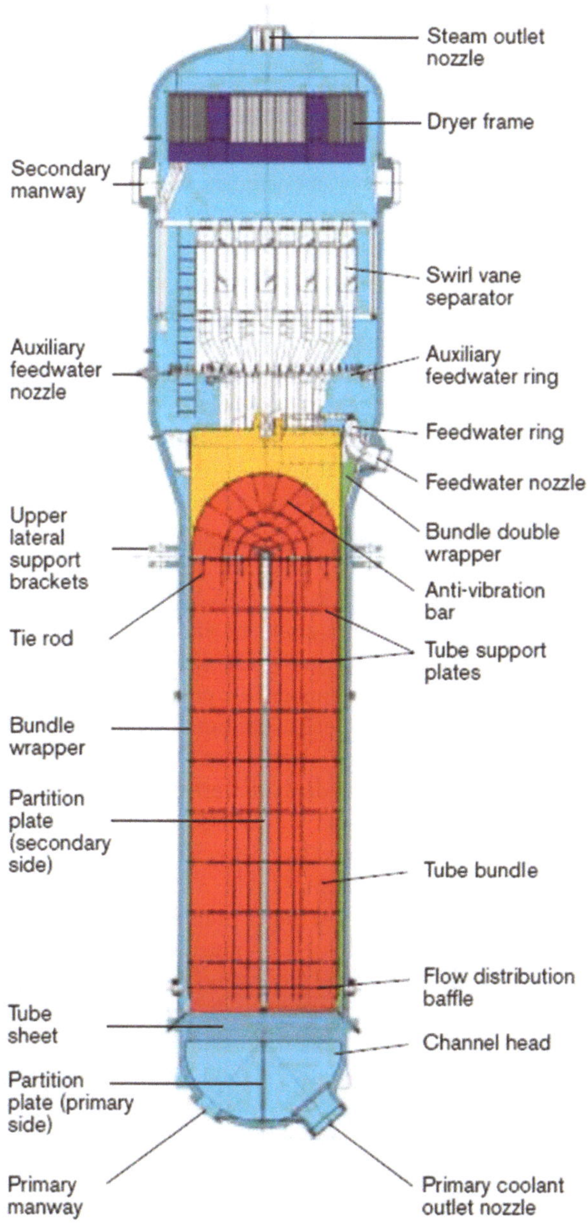

Fig. 3.7 Cutaway of steam generator for EPR (AREVA) [4]

Several supporting systems are required for operation of the coolant circuit systems. The volume control system offsets changes in volume of the reactor cooling system resulting from temperature and operational influences. It is controlled by the water level in the pressurizer (Fig. 3.6). Part of the cooling water is extracted continuously and purified in ion exchangers. At the same time, corrosion products and possibly released radioactive products are removed.

3.2.4 Containment Building

The containment building has the function to contain releases of airborne radioactivity in case of accidental conditions and shield the reactor core and cooling systems against external impacts. The containment design chosen by the manufacturers is either an inner steel containment with an outer prestressed concrete shell (KWU-PWR, AP1000) or a single (US-APWR) or a double prestressed concrete structure (EPR) with internal liner.

In case of the Kraftwerk Union PWR (Table 3.1), the reactor pressure vessel, the coolant pumps, steam generators, emergency and afterheat cooling systems as well as the vault for fresh and spent fuel elements are arranged within the reactor building, which is enclosed in a spherical double containment (Figs. 3.8 and 3.9). The double containment is made up of the inner steel containment and an outer concrete shield which is 1.8 m thick (Fig. 3.8).

The inner steel containment is held at a lower pressure than the atmospheric pressure. In this way only leakage from the outside to the interior of the containment is possible during normal operation. The spherical inner steel containment has a diameter of approx. 56 m and is designed to an internal pressure of about 0.5 MPa. If the internal design pressure of 0.5 MPa is exceeded after a core melt accident the containment atmosphere could be slowly exhausted through a valve and special aerosol filters which retain 99.9 % of the radioactive aerosols. Penetrations of the piping through the containment are equipped with vented double (Fig. 3.9) bellows and can be checked for leaks. The outer steel reinforced concrete shield protects the reactor against external impacts and shields the environment against radiation exposure in case of accidents. External impacts which are considered as a design basis for the containment are earthquakes, floods, tornados, airplane crashes, and pressure waves generated by chemical explosions.

The EPR containment (Table 3.1) is a cylindrical double containment of prestressed concrete. The inner containment has a 6 mm thick steel liner. The double concrete shield is 1.3 m (inner containment) and 1.3 m (outer containment) or in total 2.6 m thick and has a diameter of about 75 m and a height of about 60 m (Figs. 3.10 and 3.11). It protects the inner part of the containment against external events as in case of the spherical containment of the KWU-PWR. The upper part of the inner containment houses sprinkler systems supplied with water by the containment heat removal system (CHRS) for ultimate heat removal in case of a severe accident. Below the reactor pressure vessel, a so-called molten core spreading area with special cooling systems is located which can cool the hot core masses in case of a core melt accident (Figs. 3.10 and 3.11). Leak tightness of the inner containment and filter systems guarantee extremely low releases of radioactivity to the environment, even in case of severe accidents. Relocation or evacuation of the population can be avoided even in case of severe accidents (Chaps. 9 and 10).

The inner containment of modern European PWRs is required to withstand the pressure resulting from large scale hydrogen combustion processes which might occur under severe accident conditions (Chap. 8). Hydrogen can result from the

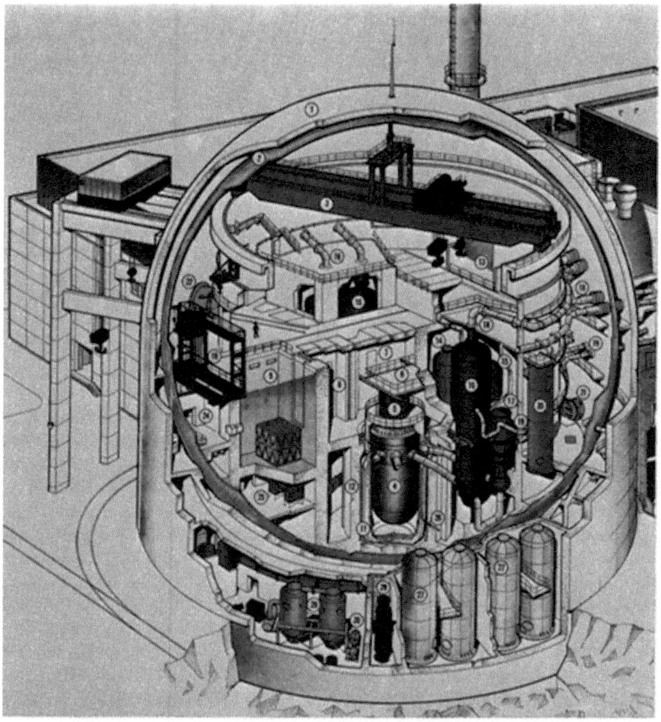

1 CONCRETE SHIELD
2 STEEL CONTAINMENT
3 REACTOR POLAR CRANE
4 REACTOR PRESSURE VESSEL
5 CONTROL ROD DRIVES
6 CABLE BRIDGE
7 REACTOR COMPARTMENT
8 CORE INTERNALS LAY-DOWN LOCATION
9 FUEL ELEMENT STORAGE POOL
10 FUELING MACHINE
11 INNER SHIELD (BIOLOGICAL SHIELD)
12 SUPPORT SHIELD (BIOLOGICAL SHIELD)
13 VESSEL HEAD LAY-DOWN LOCATION
14 PRESSURIZER
15 PRESSURIZER RELIEF TANK
16 STEAM GENERATOR
17 MAIN COOLANT PUMPS
18 MAIN STEAM PIPES
19 FEEDWATER PIPES
20 PRESSURE ACCUMULATOR
21 PERSONNEL LOCK
22 MATERIALS LOCK
23 SEMI-PORTAL STRUCTURE WITH TROLLEY
24 STORAGE FACILITY FOR FRESH FUEL ELEMENTS
25 TRANSDUCER ROOM
26 MEASURING CHAMBER GUIDE TUBES FOR EXTERNAL NEUTRON FLUX MEASUREMENT
27 FLOODING TANK
28 RESIDUAL HEAT EXCHANGER
29 COMPONENT COOLING HEAT EXCHANGERS
30 SAFETY FEED PUMPS

Fig. 3.8 Cutaway view of the containment of a 1,300 MW(e) PWR (Kraftwerk Union) [1]

chemical reaction of the fuel cladding material zircaloy with hot steam during a severe core melt accident. Also so-called hydrogen recombiners are distributed all over the containment to decrease the hydrogen concentrations in the inner containment in case of slowly developing core melt accidents. In Chap. 8 the safety aspects of rapid accident progressions with fast hydrogen generation will be discussed.

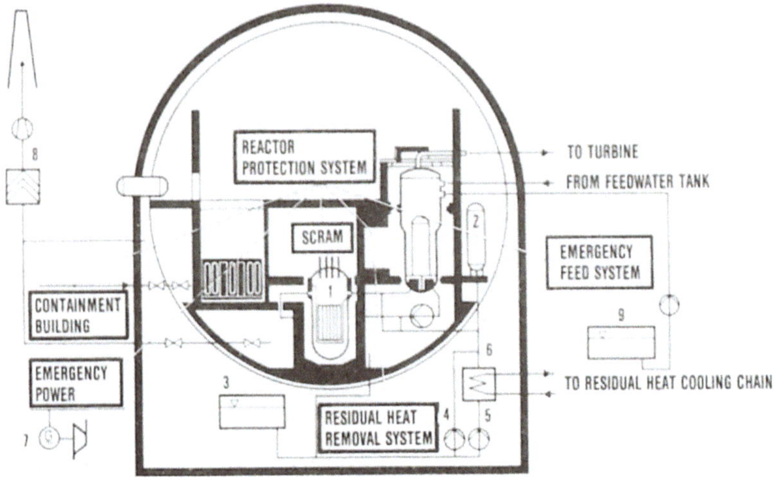

1 SCRAM SYSTEM
2 PRESSURIZER
3 FLOODING TANK
4 SAFETY FEED PUMP
5 RESIDUAL COOLING PUMP
6 RESIDUAL HEAT COOLER
7 EMERGENCY POWER SYSTEM
8 VENTILATION SYSTEMS
9 EMERGENCY FEEDWATER SYSTEM

Fig. 3.9 Reactor protection system of a PWR (Kraftwerk Union) [1]. 1. Scram system, 2. Pressurizer, 3. Flooding tank, 4. Safety feed pump, 5. Residual cooling pump, 6. Residual heat cooler, 7. Emergency power system, 8. Ventilation systems, 9. Emergency feedwater system

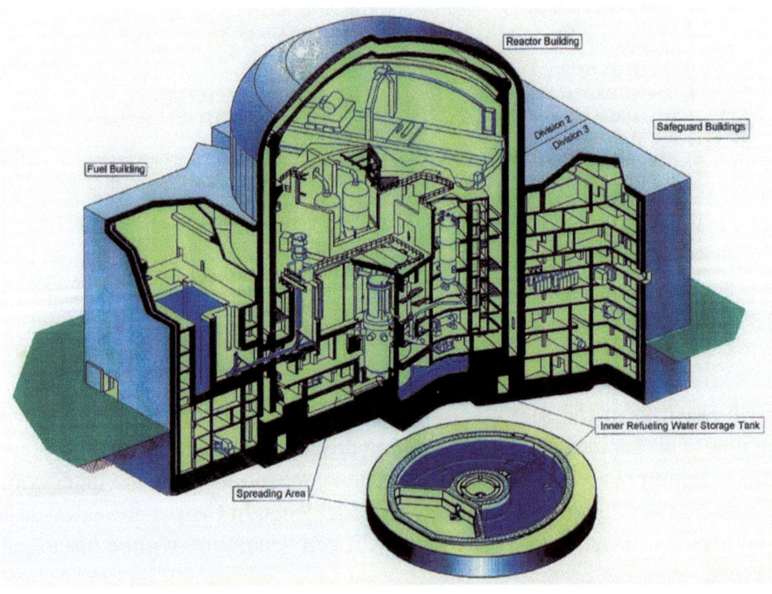

Fig. 3.10 EPR double containment with reactor pressure vessel, cooling systems and molten core spreading area (AREVA) [4, 8]

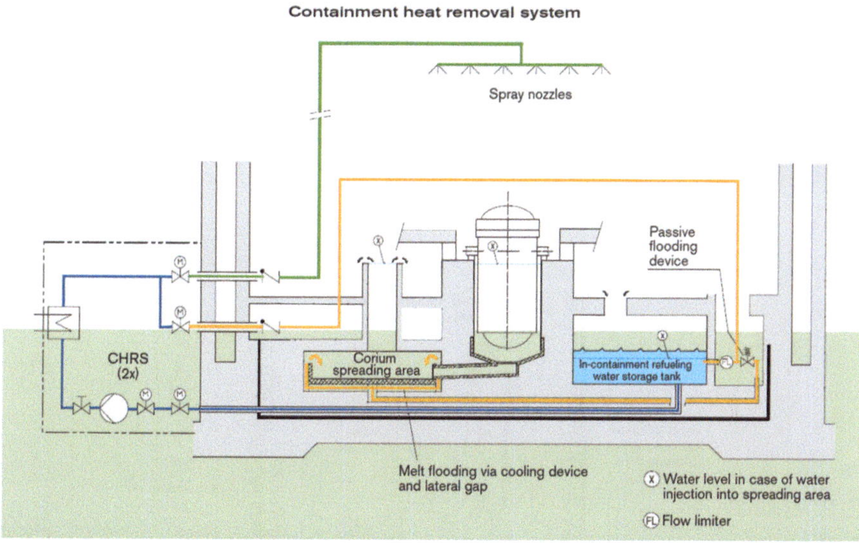

Fig. 3.11 Containment heat removal system in case of core melt for EPR (AREVA) [4, 8]

3.2.5 *AP1000 Safety Design*

The containment design of the AP1000 (Table 3.2) relies on passive safety features, e.g. natural driving forces like gravity flow, natural circulation flow and convection heat transfer, as well as compressed gas.

The AP1000 containment (Fig. 3.12) consists of an outer cylindrical reinforced concrete shield building which surrounds a cylindrical steel containment of 39.6 m diameter. During normal operation and accidental conditions the concrete shield building provides shielding against radiation from reactor components and released radioactivity. It also protects the internal reactor components from external impacts, e.g. tornados, missiles, etc. [2, 10].

In case of a core melt accident followed by heat and vapor release to the inner atmosphere of the containment the heat can be transferred by thermal conduction through the steel wall of the internal containment. Air coming into the shield containment structures flows down a baffle structure and cools the cylindrical steel wall by uprising natural convection flow (Fig. 3.12). It leaves the shield containment at the top. This cooling process can be enhanced by water flowing from the drain tank of the passive containment cooling system (PCCS) on top of the inner steel containment. This water is flowing down the outer surface of the inner steel containment. The in-containment refueling water storage tank (IRWST) with its large water volume provides a large heat sink for passive residual heat removal.

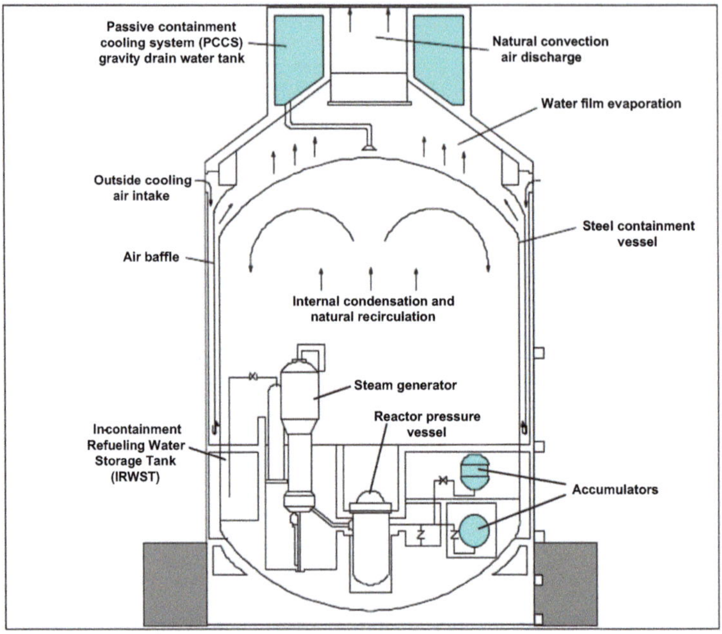

Fig. 3.12 AP1000 passive containment cooling system [2, 9]

3.2.5.1 Emergency Core Cooling System (ECCS)

The passive core cooling system (PXS) protects the AP1000 plant against core cooling accidents (leaks and ruptures at different locations of the reactor cooling system). The PXS provides safety injection of coolant water, depressurization of the primary system and residual heat (afterheat) removal.

In case of e.g. a double-ended rupture of a main reactor cooling line the safety injection and depressurization systems start to act (Fig. 3.13). The PXS uses three sources of cooling water to maintain core cooling through safety water injection. The safety water injection systems including two core water makeup tanks (CMTs) at high coolant pressure (15.5 MPa borated water), the two accumulators with borated water at medium coolant pressure (4.8 MPa) and the IRWST with borated water which starts to act at pressures below 0.18 MPa (after full depressurization). These three water injection systems are connected to separate injection nozzles at the reactor pressure vessel [2, 9]. In this way no safety injection water can be lost in case of breaks in the main cooling lines (Fig. 3.13).

In case of an accident sequence with a threat of core melting, the primary coolant system can be depressurized after reactor shutdown by the automatic depressurization system (ADS) (Fig. 3.13) into the IRWST via two depressurization stages with threefold valve lines at the pressurizer and a third and a fourth depressurization stage connected to the main coolant lines [2].

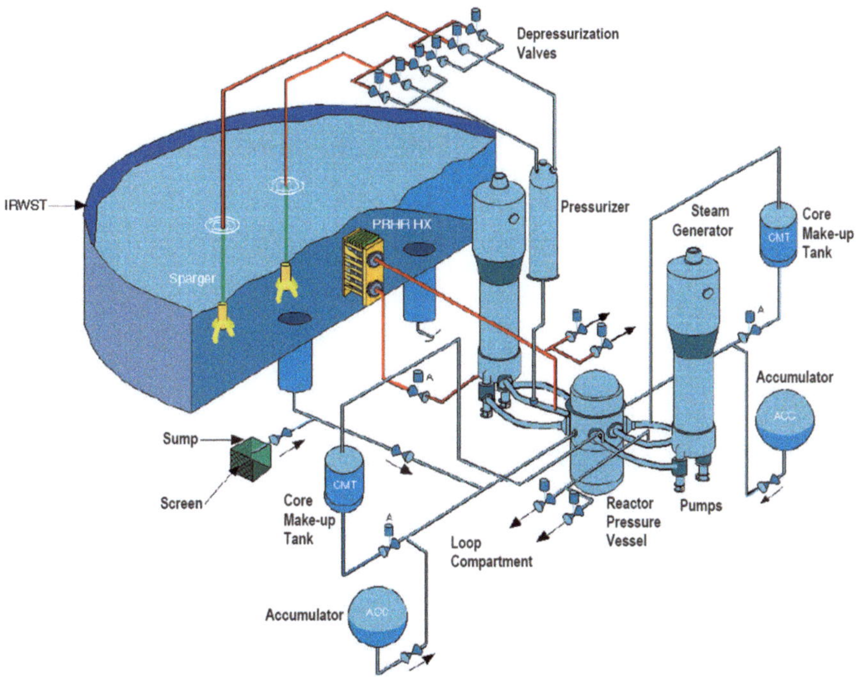

Fig. 3.13 Automatic depressurization system, safety coolant water injection and residual heat removal of AP1000 [2, 9]

3.2.5.2 Passive Residual Heat Removal

After depressurization below 0.18 MPa the IRWST provides long term cooling of the core by the opening of squib valves. For passive residual heat removal a passive residual heat removal heat exchanger (PRHR HX) is connected through inlet and outlet lines to one of the main reactor coolant loops. The heat sink for the PRHR HX is provided by the large water volume of the IRWST. In this way the PRHR HX can also counteract to temperature transients caused by imbalances on the steam generator and feedwater side.

3.2.5.3 Passive Containment Cooling

In case of boiling water in the IRWST the steam in the atmosphere of the inner steel containment condenses at the inner wall. The resulting water is collected and flowing back into the IRWST. The heat is transferred by heat conduction through the wall and removed by the air or water flowing at the outer surface. In this way all decay heat released in the molten core can be transferred to the outside atmosphere (Fig. 3.12).

In case of core melting the IRWST can be drained into the reactor cavity surrounding the reactor pressure vessel. The reactor pressure vessel is then surrounded by water. This prevents failure of the wall of the reactor pressure vessel [2, 10] (see also Chap. 10).

The design pressure of the inner containment is 0.41 MPa. The thickness of the steel wall of the inner containment is 44 mm. The thickness of the reinforced concrete wall of the outer shield building is 91.4 cm [11, 12]. The design basis for seismic events is 0.3 g ground acceleration.

The core coolant inventory in the containment for recirculation cooling and boration of the core is sufficient to last at least 30 days. Hydrogen igniters and autocatalytic recombiners shall prevent hydrogen explosions or detonations.

3.2.6 The US-APWR Containment Design

The US-APWR design with 1,700 MW(e) power output (Table 3.2) is based on the Mitsubishi (Japan) APWR design modified to meet the US-licensing requirements. It is a four loop design (Fig. 3.14). Its core has 257 fuel elements of 17 × 17 fuel rods (Fig. 3.14). The core is reloaded every 18 months [3, 5, 13] (Fig. 3.15).

The US-APWR has a single wall concrete containment and an in-containment refueling water storage tank (IRWST) with large water volume. In case of threat of core melting the steam/water mixture in the reactor pressure vessel can be released into the IRWST via drain lines after opening of depressurization valves above the pressurizer (Fig. 3.16). Water of the IRWST can be pumped into the reactor cavity. In this way the lower part of the outside pressure vessel is flooded with water to prevent core melt through to the bottom. The bottom of the reactor cavity is clad by a steel liner where the core debris shall be spread into a thin melt layer for long term cooling. In case of boiling water in the IRWST the steam would rise into the upper part of the containment. It shall be condensed there by water of a containment spray system. The containment atmosphere can additionally be cooled by an alternative containment cooling system. Hydrogen formed during a core melt accident by chemical reaction between steam and the hot zirconium cladding will be monitored in the upper part of the containment. The hot hydrogen shall be burnt by a number of igniters.

Water for the containment spray system can be fed from the IRWST by a pump through a heat exchanger (CS/RHR HX) or by a fire-water pump from an outside water tank. Similarly a pump can take water from the IRWST and feed it back to the cylindrical cavity around the reactor pressure vessel. If needed, water can be pumped to the IRWST from the outside water tank via the fire-water pump. Emergency feed water for the steam generators can be provided by emergency feed water pumps from outside water sources.

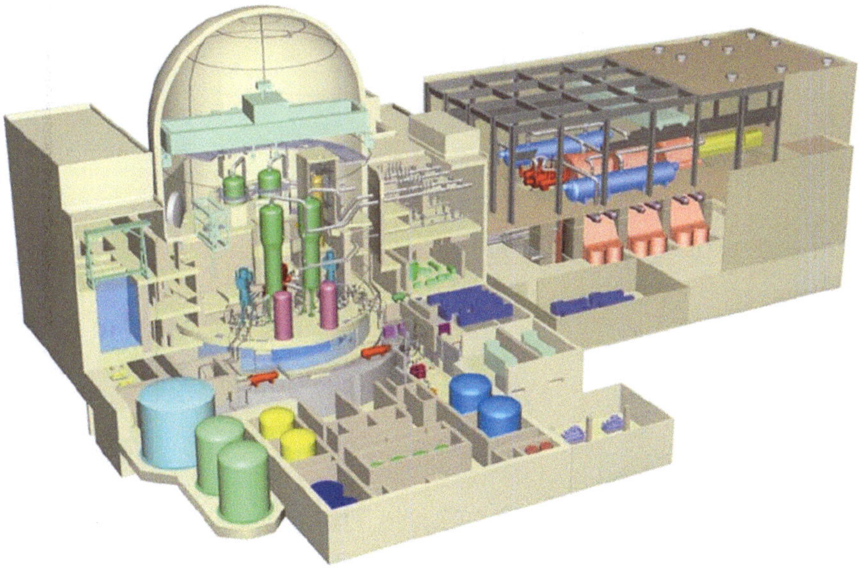

Fig. 3.14 Overview of US-APWR design [13]

3.2.7 Control Systems

At nominal power all PWRs have negative feedback coefficients of coolant temperature and power (see Chap. 2). Any reduction in core coolant temperature therefore will result in an increase in reactivity and power. If higher loads are demanded by the generator and the turbine, more heat must be extracted from the primary cooling system through the steam generators. This is done by opening the turbine governor valve which causes the primary coolant temperature in the reactor core to drop and subsequently the reactor power to rise; the power automatically balances out at a slightly higher level. However, in order to prevent the steam temperature and the steam pressure from dropping too far, the control elements are also moved up at the same time.

Reactivity changes in the core are balanced by axial movements of the control elements: slow changes of the kind brought about by fuel burnup and fission product build up are also controlled by changing the boric acid concentration of the primary coolant. Also gadolinium is mixed with the UO_2 fuel as a burnable poison if very high burnup of the fuel shall be attained.

The coolant pressure is kept at the systems pressure by the pressurizer (Fig. 3.6). The water level in the pressurizer is controlled by the volume control system. The addition of feed water to the steam generator must be matched by the feed water control system as a function of the amount of steam extracted for the turbine.

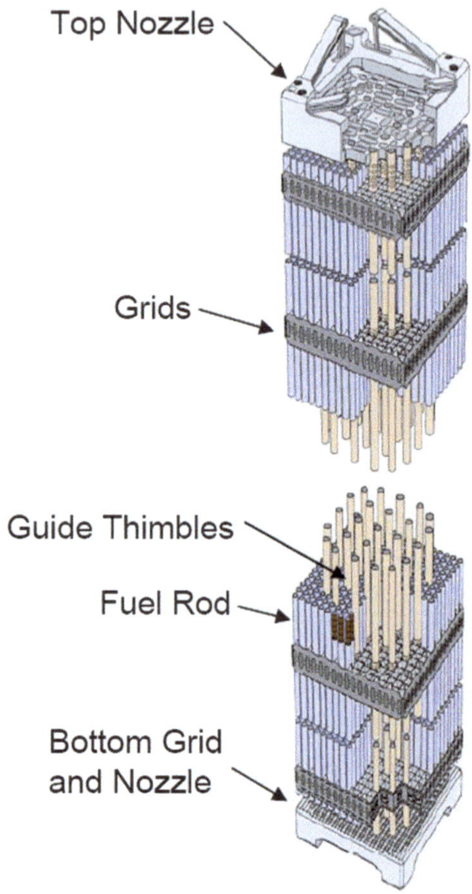

Fig. 3.15 Fuel assembly design of US-APWR [9]

3.2.8 PWR Protection System

The PWR protection system (Fig. 3.9) processes the main measured data important for plant safety, and automatically initiates, among other steps, the following actions as soon as certain set points are exceeded:

- fast shutdown of the reactor core with turbine trip (reactor scram) and separation from the electrical grid,
- emergency power supply,
- afterheat removal and emergency water injection,
- closure of reactor containment.

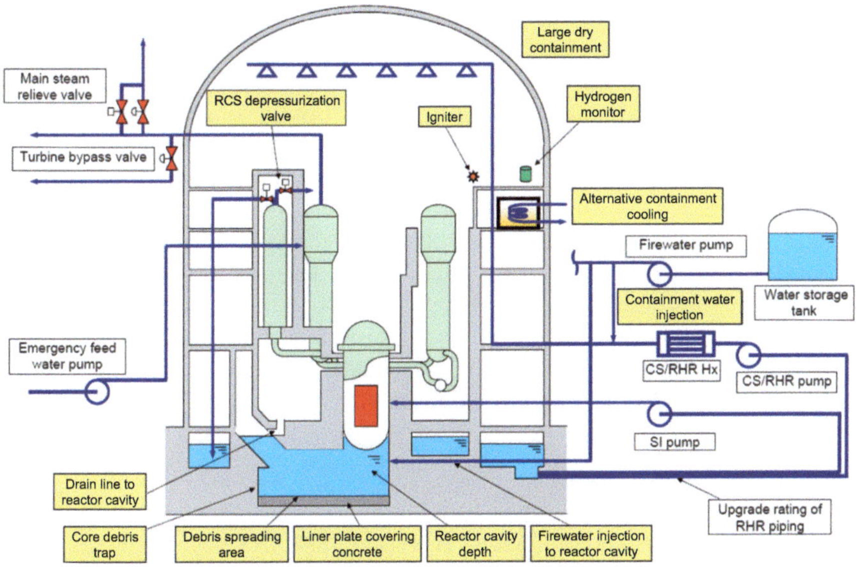

Fig. 3.16 Containment design and safety systems of US-APWR [13]

3.2.8.1 Reactor Scram

In a reactor scram, the absorber rods of the PWR are dropped into the reactor core by gravity (Fig. 3.9). The reactor goes subcritical and the reactor power drops to the level of afterheat generation (Chap. 2).

3.2.8.2 Emergency Power Supply

During normal reactor operation, the plant is connected to the public grid system. In the case of a breakdown of the public grid, the plant is disconnected from the grid and the output of the turbogenerator and the power generation of the reactor is reduced to the level of the plant requirement. The plant generates its own supply in isolated operation and is maintained operational. If also the isolated mode of operation fails, only the safety systems will be supplied by emergency power. Emergency power is supplied in a redundant layout by diesel generators, gas/combustion turbines and battery systems. If the operation room is endangered during an accident by fire, poisonous or radioactive gases or aerosols the plant can be operated from an emergency operation room (Chap. 10).

3.2.8.3 Emergency Feedwater System

The emergency feed water system (Figs. 3.9 and 3.16) supplies feed water to the steam generators, if the main feed water pumps can no longer do so. They are supplied by emergency power in case the main power supply was to fail. The emergency feed water system is equipped with fourfold redundancy (KWU-PWR, US-APWR, EPR), having borated water reserves (feed water tank and demineralized water tank) which allow the removal of afterheat to be maintained for many hours. Removal of the afterheat from the reactor core is ensured first by natural convection of the primary cooling water flowing through the core and steam generators. Natural convection is enforced by installing the steam generators at a higher level than the reactor core. After the coolant pressure in the reactor cooling system has dropped to a sufficiently low level the afterheat removal system takes over this function.

3.2.8.4 Emergency Cooling (Safety Injection) and Afterheat Removal Systems

The emergency cooling (safety injection) system of PWRs consists of the high pressure safety feed system, the flooding tank or accumulators and the low pressure safety injection system. The low pressure safety injection system is combined with the residual heat or afterheat system which serves for both operational and safety related purposes (see Figs. 3.8, 3.9, 3.13, and 3.16):

- When the power of the PWR is shut down by the protection system, the emergency cooling (safety injection) and afterheat removal system are started up automatically, after the pressure and the temperature in the primary systems have dropped to sufficiently low levels. These systems then remove the afterheat and continue to cool the reactor core and coolant circuits.
- In a loss-of-coolant accident the emergency cooling (safety injection) and afterheat removal system have to maintain the coolant level in the reactor pressure vessel and ensure cooling of the reactor core. The emergency cooling (safety injection) and afterheat removal system for AP1000 was described in Sect. 3.2.5.1. It has fourfold redundancy for the KWU-PWR, US-APWR and EPR and is supplied by emergency power. It can feed water both into the cold (inlet) and the hot (outlet) main coolant lines by means of a high pressure feed system (e.g. 15.5 MPa in case of AP1000, 11 MPa in case of KWU-PWRs, 9.2 MPa in case of EPR). In addition there is the medium pressure injection by accumulators (e.g. 4.5 MPA for AP1000 and 2.5 MPa for KWU-PWRs) and a low pressure feed system <1 MPa. Major leaks would cause the pressure in the reactor cooling system to drop so quickly that the high pressure feed system would not be started up. In that case, borated water will be injected into the primary main cooling system directly from the flooding tanks or accumulators and through the low pressure injection systems at <1 MPa from the flooding

tanks, e.g. in case of the KWU-PWRs. If the flooding tanks or accumulators have run dry, the low pressure injection system feeds water from the reactor building sump into the primary systems in case of KWU-PWR. The building sump can collect leakage water from the primary system (sump recirculation operation). In case of AP1000, US-APWR or EPR, water discharged inside the containment is collected in the in-containment refueling water storage tank (IRWST) located at the lower part of the containment. The low pressure safety injection and residual heat removal systems can take water from the IRWST (AP1000, US-APWR, EPR).

Smaller leaks cause the pressure to drop only gradually, so that initially only the high pressure feed systems will start to function. However, the pressure and temperature drop in the main coolant system is supported by a temperature drop of 100 °C/h on the secondary side.

In case of AP1000, US-APWR and EPR the primary coolant system under high pressure can also be depressurized automatically via several depressurization valve lines on top of the pressurizer into the IRWST. After the coolant pressure and temperature have dropped sufficiently below 1 MPa, the low pressure feed systems will be started up.

The afterheat is discharged into river, lake or sea water, a cooling pond or special cooling towers by way of intermediate heat exchangers (secondary parts of the emergency cooling and afterheat removal system), which are also fourfold redundant (KWU-PWRs, US-APWR, EPR) and driven by emergency power.

3.3 Boiling Water Reactors

The Boiling Water Reactors (BWRs) built today by a number of manufacturers in the USA, Europe and Japan are characterized by almost identical technical designs. This section will deal with the standard BWR-1,300 of Kraftwerk Union [14] and the SWR-1,000 (KERENA) as designed by AREVA [15]. In addition the advanced boiling water reactor (ABWR) built by General Electric (USA), TOSHIBA and HITACHI (Japan) will be presented. Figure 3.17 shows the functional design diagram of a BWR plant. In a BWR steam is directly produced in the reactor core and dried in upper steam dryer structures of the reactor pressure vessel. The steam then flows directly to the turbine generator system. The steam condenses in the condenser of the turbine system. The condenser is cooled by water from a cooling tower. Condensate water is then pumped back through a feedwater tank to the reactor core.

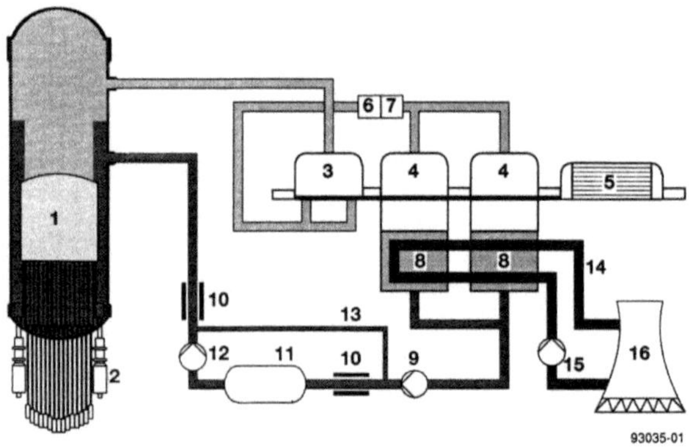

1	Reactor core	10	Pre-heater
2	Main coolant pump	11	Feedwater tank
3	High pressure part of turbine	12	Feedwater pump
4	Low pressure part of turbine	13	Interconnecting pipe
5	Generator	14	Cooling water
6	Water separator	15	Cooling-water pump
7	Superheater	16	Cooling tower
8	Condenser		
9	Condensate Pump		

Fig. 3.17 Functional diagram of a boiling water reactor [14]. 1. Reactor core, 2. Main coolant pump, 3. High pressure part of turbine, 4. Low pressure part of turbine, 5. Generator, 6. Water separator, 7. Superheater, 8. Condenser, 9. Condensate pump, 10. Pre-heater, 11. Feedwater tank, 12. Feedwater pump, 13. Interconnecting pipe, 14. Cooling water, 15. Cooling-water pump, 16. Cooling tower

3.3.1 Core, Pressure Vessel and Cooling System of a BWR

The reactor core of BWRs consists of a square array of 784 or 664 so-called ATRIUM fuel elements, about 3.74 or 3.0 m long for the BWR-1,300 of Kraftwerk Union and the SWR-1,000 of AREVA, respectively (Table 3.3). Each fuel element contains either 10×10 (ATRIUM 10) or 12×12 (ATRIUM 12) fuel rods in a closed square box of 131×131 or 155×155 mm, the so-called fuel channel (Fig. 3.18). The ATRIUM fuel elements have a 3×3 cm or 4×4 cm central water channel to achieve a relatively flat power profile across the fuel element [1, 6, 7, 14–16]. The core and fuel element data of the ABWR of TOSHIBA are given in Sect. 3.4. For moderation of the neutrons and cooling of the core, water flows through the core and is allowed to boil in the upper part of the core. Cruciform absorber plates (Fig. 3.19), containing boron carbide or hafnium as absorber material, are installed in between a set of four square fuel elements. The absorber plates are moved hydraulically into and out of the reactor core from below.

The fuel rods have claddings of Zircaloy and contain UO_2 pellets with enrichments of about 4.5 % or 5.4 % U-235. The fuel is unloaded after a maximum burnup of 55,000 or 65,000 MWd/ton. Roughly one quarter of the fuel elements are

Table 3.3 Design characteristics of the BWR power plants BWR-1,300 of Kraftwerk Union and SWR-1,000 of AREVA [15, 16]

		BWR-1,300 (Kraftwerk Union)	SWR-1,000 (KERENA) (AREVA)
Reactor power			
Thermal	MW(th)	3,840	3,370
Electrical net	MW(e)	1,284	1,254
Plant efficiency (net)	%	35	37.2
Reactor core			
Equivalent core diameter	m	4.8	5.3
Active core height	m	3.74	3.0
Specific core power	kW(th)/l	56.8	51
Density	kW(th)/kg U	28.2	24.7
Number of fuel elements		784	664
Total amount of fuel (uranium mass)	kg U	136,000	136,300
Fuel element and control element			
Fuel		UO_2	UO_2
U-235 fuel enrichment average	w/%	4	5.4
Cladding material		Zircaloy	Zircaloy (Fe combined)
Cladding outer diameter	mm	10.28	10.28
Cladding thickness	mm	0.62	0.62
Av. specific fuel rod power	W/cm	129	116
Fuel assembly array		10 × 10 ATRIUM	12 × 12 ATRIUM
Control/absorber rod type		Cruciform control element inserted from the bottom between a set of 4 fuel assemblies	Cruciform control element inserted from the bottom between a set of 4 fuel assemblies
Number of control elements		193	157
Heat transfer system			
Primary system			
Total coolant flow rate (core flow)	t/s	14.3	13.2
Coolant pressure	MPa	7.0	7.5
Feedwater temp.	°C	216	220
Core coolant inlet temp.	°C	278	278
Steam outlet temp.	°C	286	289
Steam supply system			
Steam generation	t/s	1.94	1.85
Steam pressure	MPa	7.0	7.3
Steam temperature	°C	286	289
Fuel cycle			
Average fuel burn up	MWd/ton	56,000	65,000
Refueling sequence		1/4 per 18 months	1/4 per 18 months
Average fissile fraction in spent fuel			
– U-235	%	0.8	0.8
– Pu-fiss.	%	0.7	0.7

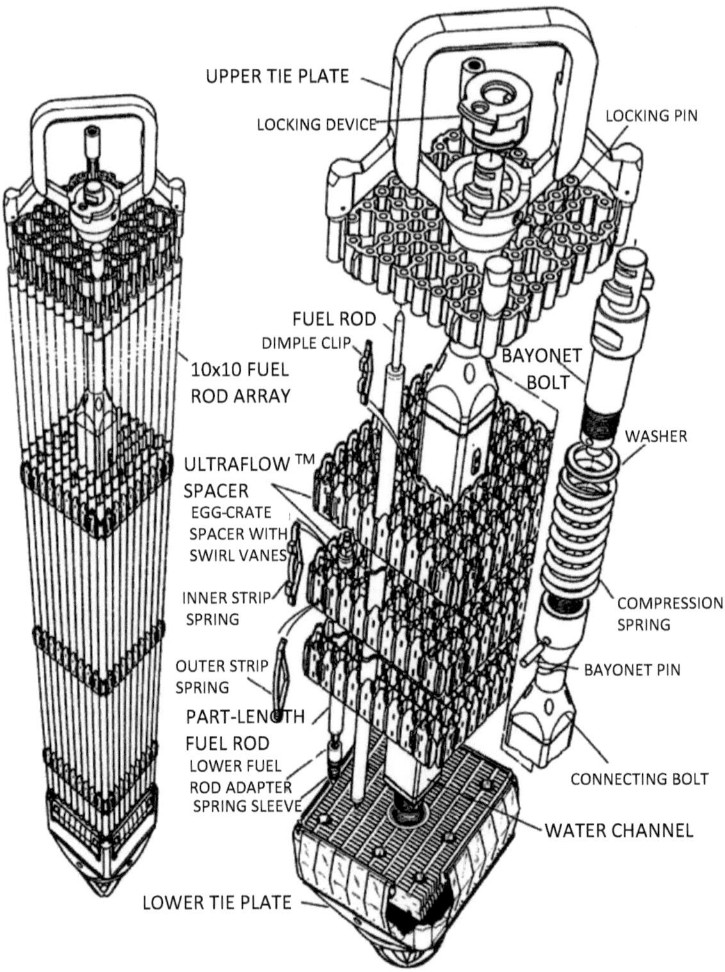

Fig. 3.18 ATRIUM-10 boiling water reactor fuel element [6]

replaced by fresh fuel elements after 18 months. Fuel elements which have not attained their maximum burnups at that time are reshuffled (moved to different radial positions of the core).

The fuel rods contain gadolinium as a burnable neutron poison to compensate for the burnup of fissile material and the build up of fission products during reactor operation. The average power density in the core is 57 or 51 kW(th)/l and 28.2 or 24.7 kW(th)/kg uranium for the BWR-1,300 of Kraftwerk Union and SWR-1,000 (KERENA) of AREVA, respectively. The water inlet temperature in the lower part of the core is 216 °C or 220 °C, the steam outlet temperature is 286 °C or 289 °C, which corresponds to a saturation steam pressure of roughly 7.0 or 7.5 MPa (Table 3.3).

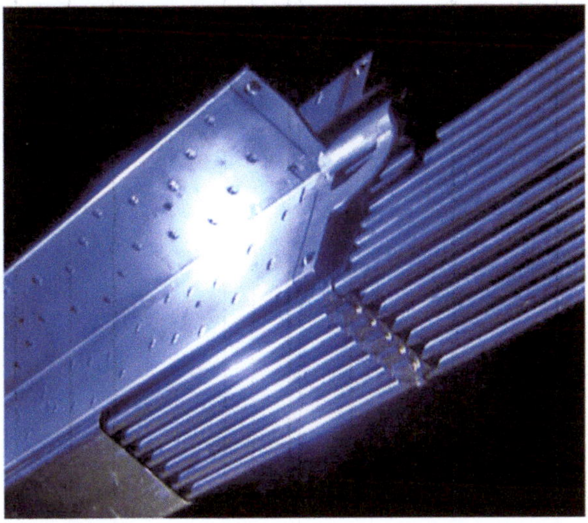

Fig. 3.19 Cruciform absorber element of a boiling water reactor core with a 7 × 7 fuel element [14]

Steam is generated by water boiling in the upper part of the reactor core. To provide sufficient core flow for ample heat transfer, BWRs employ internal recirculation pumps which increase the core flow rate. The core with the absorber plates is contained in a large steel pressure vessel (Fig. 3.20). Above the core, there are the steam separators and steam dryers. The reactor vessel head can be removed for loading and unloading of fuel elements. A BWR pressure vessel has a diameter of 6.6 or 7.1 m, a wall thickness of roughly 160 mm, and a height of 23 m. The Kraftwerk Union BWR-1,300 and the AREVA SWR-1,000 (KERENA) pressure vessels are made of, e.g. 22NiMoCr37 steel, with the inside being plated with austenitic stainless steel.

The saturated steam flows from the reactor pressure vessel directly to the turbo-generator system and is pumped back from the condenser and feedwater tank to the pressure vessel. The condenser is cooled by water from a cooling tower or from a river (see Fig. 3.17).

A question of particular interest in BWRs with direct cooling systems is the radioactivity of the cooling water. The amount of radioactivity is determined by the impurities contained in the water and by an (n,p)-reaction of oxygen producing nitrogen, N-16, with a half life of 7.2 s. Many years of experience in operating boiling water reactors have shown that, because of the short half-life of the N-16, maintenance work on the turbine, the condenser and the feed water pumps is not impaired significantly by radioactivity.

The water circulation in the reactor pressure vessel and through the core can be used for changing the reactor power. Reduction of water flow through the core will result in a higher evaporation rate and in a larger volume of bubble formation. Increasing the volume of steam in the core reduces the moderation of the neutrons, and as a consequence, the criticality or effective multiplication factor k_{eff} (see Chap. 2) and the reactor power will be reduced. In this way, changes in the water

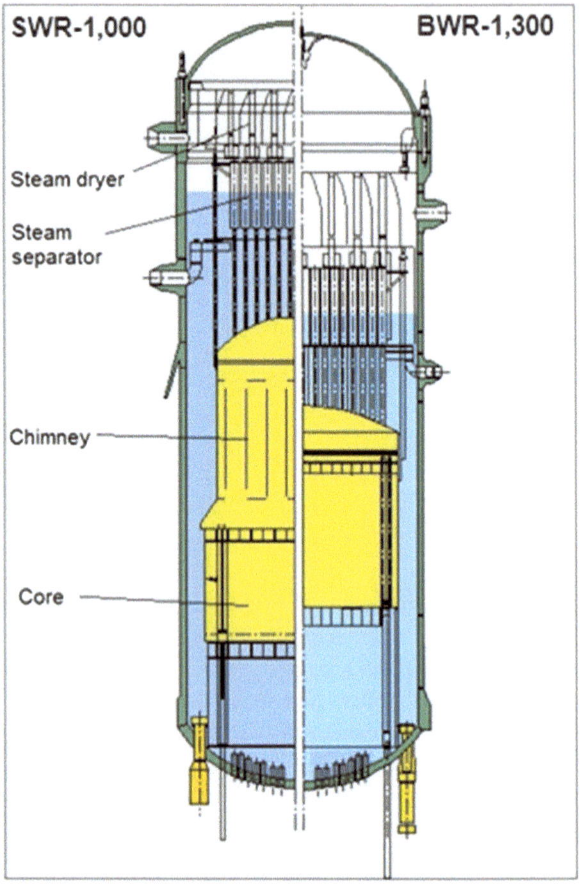

Fig. 3.20 Reactor pressure vessel of a SWR-1,000 (KERENA) or a BWR-1,300 [14–16]

flow can be used to control the reactor power in a broad range without movement of control rods. BWRs can automatically follow the load requirements of the turbine, by sensing pressure disturbances at the turbine, transmitting these signals to the recirculation flow control valve and regulating core flow and therefore reactor power.

In order to ensure high quality of the reactor feed water, all the feed water recirculated from the turbine condenser is pumped through filters (demineralizer units) and cleared of any corrosion products and other impurities.

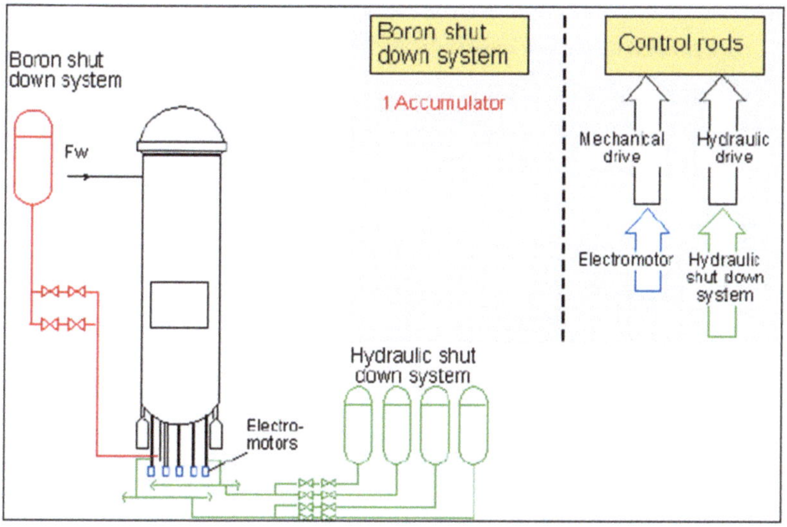

Fig. 3.21 Shutdown systems of BWR-1,300 and SWR-1,000 (KERENA) [14–16]

3.3.2 Boiling Water Reactor Safety Systems

If there are perturbations in the criticality or effective multiplication factor, k_{eff}, or losses of coolant flow, the reactor is shut down in a short time by rapid insertion of the absorber plates. As a backup shutdown system, the coolant (moderator) can be poisoned by a neutron absorbing boric acid solution and, in this way, also quench the nuclear reaction and shut down the reactor (Fig. 3.21).

Ruptures in any pipe in the primary cooling system (e.g., the recirculation system) cause the main steam pipes and the feed water pipes to be blocked by two series-connected isolation valves and by two series-connected non-return valves, respectively. This action isolates the reactor pressure vessel within the containment from the outer turbine and condenser cycles. When these isolation valves are closed, or if there is overpressure in the reactor pressure vessel, BWR safety/relief valves are automatically actuated, allowing a path for steam to be discharged from the reactor vessel. In this case, the steam is discharged into a large water pool inside the pressure suppression containment.

In case of the BWR-1,300 of Kraftwerk Union steam is discharged into a large pressure suppression pool (Fig. 3.22). In case of SWR-1,000 (KERENA) of AREVA more passive safety systems are installed as compared to BWR-1,300 and the steam is discharged into four core flooding pools. In the following part both designs are described separately.

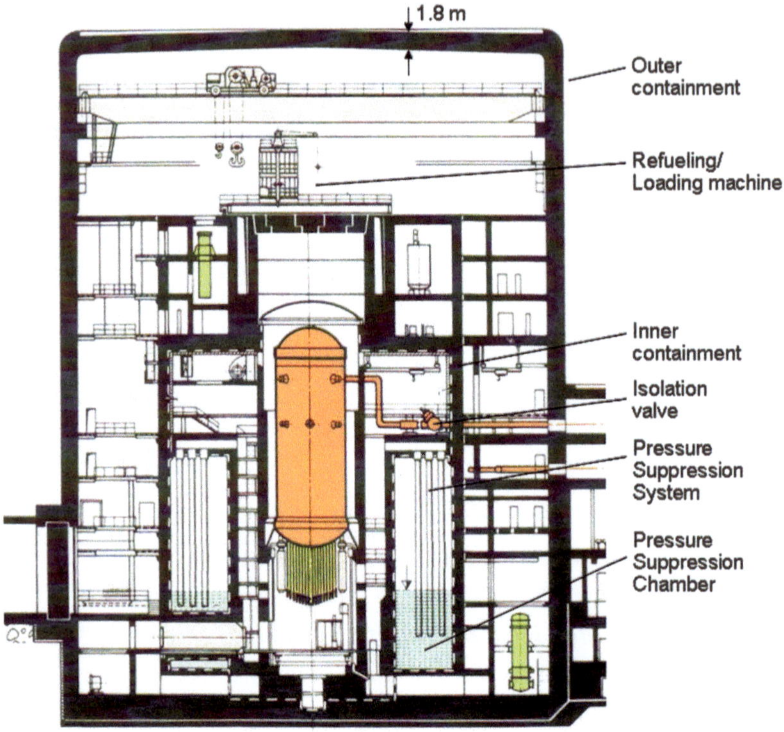

Fig. 3.22 Schematic view of the inner and outer containment of BWR-1,300 (Kraftwerk Union) [14]

3.3.2.1 Pressure Suppression System of Kraftwerk Union BWR-1,300

Figure 3.22 shows the pressure suppression containment system of the Kraftwerk Union BWR-1,300. The reactor pressure vessel, the recirculation system and the pressure relief valves of the main steam pipes are accommodated in the inner containment, also called drywell, which isolates them from the rest of the reactor containment. The drywell consists of a reinforced concrete structure with horizontal vent openings which allow communication between the pool and the drywell. The drywell is kept dry.

If primary coolant is released through a leak in the primary system pipes, steam will enter into the drywell and will be channeled into the pressure suppression pool, where it will condense. An inner containment surrounds the drywell pressure suppression system and all reactor equipment. It is designed to withstand temperatures and pressures that could be caused by a loss-of-coolant accident and to retain non-gaseous fission products which could potentially be released from the reactor system during an accident.

The inner containment is filled with nitrogen. Hydrogen recombiners are located at certain positions to burn hydrogen in case of a cladding-steam interaction or a core melt accident.

3.3.2.2 Emergency Core Cooling and Afterheat Removal Systems

If the water level in the reactor pressure vessel drops, or if there is a leak (loss-of-coolant accident), water is automatically added to the reactor pressure vessel by the following systems [17]:

- a three train high pressure core water injection system
- a three train low pressure coolant injection system and
- a three train residual core heat removal system.

The pressure pump of the high pressure water injection system initially takes water from tanks that store condensate water or, if necessary, takes water from the pressure suppression pool. The water is then directly pumped into the pressure vessel. Its function is to supply large quantities of water to the core in case of a loss of coolant accident while the reactor is still in a high pressure condition. It prevents fuel cladding damage in the event the core becomes partially or fully uncovered.

In case it becomes necessary to use the low pressure system, the reactor pressure vessel can be depressurized. This is accomplished by opening the safety relief valves and by discharging steam to the pressure suppression pool for condensation. After this has happened, the low pressure water injection system can be used by taking water from the suppression pool and feeding water directly into the inside of the reactor vessel.

The low pressure coolant injection system is actually one mode of the residual heat removal system. Two heat exchanger loops can be used to cool the suppression pool or the heat exchangers can be bypassed and water can be injected at low pressure directly into the pressure vessel to cool the core.

An additional function of the residual heat removal system is to remove and condense any steam in the drywell. Water is taken from the pressure suppression pool and sprayed into the inner containment free volume. Thus over-pressurization of the containment can be prevented (Fig. 3.23).

3.3.2.3 SWR-1,000 (KERENA) Containment of AREVA and Passive Cooling Systems

The reactor pressure vessel is housed in an inner prestressed concrete containment with a steel liner (Fig. 3.24). This inner containment is subdivided into a pressure suppression chamber, the drywell and four hydraulically connected large core flooding pools. The core flooding pools serve as a heat sink for passive heat removal from the reactor pressure vessel by emergency condensers and the safety relief valves. The drywell houses the reactor pressure vessel with the control rod drive

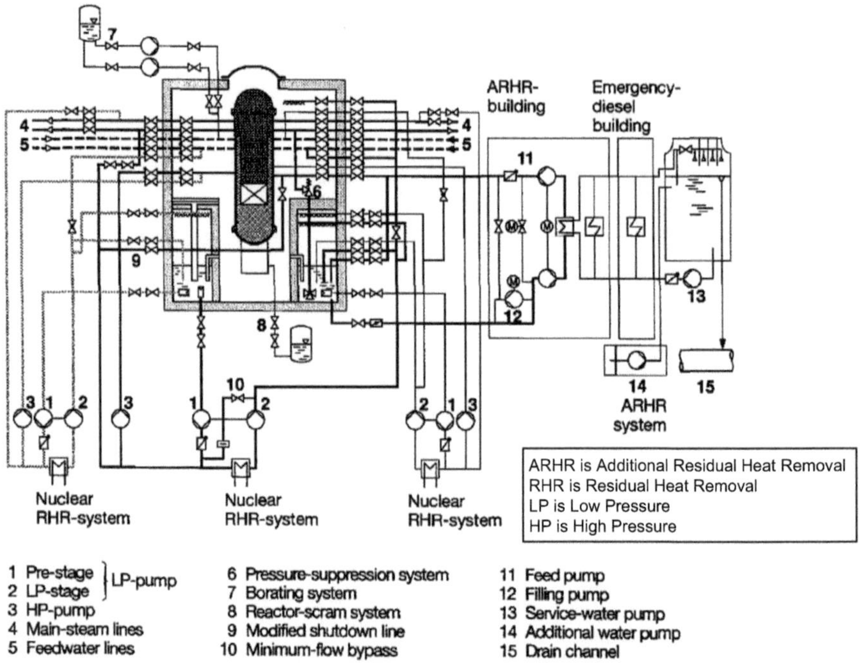

Fig. 3.23 BWR-1,300 residual-heat-removal systems of Kraftwerk Union [17]

and the control rod shutdown system, the three steam lines and two feed water lines, as well as the safety relief valves, the containment isolation valves and the pressure pulse transmitters. In addition the high pressure part of the water cleanup system and the flooding lines for passive flooding of the reactor pressure vessel and flooding of the drywell in case of a drastic loss of coolant are located in the drywell. During operation the inner containment is inertized by nitrogen, to prevent hydrogen explosions or detonations in case of a loss of coolant accident with hydrogen release. Above the inner containment the water filled shielding pool and the adjacent fuel element storage pool are located. In case of refueling the reactor pressure vessel cover is opened and the fuel elements are loaded or unloaded through the shielding water pool and transferred to the fuel element storage pool.

The main steam lines and the feed water lines are equipped with two containment isolation valves, one located inside and one outside of the inner containment wall. The residual heat removal systems are located underneath the pressure suppression chamber in non-inertized compartments which are accessible for inspection and repair from the outside [15, 16].

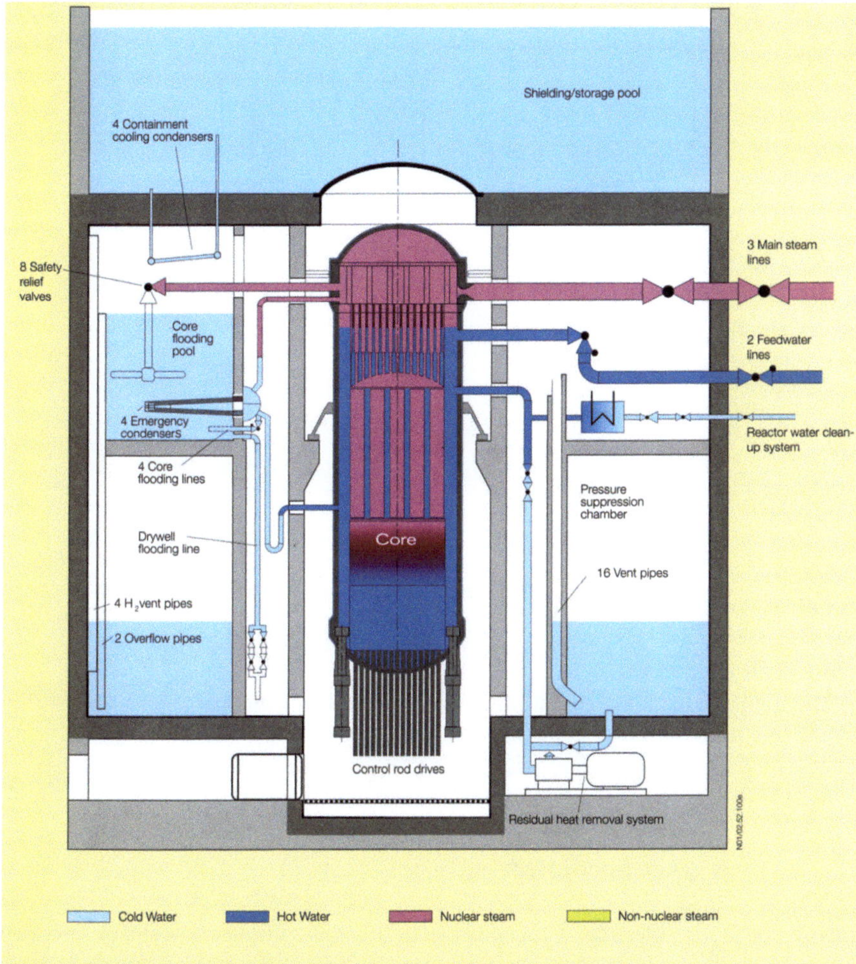

Fig. 3.24 Inner containment with safety systems of SWR-1,000 (KERENA) [15, 16]

3.3.2.4 Safety Relief Valve System

The safety relief valve system (Fig. 3.25) consists of the safety relief valves, the relief lines and steam quenchers which are located in the core flooding pools.

They act in case of:

- overpressure to protect the pressure boundaries (pressure relief)
- the water level in the reactor pressure vessel falling below specified limits (loss of coolant accident, automatic depressurization)
- turbine trip (short term removal of excess steam)
- severe accident mitigation (depressurization).

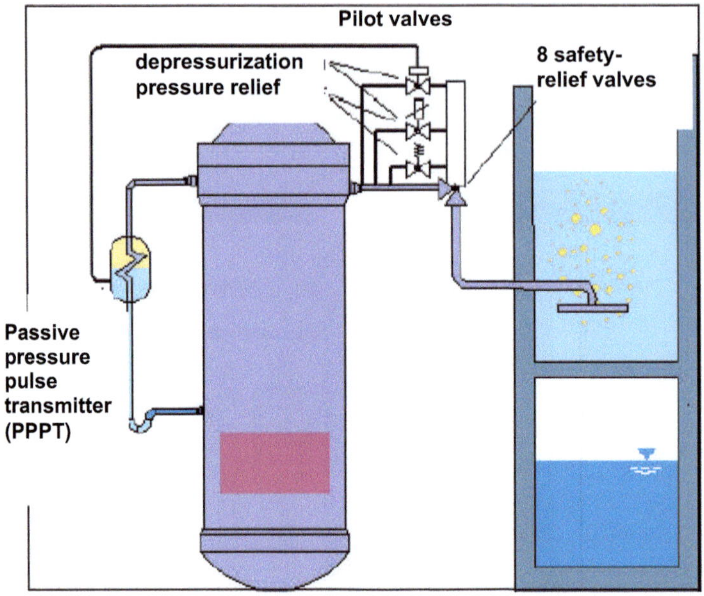

Fig. 3.25 Safety relief valve system of SWR-1,000 (KERENA) [15, 16]

The eight main safety relief valves are actuated by solenoid pilot valves or by diaphragm pilot valves via passive pressure pulse transmitters. Standard spring loaded valves are also used to initiate the pressure relief function.

3.3.2.5 Emergency Condensers

Each of the four emergency condensers (Fig. 3.26) consists of a steam line (top connection), the heat exchanger tubes and a condensate return line (lower connection). Each return line is equipped with an anti-circulation loop and a passive outflow reducer. The emergency condensers are actuated passively when the water level in the reactor pressure vessel drops to a certain level. In this case the emergency condenser tubes fill with steam which condensates. The condensate returns back to the reactor pressure vessel.

3.3.2.6 Containment Cooling Condensers

The containment cooling condensers (Fig. 3.27) remove residual heat passively from the containment atmosphere to the shielding storage pool located above the inner containment. Each containment condenser consists of tubes arranged at a slight angle to horizontal. The tubes penetrate the upper wall of the inner containment. They are open to the shielding/storage pool. The containment cooling

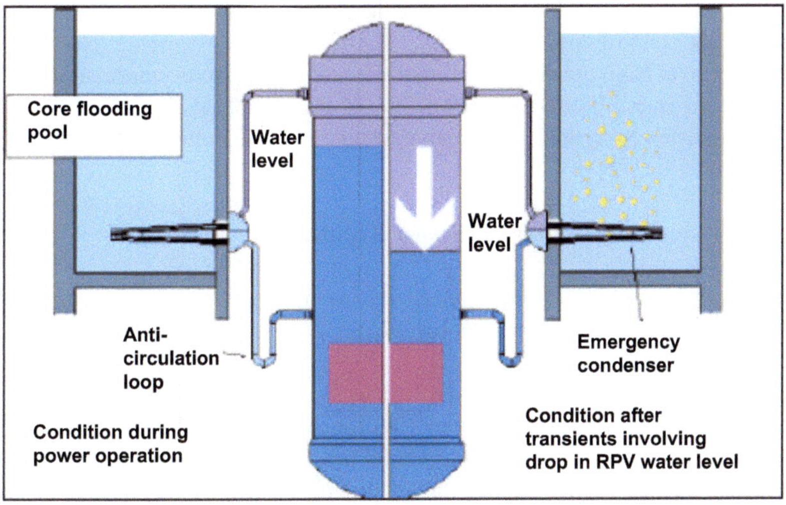

Fig. 3.26 Emergency condenser of SWR-1,000 (KERENA) [15, 16]

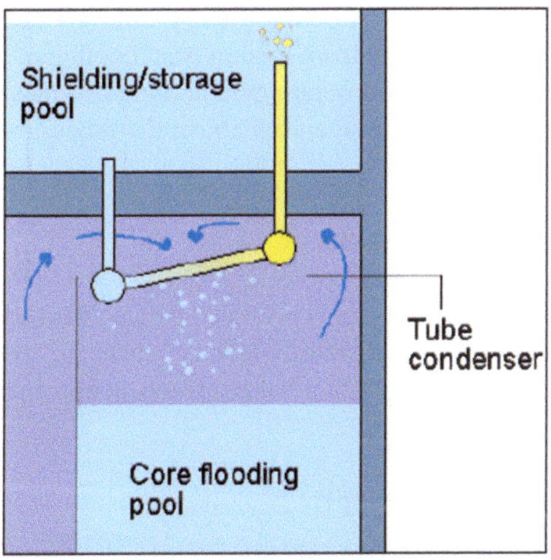

Fig. 3.27 Containment cooling condenser of SWR-1,000 (KERENA) [16, 18]

condensers are passive cooling systems. In case of high temperatures of the upper containment atmosphere the water in the tubes warms up or evaporates and the heat is transferred to the shielding/storage pool.

3.3.2.7 Residual Heat Removal and Core Flooding Systems

Two active low pressure core flooding and residual heat removal systems perform the following tasks:

- reactor core cooling during operational shutdown as well as under accident conditions (loss of coolant)
- heat removal from the core flooding pools and the pressure suppression water in case of pressure relief conditions or loss of the main heat sink
- water transfer operations prior and after core refueling operations.

3.3.2.8 Systems for Control of Severe Accidents

Because of the above described passive cooling systems, high pressure water injection systems as in other BWRs are no more needed for the SWR-1,000 (KERENA).

Severe core melt accidents could occur if all active and passive injection functions would fail. This is extremely improbable. In this case a core melt under high pressure in the reactor pressure vessel can be ruled out by the design of the depressurization system. If the reactor core melts at low pressure it could be kept in the lower part of the pressure vessel and be cooled by water from the outside by thermal conduction through the remaining steel structures in the lower part of the reactor pressure vessel (Fig. 3.28). A special flooding system can flood water from the core flooding pools to the lower part of the drywell. A water pool surrounds the lower part of the pressure vessel. Steam arising from cooling of the reactor pressure vessel would be condensed at the containment cooling condensers. The heat would be transferred to the shielding/storage pool above the inner containment.

3.3.2.9 Auxiliary Cooling Systems

The water pool for spent fuel elements is cooled by four heat exchangers operating in the natural convection mode. Reactor water and spent fuel pool cleanup systems with filters operate to keep the cooling water always at specified conditions.

3.3.2.10 Emergency Power Supply

An external grid supplies emergency power for all electrical loads which have to remain available even in the event of a loss of the normal auxiliary power supply grid. Emergency diesel generators or gasturbine-generators can take over in the emergency case and provide an independent power supply.

3.3.2.11 Containment System

A reinforced concrete containment (Figs. 3.22 and 3.24) encloses and protects the inner containment against such external events as they were described for the PWR in Sect. 3.2.4. The annulus between the outer concrete shield building and the

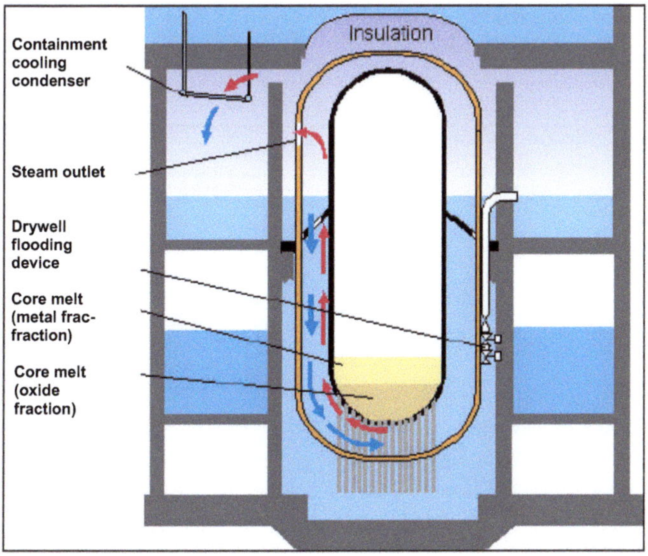

Fig. 3.28 Cooling of RPV exterior in the event of core melt of SWR-1,000 [16]

containment is maintained at a pressure lower than atmospheric so that any radioactive gases leaking into this annulus can be filtered prior to the release to the environment.

In case of the SWR-1,000 (KERENA), radioactivity releases to the outer environment would be extremely low. Even in case of severe core melt down accidents, evacuation or relocation of the population would not become necessary.

3.4 The Advanced Boiling Water Reactors

The Advanced Boiling Water Reactor (ABWR) was designed and is manufactured by General Electric (USA) as well as by TOSHIBA and HITACHI (Japan) [18–21]. ABWR plants are operating in the USA, Japan and Europe. The ABWR has many similar design features compared to the BWR-1,300 described in Sect. 3.3. The main design data of the TOSHIBA ABWR and an upgraded ABWR-II design are given in Table 3.4.

3.4.1 Core and Reactor Pressure Vessel of ABWR

The core with fuel elements and control elements as well as the pressure vessel of the ABWR are displayed in Fig. 3.29.

Table 3.4 Design characteristics of the ABWR (General Electric, HITACHI, TOSHIBA) and ABWR-II (General Electric, HITACHI, TOSHIBA) [21, 22]

		ABWR (General Electric, HITACHI, TOSHIBA)	ABWR-II (General Electric, HITACHI, TOSHIBA)
Reactor power			
Thermal	MW(th)	3,926	4,960
Electrical net	MW(e)	1,300	1,718
Plant efficiency (net)	%	33.1	33
Number of main steam lines		3	4
Reactor core			
Equivalent core diameter	m	5.16	5.41
Active core height	m	3.81	3.71
Specific core power	kW(th)/l	49.2	58.1
Density	kW(th)/kg U	25	26.1
Number of fuel elements		872	424
Total amount of fuel (uranium mass)	t U	157	190
Fuel element and control element			
Fuel		UO_2	UO_2
U-235 fuel enrichment average	w/%	4	5.2
Cladding material		Zircaloy (annealed)	Zircaloy (annealed)
Cladding outer diameter	mm	10.3	10.3
Cladding thickness	mm	0.66	0.66
Av. specific fuel rod power	W/cm	133	116
Fuel assembly array		10 × 10 (square lattice)	8 × 8 × 4[a] (square lattice)
Number of control rods		205	197
Absorber material		B_4C, Hafnium	B_4C, Hafnium
Soluble poison		Boron	Boron
Number of control elements		193	157
Heat transfer system			
Primary system			
Total coolant flow rate (core flow)	t/s	14.5	15.7
Coolant pressure	MPa	7.07	7.17
Feedwater temp.	°C	216	216
Core coolant inlet temp.	°C	278	277
Steam outlet temp.	°C	288	288
Steam supply system			
Steam generation	t/s	2.12	2.68

(continued)

Table 3.4 (continued)

		ABWR (General Electric, HITACHI, TOSHIBA)	ABWR-II (General Electric, HITACHI, TOSHIBA)
Steam pressure	MPa	7.0	7.0
Steam temperature	°C	288	288
Fuel cycle			
Average fuel burn up	MWd/ton	>50,000	>60,000
Refueling sequence		24 Months	18 Months
Average fissile fraction in spent fuel			
– U-235	%	0.8	0.8
– Pu-fiss.	%	0.7	0.7

[a]See Fig. 3.33

Fig. 3.29 Core and primary system of the ABWR [18]

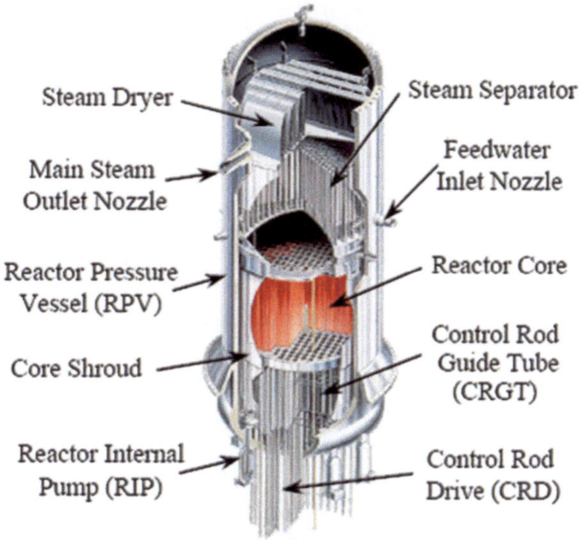

The steam is generated by water boiling in the upper part of the reactor core. The core flow is driven by ten internal recirculation pumps. The feedwater coming from the feedwater tank mixes with the water coming from steam separators and dryers and flows downward in an annular ring between the reactor pressure vessel and the core shroud. Then it changes its flow direction and flows upwards through the core. Steam is separated from water droplets in separators and dried in the dryers. It leaves the reactor pressure vessel via three or four outlet steam pipes. The control rods are vertically inserted and withdrawn from below the core. The control rod guide tubes are mounted on the reactor pressure vessel bottom.

The saturated steam flows from the reactor pressure vessel directly to the turbine and is pumped back from the condenser and feedwater tank to the pressure vessel. The condenser is cooled by water from a cooling tower or from a river (Fig. 3.17).

3.4.2 The ABWR Safety and Depressurization Systems

The reactor shutdown systems and the backup shutdown systems are similar to those described in Sect. 3.3.2 and Fig. 3.21. Instead of boric acid also sodium pentaborate is used [18].

The depressurization chamber (wet well), the dry well part of the inner containment and the outer secondary containment of the ABWR are shown in Fig. 3.30. The inner containment is a prestressed concrete building clad by steel plates and filled with nitrogen.

Ruptures of any pipe in the primary coolant system cause the main steam pipes and the main feedwater pipes to be blocked by two series-connected main steam isolation valves (MSIV) and by series-connected non-return valves. In this way the reactor pressure vessel within the inner containment is isolated from the turbine-condenser system. Under accidental conditions with overpressure in the reactor pressure vessel, the safety relief valves (SRV) are actuated automatically (Automatic depressurization system) (Fig. 3.31). Steam is discharged into the large water pool in the pressure suppression chamber (wet well) (Figs. 3.30 and 3.31). If in case of a core melt also fission products and hydrogen are streaming into the pressure suppression pool, the gaseous fission products and hydrogen will collect in the upper dry well part of the inner containment. The inner containment is inertized by nitrogen. This avoids hydrogen detonations. In addition hydrogen recombiners will decrease the content of hydrogen there.

3.4.3 Emergency Cooling and Afterheat Removal System of the ABWR

In case the water level in the reactor pressure vessel drops during a loss of coolant accident, water is automatically fed into the reactor pressure vessel by the following systems:

- two automatic high pressure core water flooding systems. Their pumps are motor-driven. The water is taken from the large suppression pool or from the condensate storage tank (Fig. 3.31).
- one automatic reactor core isolation cooling system. The pump of this system is driven by a steam turbine, the steam is coming from the reactor pressure vessel. The flooding water is taken from the suppression pool or from the condensate storage tank.

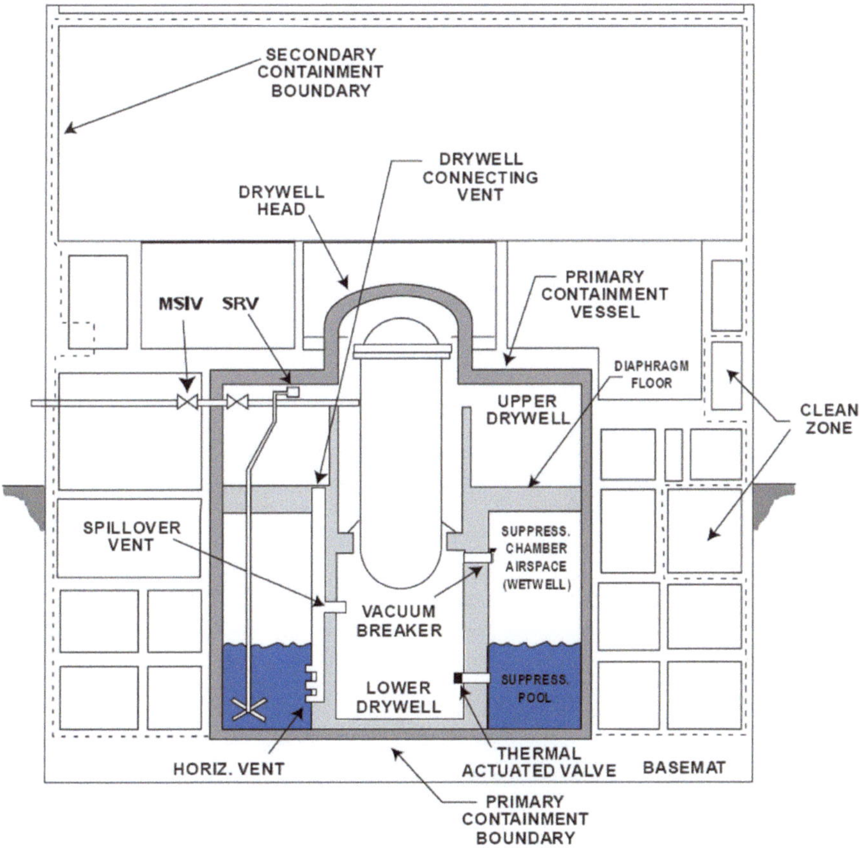

Fig. 3.30 ABWR—Containment structure [22]

– three automatic low pressure core flooding systems. These pumps are motor-driven. The water is taken from the large suppression pool. The three low pressure core flooding systems are interconnected by valves with the three residual heat removal systems. In this case the water is cooled in heat exchangers. The residual reactor heat removal systems are applied for cooling of the core after reactor shutdown (residual heat conditions).
– Suppression pool cooling
  The three residual heat removal systems are also used to cool the larger suppression pool in case its water overheats as a consequence of core melt accidents.
– Containment spray cooling system
  Two of the three residual heat removal systems can also supply the water for spray systems in the wet well and in the dry well part of the inner containment. This spray water can condense steam in the wet well or in the dry well part of the inner containment.

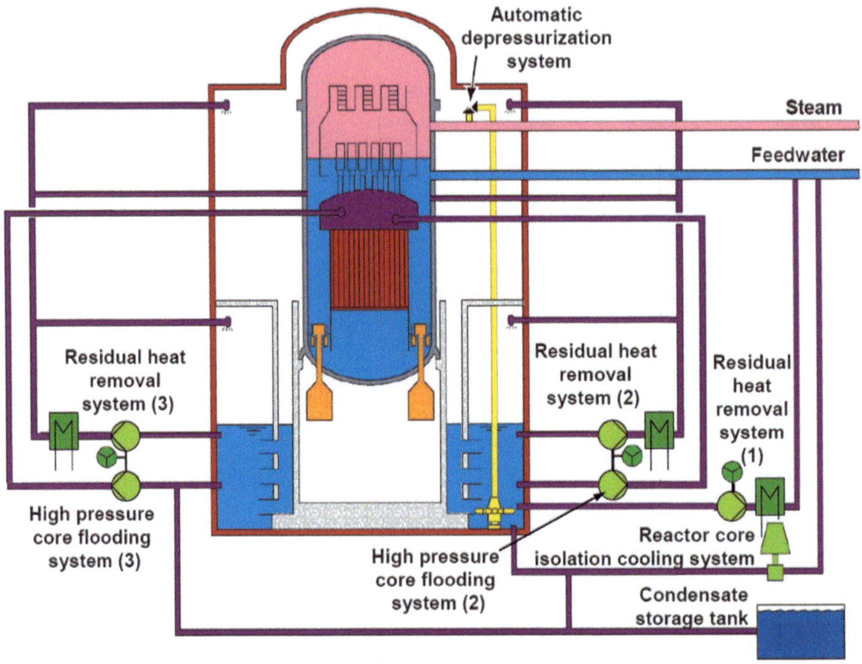

Fig. 3.31 ABWR Emergency core cooling system [19, 22]

3.4.4 Emergency Power Supply of ABWR

In case of loss of power supply for the reactor plant the emergency power generators will start up automatically. The ABWR is equipped with three safety class diesel-generators of 7 MW(e) each. In addition there is one combustion turbine generator installed with a power of 20 MW(e) as an alternate diverse emergency power source.

The reactor core isolation cooling system is driven by the steam turbine with high pressure steam from the reactor pressure vessel. Batteries supply electricity for the plant instrumentation and for other parts of the electrical emergency cooling equipment.

3.4.5 The ABWR-II Design

The ABWR-II with about 1,700 MW electrical power is an advanced BWR design of General Electric, HITACHI and TOSHIBA which evolved from the ABWR. The objective is to further improve the ABWR safety design (Fig. 3.32).

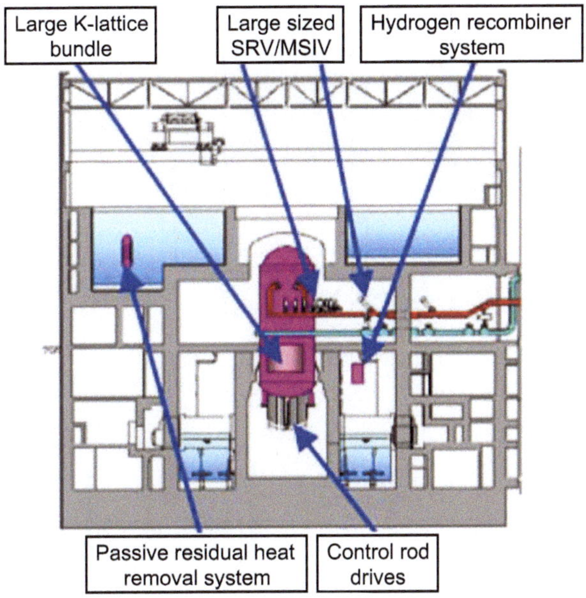

Fig. 3.32 Main features of ABWR-II Plant-System [22]

The ABWR-II core has a larger fuel element, a somewhat modified emergency core cooling system (ECCS) and additional passive heat removal systems for accidental (beyond design basis) conditions.

3.4.5.1 ABWR-II Core Design and Coolant System

The larger ABWR-II fuel element box consists of four subbundles (Fig. 3.33). Each of the four subbundles has an 8×8 fuel rod arrangement. A so-called K-lattice control rod concept (Fig. 3.33) allows the number of control rods in the core to be increased. As a consequence the ABWR-II core has only 424 fuel element boxes and 197 control elements. The dimensions of the fuel rods and the cladding materials are equal to those of the ABWR.

The control rod drive system consists of the fine motion control rod drive mechanisms, the hydraulic control unit and the control rod drive hydraulic subsystems. Compared to the ABWR design an improvement was made by introducing a magnetic coupling which transmits torque between the control rod drive motor and the control rod drive shaft through the pressure boundary at the bottom of the reactor pressure vessel.

The ABWR-II has four main steam lines and 14 safety/relief valves in the main steam lines. The discharge capacity during depressurization was increased by 70 % compared to the ABWR design.

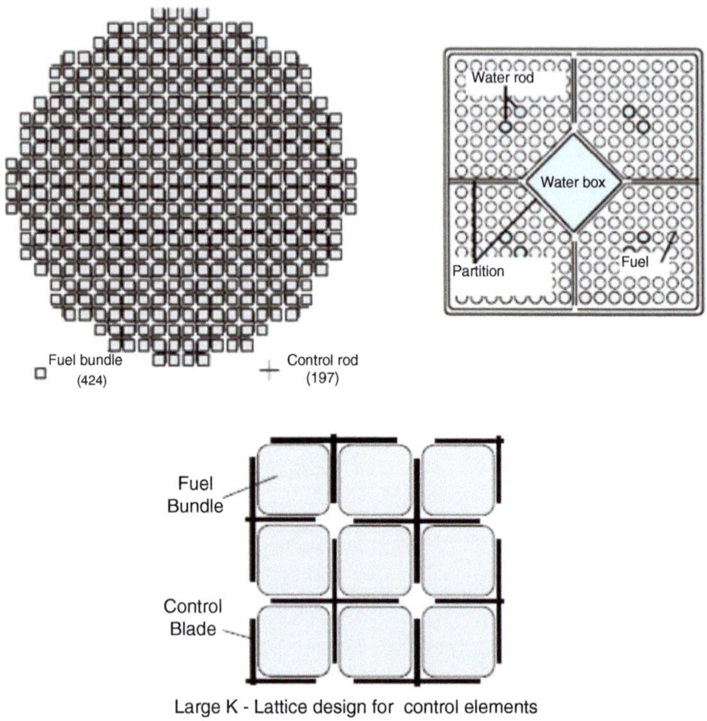

Fig. 3.33 Core configuration, fuel element box and K-lattice design for control element arrangement of ABWR-II [22]

3.4.5.2 Active and Passive Safety Systems

Active Emergency Core Cooling Systems

An advanced reactor core isolation cooling system with steam turbine equipped with a generator has the capacity of long term operation and power supply beyond battery capacity during a station black out situation. The high pressure coolant injection system is, therefore, arranged in two trains. The low pressure coolant injection system and residual heat removal system are operating in four trains. The emergency power is supplied by two diesel generators and gas turbines.

ABWR-II Passive Heat Removal Systems

In the ABWR-II design the wet well part and the dry well part of the prestressed inner containment are separated from each other. The passive heat removal system of ABWR-II consists of

- a passive reactor core cooling system
- a passive containment cooling system together with a common heat sink pool above the inner containment.

These passive heat removal systems provide a backup for residual heat removal (afterheat) and for heat removal during core melt down situations (Fig. 3.32) [23, 24].

The large water pool above the inner containment houses an intermediate heat exchanger (ultimate condenser) with pipe connections down to the wet well part of the inner containment, into the steam suppression pool and into the cavity below the pressure vessel.

In a severe accident situation, when the water in the pressure suppression chamber (wet well) would start to boil, the steam would flow to the heat exchanger (ultimate condenser) in the upper water pool and condense there. The condensate would then flow back to the pressure suppression chamber.

In case of reactor pressure vessel bottom melt through by the molten core, the debris would fall onto the steel plated cavity underneath the reactor pressure vessel. Openings with fusible valves in the cavity wall would open, water of the pressure suppression pool would cover the molten core debris and the steam produced would flow to the heat exchanger (ultimate condenser), condense there and the water would flow back in the pressure suppression chamber. This water can also be used for water spray systems in the inner containment to condense steam.

Hydrogen recombiners will lower the hydrogen concentration after a core melt accident. This will also lower the pressure in the inner containment. Thus, the need for containment venting in case of overpressure can be avoided [23, 24].

References

1. Kessler G (2012) Sustainable and safe nuclear fission energy. Springer, Heidelberg
2. Cummings WE et al (2013) Westinghouse AP1000 advanced passive plant. In: Proceedings of ICAPP 2003, Paper 3235, Cordoba, Spain
3. US-APWR Design. http://www.mnes-us.com/htm/usapwrdesign.htm
4. EPR – European pressurized water reactor, the 1600 MWe reactor. http://www.areva-np.com/common/liblocal/docs/Brochure/EPR_US_%20May%202005.pdf
5. US-APWR. http://pbadupws.nrc.gov/docs/ML1109/ML110980212.pdf
6. Strasser A et al (2005) Fuel fabrication process handbook. Advanced Nuclear Technology International, Surahammar
7. Güldner R et al (1999) Contribution of advanced fuel technologies to improved nuclear power plant operation. In: The Uranium Institute 24th annual symposium, London
8. Czech J et al (1999) European pressurized water reactor: safety objectives and principles. Nucl Eng Design 187:25–32
9. Foulke L (2011) Safety systems for pressurized water reactors. In: IAEA workshop 2011. http://www.iaea.org/NuclearPower/Downloads/Technology/meetings/2011-Oct-Simulators-WS/Foulke.3-Passive.Safety-1.pdf

10. The Westinghouse AP1000 Advanced Nuclear Power Plant, Plant description. http://www.ap1000.westinghousenuclear.com/ and http://www.westinghousenuclear.com/docs/AP1000_brochure.pdf
11. Nuclear Monitor (2009) http://www.nirs.org/mononline/nm697.pdf
12. Westinghouse Electric Company (2011) New nuclear reactors: generic design assessment. Westinghouse Electric Company LLC AP1000® Nuclear Reactor. http://www.hse.gov.uk/newreactors/reports/step-four/technical-assessment/ap1000-onr-gda-sr-11-002-rev-0.pdf
13. Mitsubishi US-APWR. http://www.iaea.org/NuclearPower/Downloadable/Meetings/2011/2011-07-04-07-08-WS-NPTD/3_JAPAN_APWR_Mitsubishi_Nojiri.pdf
14. (2000) Kernkraftwerk Gundremmingen 1280 MWe BWR. www.kkw-gundremmingen.de
15. (2007) SWR-1,000, an advanced boiling water reactor with passive safety features. http://pbadupws.nrc.gov/docs/ML0401/ML040150573.pdf
16. Stosic ZV et al (2008) Boiling water reactor with innovative safety concept, the generation III + SWR-1000. Nucl Eng Design 238:1863–1901
17. Kersting E et al (1993), Safety analysis for boiling water reactors, a summary. Gesellschaft für Anlagen- und Reaktorsicherheit, GRS-98
18. Yamada K et al (2003) ABWR design and its evolution - primary system design of ABWR and ABWR-II. In: GENES4/ANP2003, Kyoto, 15–19 Sept 2003, Paper 1161
19. Beard JA (2007) ABWR-overview. https://docs.google.com/file/d/0B744LXrbu_Y3ODYyNWNmYmQtYjVlZi00ZWY4LWFlNjAtNDI4MWI2Y2NiMDBm/edit?hl=en&pli=1
20. Hitachi, Ltd. (2007) Design features, construction and operating experiences of abwr, improvement of safety, economics and reliability. http://www.reak.bme.hu/MTAEB/files/konferencia_20070308/tpresent/Hitachi_Design+Features+of+ABWR.pdf
21. Toshiba Corporation (2011) US-ABWR and EU-ABWR Design, safety technology, operability features and their current deployment. http://www.iaea.org/NuclearPower/Downloadable/Meetings/2011/2011-07-04-07-08-WS-NPTD/4_JAPAN_ABWR_Toshiba_Ishibashi.pdf
22. IAEA (2004) Status of advanced light water reactor designs. IAEA Tecdoc-1391. http://www-pub.iaea.org/MTCD/publications/PDF/te_1391_web.pdf
23. Sato T (1999) Study of advanced reinforced concrete containment for the next generation boiling water reactors. In: 7th International conference of nuclear engineering (ICONE 7330), Tokyo
24. Sato T (2004) Basic concept of a new future BWR. Nucl Eng Design 230:181–193

Chapter 4
Radioactive Releases from Nuclear Power Plants During Normal Operation

Abstract This chapter lists the different main radioactive isotopes produced in the fuel of nuclear plants which are released in very small amounts into the environment. It then explains the different pathways leading to radioactive exposure of the human body. Several containment barriers in a nuclear power plant lead to extremely low leak rates of radioactive substances. This is followed by the definition of the radiation dose, the radiation weighting factors, the tissue weighting factors, the equivalent radiation dose and the natural background radiation dose. Radiation exposure from man-made sources is strictly controlled by governmental agencies. Permissible radiation dose limits were set by governments for radioactive releases from nuclear installations for the population and for employees who might receive enhanced radiation during their occupation. This also holds for rescue operation teams after a severe nuclear accident. Finally the radioactive effluents of LWRs and the effective doses to the public for airborne and liquid effluents of LWRs are presented. This is compared with the release of radioactive nuclides from coal fired plants.

4.1 Radioactive Releases and Exposure Pathways

During normal operation of nuclear power plants and other facilities of the nuclear fuel cycle, small amounts of radioactivity are released into the environment at a monitored and controlled rate. Airborne radioactivity includes the radioisotopes of the noble gases krypton, xenon, radon, of tritium, C-14, and also of fission products and fuel aerosols. Liquid effluents released into rivers, large lakes or the ocean contain tritium, fission products and other radioactive substances [1, 2]. As a consequence, man may be exposed to ionizing radiation through various exposure pathways (Fig. 4.1):

G. Kessler et al., *The Risks of Nuclear Energy Technology*, Science Policy Reports,
DOI 10.1007/978-3-642-55116-1_4,

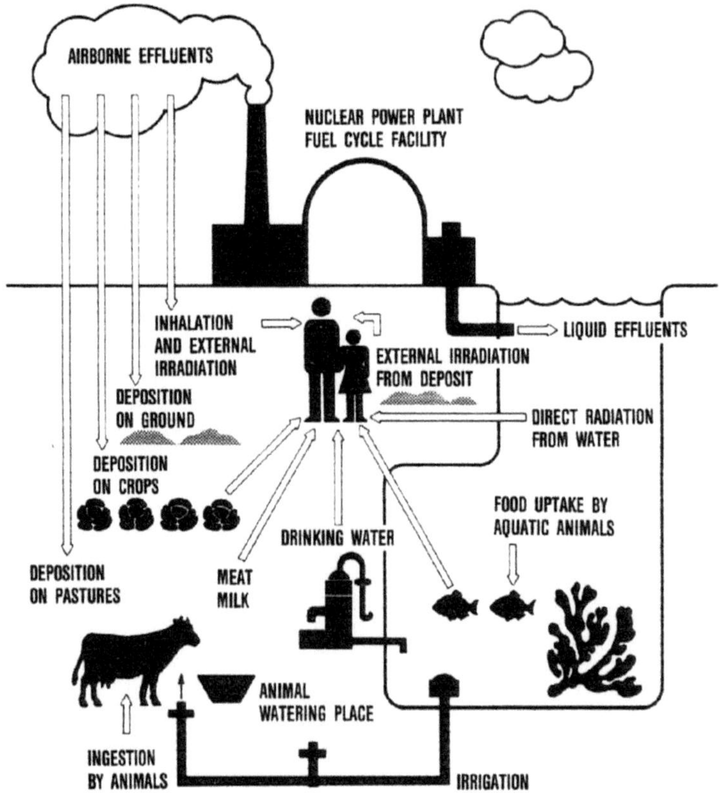

Fig. 4.1 Possible exposure pathways to man from the nuclear fuel cycle [3]

- external β- and γ-radiation of the gaseous radioactive nuclides in the atmosphere (β- and γ-submersion) or by immersion in water (swimming),
- radiation from aerosol particles deposited on the ground (soil radiation),
- internal exposure following inhalation of radioactive nuclides (inhalation),
- internal exposure as a result of the intake of contaminated food or water (ingestion).

The release rate of radioactive nuclides into the environment depends on the retention mechanisms incorporated in the engineered safety design of a fuel cycle facility. Series connection of several containment barriers with low leak rates and other technical measures allow very high retention factors or low release factors to be attained.

Gaseous nuclides or aerosols escaping from the plant or discharged in a controlled way through a stack are diluted in the ambient atmosphere. Dilution is a function of the height of the exhaust stack, the turbulence conditions of the atmosphere, and the distance from the plant. Radionuclides are also deposited on the ground by dry and wet disposal. Aqueous radioactive discharges are diluted as a function of the quantitative relationships between the liquid effluents and the ambient water volume.

The individual radioactive nuclides can enter the human body on various pathways, where they again may have different radiological impacts.

4.1.1 Exposure Pathways of Significant Radionuclides

4.1.1.1 Tritium, Carbon-14 and Krypton

Tritium (half-life 12.4 year) is produced in the reactor core mainly by ternary fission; on the average, about 1 in 10^4 fissions of U-235 is accompanied by the formation of tritium. About twice as many tritium nuclei are formed in the fission of Pu-239. In addition, tritium is generated in the coolant by neutron capture in deuterium atoms (deuterium has an abundance in natural water of 0.015 at.%), and by the interaction of neutrons with the boron control material. It is released from nuclear reactors and reprocessing plants as HT gas or as tritiated water (HTO), either into the atmosphere or into, e.g., a river or lake or into the ocean. Gaseous tritium, HT, is very soon oxidized into HTO. Ultimately, any tritium escaping or being released in a controlled manner thus will be present as tritiated water. Plants and animals may contain HTO/H_2O ratios close to those existing in the environment. Radioactive exposure of the human body then occurs as a result of the ingestion of food and drinking water. Moreover, tritiated water (HTO) can be absorbed by inhalation and through the skin of the human body. In this way, the β-radiation (maximum energy 18 keV) of tritium causes a whole body exposure.

C-14 (half-life 5,730 year) is built up in the reactor core by (n,p)-reactions with N-14, (n,α)-reactions with O-17, and (n,γ)-reactions with C-13. C-14 emits β-radiation (maximum energy 156 keV). In plants and animals, $^{14}CO_2/^{12}CO_2$ ratios may be established which are very close to those in the atmosphere. Radioactive exposure of the human body then occurs mainly as a result of the ingestion of food (milk, vegetables, meat). Direct inhalation and exposure from the ambient atmosphere only play minor roles.

Kr-85 (half-life 10.7 year) is a fission product. Kr-85 emissions from a LWR are diluted in the atmosphere. Approximately 99.6 % of the Kr-85 nuclei decay by emitting β-particles with a maximum energy of 0.67 MeV. Only 0.4 % of the Kr-85 nuclei decay by emitting a β-particle (maximum energy, 0.16 MeV) and γ-radiation (0.51 MeV). There is no reduction in the airborne concentration as a result of deposition or washout. Kr-85 is only sparingly soluble in water. Its main radiological impact on the human body is due to the exposure to the skin. The inhalation of Kr-85 plays a smaller role.

4.1.1.2 Radioisotopes of Iodine

For releases from nuclear reactors the shortlived radioactive isotopes of iodine, I-131 (half-life 8 days) and I-133 (half-life 20 h), are important fission products. I-129 (half-life 1.7×10^7 year) remains an important radioisotope. Radioiodines in the airborne effluents from a nuclear power plant occur partly as elemental iodine and partly as an organic compound (e.g. methyl iodide). Airborne iodine is deposited on the surfaces of grass or vegetables. If liquid effluents containing radioiodine are discharged into rivers or lakes etc., a possible major pathway will be their accumulation in fish or plants. The human body can take up radioiodine with the inhalation of air, ingestion of vegetables or fish, and by drinking milk. The radioiodine absorbed by the human body is concentrated mainly in the thyroid. Radioiodine emits both β- and γ-radiations.

Iodine aerosols or organic iodine compounds may, in particular under abnormal operating conditions, deposit on components and concrete walls in the reactor building. These are usually covered by suitable paintings. Nevertheless this may impair decommissioning of these parts of the reactor plant [4].

4.1.1.3 Strontium and Cesium

The fission product Sr-90 (half-life 29.1 year) can be released into the atmosphere as aerosols during severe nuclear accidents or with liquid effluents into rivers during nuclear accidents. Through the food chain (milk, vegetables, fish, meat and drinking water), Sr-90 enters the human body. Like calcium, Sr-90 is deposited preferably in bones, representing a major burden on the blood forming organs because of the long biological residence times of 18 years and the β-radiation of 2.3 MeV maximum energy of the daughter product, Y-90 (half-life 2.7 days).

Releases of radioactive cesium by way of gaseous and liquid effluents during severe nuclear accidents also cause radiological exposures of the human body through the uptake of food, as in the case of strontium. Cs-134 (half-life 2.1 year) and Cs-137 (half-life 30 year) also emit γ-radiation in addition to β-radiation. Cesium can largely replace potassium in living organisms and, like the latter, is distributed throughout the body in highly soluble compounds.

4.1.1.4 Plutonium Isotopes

Plutonium isotopes may be discharged into the atmosphere during severe nuclear accidents with gaseous effluents as aerosols of PuO_2 or into rivers together with liquid effluents. The following plutonium isotopes are of main interest: Pu-238 (half-life 87.8 year): Pu-239 (half-life 24,100 year); Pu-240 (half-life 6,450 year); Pu-241 (half-life 14.4 year); Pu-242 (half-life 3.9×10^5 year). The highest burden results from inhalation, in which case plutonium is deposited in the lung. Moreover, plutonium may be ingested together with vegetables, milk, meat, fish and drinking water. Plutonium taken up by way of ingestion is deposited preferably in bone tissue.

4.2 Radiation Dose

The measure of radiation absorbed is the energy dose [1, 5, 6]. Historically the unit was called rad (radiation absorbed dose). It was originally defined as a radiation dose that deposits 100 erg of energy per gram of absorbing material:

$$100\ \mathrm{erg/g} \triangleq 10^{-5}\,\mathrm{Joule/g} \triangleq 1\ \mathrm{rad}$$

To-day, the *absorbed dose* is measured

$$\mathbf{1\ Gray\ (Gy)} = 100\ \mathrm{rad} = 10^{-3}\,\mathrm{Joule/g} = 1\ \mathrm{Joule/kg}$$

The biological effects of the radiation dose absorbed also depend on the energy and type of radiation (γ-radiation, β-radiation, α-particles, neutrons, protons). To take these differences into account, an equivalent dose was defined. The unit was called, historically, rem (roentgen equivalent man) and is Sievert (Sv) now

equivalent dose

$$\mathbf{1\ Sievert\ (Sv)} \triangleq 100\ \mathrm{rem}$$

The absorbed dose and the equivalent dose are related as follows:

$$\text{equivalent dose H (Sv)} = \text{absorbed energy dose (Gy)} \times \text{radiation weighting factor } (w_R)$$

The radiation weighting factor, w_R, (Table 4.1) is a measure of the relative effects of the nuclear particles in producing damage for a given energy deposition. It is defined by the International Commission on Radiation Protection (ICRP). For most of the γ- and β-radiations (photons or electrons) of fission products the radiation weighting factor can be taken as unity. For neutrons the radiation weighting factor depends on the neutron kinetic energy and varies between 2.5 and 20. (The radiation weighting factors for neutrons vary slightly between ICRP 60 (1991) [5] (step function) and ICRP 103 (2007) [7] (piecewise linear functions). For protons with an energy of >2 MeV the radiation weight factor is 5 in ICRP 60 [5] and 2 in ICRP 103 [7]. For alpha-particles entering the body and accumulating in certain tissues the radiation weighting factor is 20 according to both ICRP Publication 60 (1991) [5] and ICRP 103 (2007) [7].

To express the different sensitivities of organs, so-called tissue weighting factors, w_T, have been introduced (see Table 4.2). This gives rise to the "effective

Table 4.1 Radiation weighting factor, w_R, according to ICRP 60 [5] and ICRP 103 [7]

Type and energy of radiation	ICRP 60 [5] Radiation weighting factor (w_R)	ICRP 103 [7] Radiation weighting factor (w_R)
Photons (all energies)	1	1
Electrons, Muons (all energies)		
Neutrons	1 (Multiple step function)	1 (Continuous function) (See [7])
<10 keV	5	
10 keV to 100 eV	10	
>100 keV to 2 MeV	20	
>2 MeV to 20 MeV	10	
>20 MeV	5	
Protons >2 MeV	5	2
Alpha particles, fission fragments, heavy nuclei	20	20

dose". The "effective dose" is the sum total of the equivalent doses in all organs (tissues) of the body weighted by the tissue weighting factors,

$$\text{effective dose} = \sum_T w_T \cdot H_T$$

where

$H_T \triangleq$ equivalent dose in the organ (tissue),
$w_T =$ organ weighting factor for tissue, T.

4.3 Natural Background Radiation

All individuals are exposed to natural background radiation, which consists of cosmic radiation, external terrestrial radiation from naturally radioactive isotopes in the soil and rocks or houses and internal radiation after inhalation and ingestion of naturally radioactive isotopes in the human body.

The mean natural background whole body dose on earth is 2.4 mSv/year. Accordingly, the average lifetime dose (70 years) is about 170 mSv.

The average background whole body dose in the USA is about 3.7 mSv/year (medical X-rays excluded). It varies with altitude, geographic location, etc. In Kerala, India, or in Brazil with large monazite reserves (thorium), the background

Table 4.2 Tissue weighting factors w_T for determination of the effective dose according to ICRP 60 (1991) [5] and ICRP 103 (2007) [7]

Issue	ICRP 60	ICRP 103
Tissue organ weighting factors, w_T		
Gonads	0.20	0.08
Breast	0.05	0.12
Red bone marrow	0.12	0.12
Lung	0.12	0.12
Thyroid	0.05	0.04
Bone surfaces	0.01	0.01
Colon	0.12	0.12
Stomach	0.12	0.12
Bladder	0.05	0.04
Oesophagus	0.05	0.04
Liver	0.05	0.04
Brain	++	0.01
Kidney	++	++
Salivary glands	++	0.01
Skin	0.01	0.01
Remainder	0.05^+	0.12*

$^+$see ICRP 60, *see ICRP 103, $^{++}$included in Remainder

radiation dose is roughly a factor of 8–100 higher than the mean background dose on earth of 2.4 mSv/year [8–11].

In the Federal Republic of Germany, – given here as an example—, the corresponding average background whole body dose is 2.1 mSv/year (Table 4.3) [12, 13].

4.3.1 Natural Background Exposure from Natural Sources in Germany

A major source of external radiation exposure consists of both cosmic (0.3 mSv/year) and external terrestrial radiation (0.4 mSv/year) from the natural radionuclide K-40 (half-life 1.28×10^9 year) together with the radionuclides of the natural decay chains of U-238 and Th-232. The internal component of radiation exposure is largely caused by the inhalation of the natural noble gas radon and its daughter nuclides (1.1 mSv/year), and partially also by the intake of natural radioactive substances with drinking water and food (0.3 mSv/year).

The annual mean value of the radon activity concentration in occupied spaces is about 50 Bq/m^3, which corresponds to a mean annual effective dose of about 0.9 mSv/year. Outdoors inhalation of radon and its progeny leads to about 0.2 mSv/year [12, 13]. The natural background radiation in Germany is given as an example in Table 4.3; in other countries the natural background exposure is similar.

Table 4.3 Mean effective radiation dose to the population from natural background radiation in the Federal Republic of Germany during the year 2008 [12, 13]

Mean effective dose		
1. Radiation exposure from natural sources		mSv/year
1.1	Cosmic radiation (at sea level)	~0.3
1.2	External terrestrial radiation	~0.4
	Outdoors (5 h/day)	~0.1
	Indoors (19 h/day)	~0.3
1.3	Inhalation of radon and its progeny	~1.1
	Outdoors (5 h/day)	~0.2
	Indoors (19 h/day)	~0.9
1.4	Ingestion of natural radioactive substances	~0.3
Total natural background radiation		~2.1

4.4 Radiation Exposure from Man-Made Sources

Table 4.4 shows data for the radioactive exposure of man-made radioactivity sources. These man-made radioactive exposures sum up to 1.9 mSv/year [12, 13].

4.4.1 Nuclear Weapons Tests

Numerous atmospheric nuclear weapons tests were carried out from 1945 to 1980, but since 1981 only underground tests have been performed. The general level of environmental radioactivity due to former tests in the atmosphere has steadily decreased since the Comprehensive Nuclear Test-Ban Treaty from 1964. At present its contribution to the total of human radiation exposure is less than 0.01 mSv/year [12, 13].

4.4.2 Chernobyl Reactor Accident

In April 1986, a reactor accident occurred in the Chernobyl nuclear power plant which has had serious consequences in Europe so far. In the days following that accident, large amounts of radionuclides were released into the atmosphere and distributed all over Europe. In 2008, the associated mean effective dose in Germany was less than, e.g. 0.012 mSv/year. It amounts to less than 1 % of the natural background radiation exposure. About 90 % of this radiation is caused by Cs-137 deposited on the ground. Locally also higher exposure values were found. In addition, higher concentrations of radioactive isotopes were found in mushrooms and wild animals [12, 13].

Table 4.4 Mean effective radiation dose to the population from man-made radiation sources, e.g. in the Federal Republic of Germany during the year 2008 [12, 13]

2. Radiation exposure from man-made sources	mSv/year
2.1 Fallout from nuclear weapons tests	<0.01
2.2 Effects from the accident at Chernobyl	0.012
2.3 Nuclear installations	<0.01
2.4 Use of radioactive substances and ionizing radiation in medicine	~1.9
Diagnostic nuclear medicine	~0.1
2.5 Use of radioactive substances and ionizing radiation in research and technology	<0.01
Total of man-made radiation exposures	~1.9

4.4.3 Nuclear Installations

The emission of radioactive substances from nuclear power plants, uranium enrichment plants and fuel fabrication plants, e.g. in Germany contributes only insignificantly to the radiation exposure of the population. The upper values for exposures to individuals, calculated in accordance with the "General Administrative Guideline relating to §47 of the German Radiation Protection Ordinance" [14] are clearly below the required limits (Sect. 4.6.1). The annual contribution from domestic nuclear installations and other installations located close to the German borders to the mean effective dose to the population of the Federal Republic of Germany remained below 0.01 mSv in 2008 and 2009 [12, 13].

Radiation exposures by spent fuel reprocessing plants in Europe are given in [3].

4.4.4 Medical Applications

The major part of the mean effective population dose from man-made radiation exposure is caused by medical applications of radioactive substances and ionizing radiation (thyroid and skeletal scintigraphy, X-ray diagnostics, computer tomography, positron emission-tomography). It amounted to ~1.9 mSv/year, e.g. in Germany in 2008 [12, 13].

4.4.5 The Handling of Radioactive Substances in Research and Technology

The use of ionizing radiation and radioactive substances for technological and research purposes lead to a mean contribution to the effective population exposure, e.g. in Germany of less than 0.01 mSv in 2008 [12, 13].

4.4.6 *Occupational Radiation Exposure*

All employees who might receive enhanced radiation doses during their occupation are subject to radiation protection monitoring. These persons are monitored through personal dosimeters. The average individual dose of 324,000 monitored employees, e.g. in Germany was 0.14 mSv/year in 2008. Only 57,000 monitored employees out of these 324,000 employees received an average individual dose of 0.8 mSv/year in 2008 [13].

Aircrews received an average effective dose of at least 2.4 mSv/year from cosmic radiation during the flight in 2009 [13].

4.5 Radiobiological Effects

When ionizing radiation (α-,β-,γ-radiation, neutrons, protons, ions) hits a biological cell and penetrates it, this gives rise to ionization of the atoms in various molecules of the cell. This may alter these molecules. Especially alterations (separations) of the DNA containing the hereditary information produce radiation consequences. The results may be

- mutation of the cell,
- death of the cell.

Each cell has a high repair potential [9, 11, 15, 16]. As a result, most molecular alterations will have no consequences. However, it may also happen that a mutant cell is produced which passes its modified genetic function on. A cell changed in this way may cause carcinoma or leukemia (somatic effect). When the mutation in a gonad cell is passed on to a descendant, this is called a genetic effect.

It is commonly assumed, although not unchallenged [9], that there is no radiation dose threshold for these mutant effects. Under this assumption, the radiation dose/effect relation begins at the zero point, rises linearly [5, 7], assumes a quadratic curve shape at higher radiation doses, and then levels off again at very high radiation doses when cell death occurs.

In the lower radiation dose range, the biological effect of radiation can be measured only on a statistical basis, and is therefore referred to as a stochastic effect. When significant numbers of cells are damaged or die in the higher radiation dose range, this is called a deterministic effect of radiation.

4.5.1 *Stochastic Effect*

Assessing the stochastic effect quantitatively is not easy, as it is impossible, at the present state of knowledge, to recognize whether a tumor developed as a result of an exposure to ionizing radiation or for some other reason. Epidemiological studies of large populations (atomic bomb victims of Hiroshima and Nagasaki) have been used so far in attempts to determine the number of deaths from cancer exceeding the number of deaths from cancer occurring from natural background radiation. The results of the evaluation of the data from Hiroshima and Nagasaki can be related to the dose of the preceding radiation exposure, and result in the dose/effect relationship. However, the data calculated in this way do not constitute an immutable quantity. For instance, the number of deaths from cancer increases with the age of the Hiroshima-Nagasaki population under study. The International Commission on Radiological Protection, in its publication ICRP 60 [5], for the first time took into account data of the Hiroshima-Nagasaki population more than 40 years after the atomic bombs were dropped, and extrapolated these data to the entire lifespan [9, 15].

For determination of the risk associated with low doses, the roughly linear dependence found at high radiation doses is now extrapolated back to the range of low radiation doses. In this way, the entire cancer risk is rather overestimated in a conservative sense.

Figure 4.2 reflects the hypotheses which can be extrapolated back from the data of Nagasaki-Hiroshima. Curve A, which is favored by most scientists, is based on a threshold level. Curve B, for which also a number of scientists argue, is based even on positive healing health effects (hormesis) of a low radiation dose [9, 15, 16].

The linear dose-effect relation is recommended in ICRP 60 and ICRP 103 as a conservative proposal.

4.5.2 *Deterministic Effects of Radiation*

Deterministic effects of radiation arise when a large number of cells are damaged considerably by a high radiation dose such that regeneration is not possible or the cells die. There are various threshold doses of radiation for the deterministic effect in various organs of the body.

At a radiation dose of less than 1 Gray, most tissues produce no clinical symptoms of disease (ICRP 60 [7]). However, there are exceptions:

- the male gonads:
  >0.15 Gray causes temporary sterility,
  permanent sterility results at >3 Gray;
- the bone marrow reacts with disorders of the blood at >0.5 Gray.

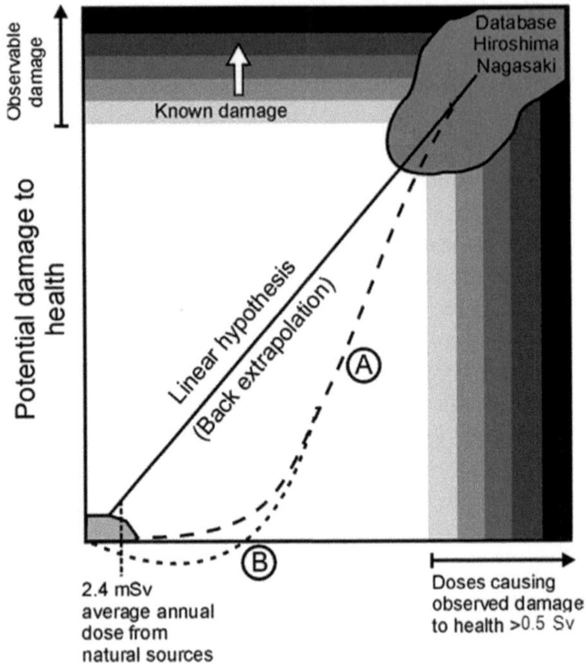

Fig. 4.2 Concept of linear dose-effect relation [8, 15]

4.5.3 Acute Radiation Syndrome

Damage to the bone marrow occurs between 1 and 20 Gray. The extent of the disorder, and the therapy employed, ultimately decide whether the irradiated accident victim survives.

Gastro-intestinal disorders are produced at >2 Gray (nausea, vomiting, hyposthenia).

At more than 6–8 Gray, there are practically no chances of survival [9, 17, 18].

4.6 Permissible Exposure Limits for Radiation Exposures

The International Commission on Radiological Protection (ICRP) has published recommendations about permissible radiation exposures to man from nuclear installations (ICRP 60 [5] and ICRP 103 [7]). In those recommendations, a distinction is made between occupationally exposed persons and individual members of the public. Considerable lower levels for exposure limits of the general population were set.

4.6.1 Limits of Effective Radiation Dose from Nuclear Installations in Normal Operation

At present, the German Radiation Protection Ordinance of 2008 [14] is valid in Germany. It is based on ICRP 60 [5] and the Euratom directive of 1996 [19]. In the USA the environmental radiation protection standards of USEPA are to be obeyed [20].

4.6.2 Radiation Exposure Limit for the Population

According to the German Radiation Protection Ordinance [14], as an example, the limit for the annual effective radiation dose due to radioactive emissions from nuclear installations to the population permissible on top of the natural background radiation is 1 mSv.

Table 4.5 shows a comparison of ICRP 60 [5] with the Euratom directive of 1996 [19] and US NRC 10 CFR 20 [20, 21].

4.6.3 Exposure Limits for Persons Occupationally Exposed to Radiation

The exposure limit per annum of the effective radiation dose for persons occupationally exposed to radiation is given in Table 4.6.

The limit for persons occupationally exposed to radiation is 20 mSv/year. This limit can be increased to 50 mSv/year as long as the sum of exposures does not exceed 100 mSv over a time period of 5 years. This is over and above background exposure and excludes medical exposure.

The exposure doses for each organ (tissue) must be determined by special tables which are not listed here in detail [5, 7, 12, 14, 19].

4.6.4 Exposure Limits for Persons of Rescue Operation Teams During a Reactor Catastrophe

After severe reactor accidents rescue operation teams working in contaminated areas inside or outside of the reactor in high radiation fields can receive a radiation exposure up to a limit of [14, 22, 23]

- 100 mSv per rescue operation and year or
- 250 mSv once in their life

Table 4.5 Exposure limit for the annual effective radiation dose from nuclear installations to the population permissible on top of the natural background radiation

Effective dose	ICRP 60 (1991)	EU (1996)	US NRC (10 CFR 20)
Limit (population) per annum	1 mSv/year	1 mSv/year	1 mSv/year

Table 4.6 Limits for persons occupationally exposed to radiation

	ICRP 60 (1991)	EU (1996)	German Radiation Prot. Ord. 2008
Limit for persons occupationally exposed to radiation	20 mSv/year	20 mSv[a]/year	20 mSv/year

[a]100 mSv in 5 consecutive years, but not more than 50 mS/year in 1 year

4.6.5 Life Time Occupational Exposure Limit

According to §56 of the German Radiation Protection Ordinance [14] an occupationally exposed person can receive a maximum of 400 mSv during its professional life. This limit can only be exceeded by 10 mSv/year if both, medical authorities and the person having reached the professional lifetime limit of 400 mSv, agree [14].

4.6.6 The ALARA Principle

Above and beyond these standards defined by ICRP and national regulations, the “as low as reasonably achievable” (ALARA) principle must be applied to all emissions of nuclear plants. This means that practically all facilities of the fuel cycle must keep below those standards [5, 7, 12, 19, 20].

4.7 Nuclear Power Plants

Most of the radioactive inventory in Bq of a nuclear power plant is made up of fission products. Gaseous fission products, such as noble gases (especially krypton and xenon) and tritium, can enter the coolant through leaks in the claddings of fuel rods. They are passed through the primary coolant purification system and the exhaust air system into carbon filter lines and into the exhaust air stack from where they are released into the environment. Emissions of shortlived isotopes, such as Kr-88 (half-life 2.8 h), can be minimized by adequate holdup of gaseous effluents in storage and decay tanks before release.

The tritium produced in the core migrates along grain boundaries and cracks of the fuel into gaps between fuel pellets and the cladding and is collected in the fission product gas plenum of the fuel rod. The zircaloy tubes of LWR fuel rods bind some 60 % of the tritium inventory. Moreover, an oxide layer building up on the outer wall of the zircaloy cladding tube acts as a diffusion barrier to the tritium. As a consequence, more than 99.9 % of the tritium formed is retained in the LWR fuel rod. Only if cladding tube failures occur, releases of tritium into the cooling water will be increased. Some of the tritium is discharged with the gaseous effluents. In water cooled reactors most of it remains in the coolant as tritiated water. Some of the tritiated water is released at a controlled rate. Improved methods under study are the concentration of tritiated water by evaporation and its prolonged storage in decay tanks. With a half-life of tritium of about 12 years, some 90 % will have decayed after 40 years.

Besides the radioactive noble gases and tritium, also such elements as rubidium, strontium, technetium, ruthenium, silver, tellurium, antimony, iodine, cesium, barium, rubidium, lanthanum and cerium are radiologically significant. Except for iodine, cesium and rubidium, these elements have only low volatilities. They may enter the primary coolant through defective fuel rod claddings. Non-volatile fission products can enter the liquid effluent only through the primary coolant purification system.

4.7.1 Radioactive Effluents from PWRs and BWRs

Radioactive effluents from PWRs and BWRs are given here for two countries: Germany and the USA.

4.7.1.1 Radioactive Effluents from PWRs and BWRs in Germany

Table 4.7 shows emission data of typical German PWRs and BWRs (data are normalized to 1 GW(e)·year). This set of data was reported in the annual report for 2008 on radioactive releases and radiation exposure from German nuclear installations [12, 13]. The data collected for 2009 differ slightly [13]. The PWR data are given for PWR plants at the site of Neckarwestheim (Germany). The BWR data are valid for two BWR plants with a total power output of 2.7 GW(e) at the site of Gundremmingen (Germany). For comparison also the half-lives of the different emitted isotopes are given.

Table 4.7 Radioactive emissions (airborne and liquid) from a PWR- and a BWR-plant site with two reactors [12]

Isotope	Half-life	PWR plant, Neckarwestheim, Germany	BWR-plant, Gundremmingen, Germany
Airborne effluents Bq/(GW(e)·year)			
Tritium	12.3 year	2.1×10^{11}	2.0×10^{11}
C-14	5,730 year	0.5×10^{10}	3.0×10^{11}
Ar-41	1.8 h	2.3×10^{11}	1.0×10^{11}
Co-60	5.3 year	2.4×10^{4}	
Kr-85m	4.5 h		2.5×10^{9}
Kr-85	10.8 year	1.0×10^{11}	1.5×10^{11}
Kr-88	2.8 h	0.7×10^{8}	0.8×10^{9}
I-131	8.0 day		0.4×10^{7}
Xe-131m	11.9 day	0.9×10^{10}	3.3×10^{10}
Xe-133m	2.2 day	0.7×10^{8}	1.2×10^{9}
Xe-133	5.2 day	1.8×10^{9}	5.19×10^{10}
Xe-135	9.1 h	1.1×10^{10}	1.07×10^{11}
Xe-137	3.8 min	2.2×10^{7}	2.26×10^{11}
Liquid effluents in Bq/(GW(e)·year)			
Fission + activation products		0.7×10^{6}	3.2×10^{8}
Tritium	12.3 year	1.2×10^{13}	1.3×10^{12}
α-Emitters		Below measurement limit	Below measurement limit

4.7.1.2 Radioactive Effluents from PWRs and BWRs in the US

Data for radioactive effluents from nuclear reactors were collected between 1994 and 2009 in the USA for 104 Light Water Reactors [24]. The gaseous radioactive effluents of these 69 BWRs and 35 PWRs are about equal to the German data (previous Section) if compared on a GW(e)·year basis. This also holds for a comparison of the liquid effluents of PWRs and BWRs on a GW(e)·year basis. The reason is the very similar safety design of US and German or European PWR and BWR plants (Sects. 3.1 and 3.2).

Although there are inherent design differences between LWRs and CANDU-PHWRs or HTGRs also these types of power reactors have similar radioactivity releases below the radiation limits described in Sect. 4.6. Nuclear power plants are equipped with instruments to measure continuously the amounts of gaseous and liquid radioactive effluents. These data must be reported to national environmental protection agencies or nuclear regulatory commissions. Such reports are made available to the public.

4.7.2 Occupational Radiation Exposure of Workers in Nuclear Power Plants

The average occupational exposure of 30,238 workers in nuclear power plants (PWRs and BWRs) was 0.5 mSv/year in Germany in 2009. This must be compared to the admissable exposure limit of 20 mSv/year of Sect. 4.6.3. This 0.5 mSv/year average radiation exposure is much lower than the occupational exposure of 2.4 mSv/year of aircraft personnel (Sect. 4.4.6) [12, 13].

4.7.3 Radiation Exposures Caused by Radioactive Emission from Light Water Reactors

Based on the radioactive effluents presented in the previous section the radiation exposure of the public will be discussed for two countries: Germany and the USA.

4.7.3.1 Radioactive Exposure from PWRs and BWRs in Germany

In calculating radiation exposures in Germany [12] it is assumed that gaseous effluents are released into the environment from a stack of 100 m height. Moreover, liquid effluents are introduced into the cooling water of a nuclear power plant and further diluted in the main canal with a water flow of 250 m^3/s. Taking into account statistical data about the weather conditions and following the different exposure pathways, it is possible to determine the radiation exposure in the specific environment of a plant. The exposure results in Fig. 4.3 were obtained on the basis of German rules and regulations. In this respect, it is assumed conservatively that a person stays in the same place throughout the year and ingests both drinking water and food from the immediate environment. Figure 4.3 presents the exposure data of German PWRs and BWRs for gaseous and liquid radioactive effluents [12].

For PWRs the annual effective dose is well below 1 μSv. For BWRs—due to the fact that the steam produced in the reactor core goes directly to the steam turbine—the effective annual dose is somewhat higher in case of airborne radioactive effluents. However, it is still more than a factor 1,000 lower on a GW(e)/year basis than the permissible limit of 1 mSv/year given in Sect. 4.6.2. If several plants operate at one site with several GW(e) power output the radiation exposure is higher accordingly.

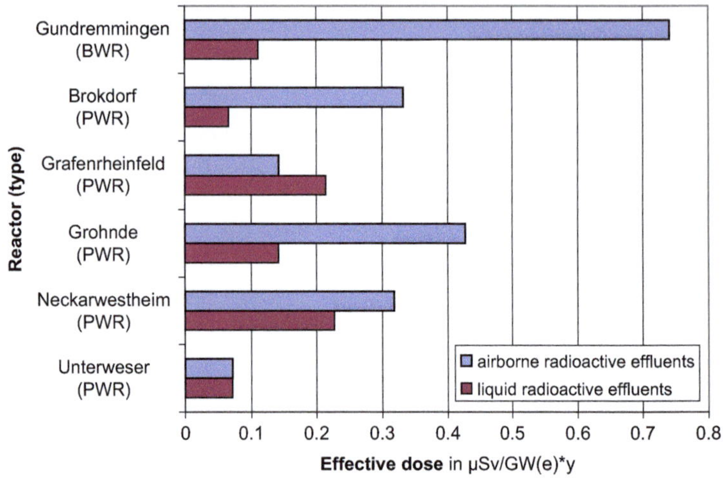

Fig. 4.3 Effective dose (μSv/(GW(e)·year)) from airborne and liquid radioactive effluents of German PWRs and BWRs in 2008 [12]

4.7.3.2 Radioactive Exposures from Light Water Reactor Effluents in the US

The gaseous and liquid radioactive effluents are about equal for PWR and BWR plants in Europe and the USA. If instead of the very conservative rules described in the previous section the radioactive effluents are distributed over the whole USA and if the radiation exposure is averaged over the whole population as described in [24] then the radiation exposures shown in Fig. 4.3 drop down by orders of magnitude.

4.7.4 Comparison with Emissions of Radioactive Nuclides from a Coal Fired Plant

A coal fired power plant for electricity production burns somewhat more than two million tons of coal per GW(e) and year. This coal contains about 1 ppm of U-238 with its daughter products, about 2 ppm Th-232 with its daughter products as well as the isotope K-40. These radioactive impurities are emitted together with the combustion gases of the coal fired plant into the environment [25]. These radioactive emissions (Bq/GW(e)·year) are shown in Table 4.8.

A comparison of the radioactive exposures of these airborne radioactive emissions of coal fired plants (Table 4.8) with those of PWRs or BWRs (Table 4.7) on a GW(e)·year basis leads to the difficulty that both power generating systems emit different radioactive nuclides. Each of these radioisotopes has different

Table 4.8 Radioactive emissions (airborne) of coal fired plants (Bq/GW(e)·year) [25]

K-40	0.8×10^9
U-238 natural radioactive family (Bq/GW(e)·year)	
Rn-222	3.2×10^{10}
Other daughters	2×10^9
Th-232 natural radioactive family (Bq/GW(e)·year)	
Rn-220	2.1×10^{10}
Other daughters	0.6×10^9

radiotoxicity (Sv/Bq). In addition the whole fuel cycle for nuclear energy must be included. Therefore a comparison is only possible on the basis of collective dose equivalents. On this basis it was shown that coal fired plants on a 1 GW(e)·year basis cause about equal collective doses compared to the whole nuclear fuel cycle of Light Water Reactors (including uranium mining, uranium fuel fabrication, uranium enrichment, the nuclear power plant and spent fuel reprocessing) [25].

References

1. Glasstone S et al (1980) Nuclear power and its environmental effects. American Nuclear Society, LaGrange Park, IL
2. La dispersion des polluants dans l'atmosphère, CLEFS CEA (1999) No 42, p 2. Commissariat à l'énergie atomique, Paris
3. Kessler G (2012) Sustainable and safe nuclear fission energy. Springer, Heidelberg
4. Nuclear Energy Agency, Committee on the Safety of Nuclear Installations (2007) State of the art report on iodine chemistry. NEA/CSNI/R(2007)1
5. International Commission on Radiation Protection (ICRP) (1991) Annals of the ICRP: ICRP Publication 60 – 1990 recommendations of the international commission on radiological protection. Pergamon, Oxford
6. Krieger H (2007) Grundlagen der Strahlungsphysik und des Strahlenschutzes. Vieweg +Teubner Verlag
7. International Commission on Radiation Protection (ICRP) (2007) Annals of the ICRP: ICRP Publication 103 – The 2007 recommendations of the international commission on radiological protection. Elsevier, Amsterdam
8. Cohen BL (1983) Before it's too late. Plenum, New York, NY, p 33
9. Jaworowski Z (1999) Radiation risks and ethics. Phys Today 52:24–29
10. Sohrabi M et al (1990) High levels of natural radiation. International Atomic Energy Agency, Vienna, p 39
11. Kesavan PL (1996) In: Wie L, Sugahara T, Tao Z (eds) High levels of natural radiation. Elsevier, Amsterdam, p 111
12. Umweltradioaktivität und Strahlenbelastung, Jahresbericht 2008, Bundesministerium für Umwelt, Naturschutz und Reaktorsicherheit (BMU), Bonn, Germany
13. Umweltradioaktivität und Strahlenbelastung, Jahresbericht 2009, Bundesministerium für Umwelt, Naturschutz und Reaktorsicherheit (BMU), Bonn, Germany

14. General Administration Guideline relating to §47 of the Radiation Protection Ordinance (Strahlenschutzverordnung) (2008) Bundesministerium für Umwelt, Naturschutz und Reaktorsicherheit, Bonn
15. Les rayonnements, l'ADN et la cellule (2000) Clefs CEA, No 43, p 6. Commissariat à l'énergie atomique, Paris
16. Zamboglou N et al (1981) Low dose effect of ionizing radiation on incorporation of iododeoxyuridine into bone marrow cells. Int J Radiat Biol 39:83–93
17. UNSCEAR (1998) Sources, effects and risks of ionizing radiation. United Nations Scientific Committee on the effects of Atomic Radiation, Report to the General Assembly, with Annexes. United Nations Printing Office, Vienna, Austria
18. United Nations Scientific Committee on the Effects of Atomic Radiation (2001) Report to the General Assembly, with Scientific Annex. Hereditary effects of radiation. New York
19. Directive 96/29 Euratom-Ionizing Radiation (2011) http://osha.europa.eu/en/legislation/directives/exposure-to-physical-hazards/osh-directives/73
20. USEPA (1977) Environmental radiation protection standards for nuclear power operations. 40 CFR 190, 42 FR 2860, 13 January 1977. GPO, Washington, DC
21. USNRC (1991) Standards for protection against radiation. 10 CFR 20, 56 FR 23361, 21 May 1991. GPO, Washington, DC
22. Iter Consult (2011) Independent technical evaluation and review, Fukushima Daiichi nuclear accident first considerations, preliminary report. http://www.iter-consult.it/ITER_Report_Fukushima_Accident.pdf
23. Feuerwehr-Dienstvorschrift FWDW 500 "Einheiten im ABC-Einsatz, Ausgabe August 2004. http://www.bfs.de/de/bfs/recht/rsh/volltext/4_Relevante_Vorschriften/4_5_fwdv500_erl_2004.pdf
24. Harris JT (2011) Radiological releases and environmental monitoring at commercial nuclear power plants, nuclear power – operation, safety and environment. Dr. Pavel Tsvetkov (ed), ISBN: 978-953-307-507-5, InTech, DOI:10.5772/17668. http://www.intechopen.com/books/nuclear-power-operation-safety-and-environment/radiological-releases-and-environmental-monitoring-at-commercial-nuclear-power-plants
25. Halbritter G et al (1982) Beitrag zu einer vergleichenden Umweltbelastungsanalyse am Beispiel der Strahlenexposition beim Einsatz von Kohle und Kernenergie zur Stromerzeugung, Kernforschungszentrum Karlsruhe, KfK 3266

Chapter 5
Safety and Risk of Light Water Reactors

Abstract This chapter starts with the goals of protection for LWR plants. These are: safe shutdown of the LWR plant, assurance of core cooling and safe and intact containment structures. The safety concept of LWRs is based on multiple containment structures and engineered safeguard components. The staggered in depth safety concept relies on accident prevention, accident limitation or accident mitigation and severe accident management. LWR plants must be designed and built on the design basis concept. Sequences of events exceeding the design basis must be counteracted by beyond design accident management measures. Probabilistic safety analyses supplements these guidelines and assures frequencies of occurrences per year for a severe accident with core melt of 10^{-5} to 10^{-7} per year. Most countries issued an Atomic Energy Act establishing the legal frame for the peaceful utilization of nuclear power. The chapter continues by describing thermodynamic and neutron physics design of a LWR core as well as the stable behavior of PWRs when controlled by movement of control elements or of BWRs when controlled by the speed of the recirculation pumps, and moving control elements. The mechanical design of the pressure vessel is of high importance. It follows the guidelines of the ASME Boiler and Pressure Vessel Code. This is accompanied by quality assurance and in-service inspections. Also the mechanical design of the reactor containment follows similar guidelines. The chapter ends by discussing the different design basis accidents which must be analyzed prior to licensing of the LWR plant.

5.1 Introduction

The purpose of reactor safety is the protection of personnel in nuclear reactors as well as the protection of the environment of these plants and of the population. Failures leading to radioactivity releases must be excluded in the design of the

G. Kessler et al., *The Risks of Nuclear Energy Technology*, Science Policy Reports,
DOI 10.1007/978-3-642-55116-1_5,

nuclear plant as far as possible or, should defects occur nevertheless, their consequences must be limited reliably.

The radioactive inventory of the core of a 1 GW(e) reactor with high burnup fuel is approx. 10^{21} Bq. It arises mainly from the fission products present in the fuel elements during reactor operation. Most of the fission products are contained in the fuel elements (fuel matrix and fuel cladding). This does not apply to some fission product gases which are accumulated in the fission gas plenum of the fuel rods.

The fuel elements can only be destroyed by overtemperatures, e.g. melting of the fuel or rupture of the fuel rod cladding due to overpressure. At the beginning of the accident this causes fission product gases to be released, e.g. tritium, carbon-14, argon, krypton, xenon, and then also highly volatile fission products, such as I-131, Cs-137, Sr-90 etc.

Overtemperatures arise from an imbalance between the heat production and the heat removal in the reactor core during reactor operation.

However, imbalances between the heat produced and the heat removed can arise also when the reactor is shut down, as the radioactive substances generate heat by radioactive decay. This decay heat power (afterheat) (Sect. 2.9) is roughly 6 % of the nominal reactor power shortly after shutdown of the reactor, and 1 % after approx. 6 h, 0.3 % after 1 week, 0.1 % after 3 months and 0.04 % after 1 year and 0.006 % after 3 years for a burnup of 50 MWd/kg. This decay heat power is slightly dependent on burnup and increases somewhat with higher burnup for cooling time periods up to about 100 years [1–3, 37–38].

Against the background of these considerations, the goals of protection listed below are required which should be ensured in a nuclear reactor under all conditions.

5.2 Goals of Protection for Nuclear Reactors and Fuel Cycle Facilities

In case of a disturbance during power operation, controlling, limiting and safety shut down systems intervene as foreseen in the plant design and reduce the power level or shut down the nuclear power plant. However, even after shutdown of the nuclear chain reaction, the reactor core needs to be cooled because of the decay heat (afterheat) produced. Based on these reactor physics characteristics arise the following basic engineered safeguards requirements (goals of protection) which must be fulfilled at all times:

- Safe shutdown of the nuclear power plant: It must be possible to shut the reactor core down safely at any time and hold it in this shut down condition.
- Core cooling: The reactor must be cooled sufficiently at all times during operation and after shutdown

- Safe containment structures, i.e. protection from malfunction-induced releases of radioactivity; limitation of radiation exposure to workers inside the reactor containment and of the population outside.

5.3 Safety Concept of Nuclear Reactor Plants

The safety concept of nuclear reactor plants is based on multiple containment structures around the radioactive materials in the reactor core (multi-barrier concept) as well as on engineered safeguards components and measures ensuring such containment.

5.3.1 Containment by Radioactivity Enclosures

The radioactive substances are enclosed in several radioactivity enclosures (see also Chap. 3). In a German PWR shown by way of example (Fig. 5.1) these are

- the oxide crystal lattice of the ceramic fuel pellets (UO_2 or mixed $UO_2(PuO_2)$ oxide),
- the zircaloy cladding tubes of the fuel (welded gastight),
- the reactor pressure vessel with the closed cooling system,
- the gastight and pressure resistant steel containment (in other reactor designs a prestressed concrete containment) enclosing the cooling systems, and the concrete structures shielding against radiation,
- the outer shell of steel-reinforced concrete. It has a limited sealing function. It also protects the plant against external impacts.

5.3.2 Multiple Level Safety Principle

In addition, the safety of a nuclear power plant is ensured by multiple levels of safety superimposed upon each other (safety concept staggered in depth) (Fig. 5.2).

5.3.2.1 First Safety Level: Reactor Physics Design, Basic Safety, Quality Assurance

Nuclear reactor plants must be designed to be safe in terms of reactor physics. This includes, e.g., safety coefficients, such as the negative power coefficient, the negative Doppler coefficient, the negative coolant temperature or void coefficient, and the proper setting of the range of the effective multiplication constant k_{eff} for

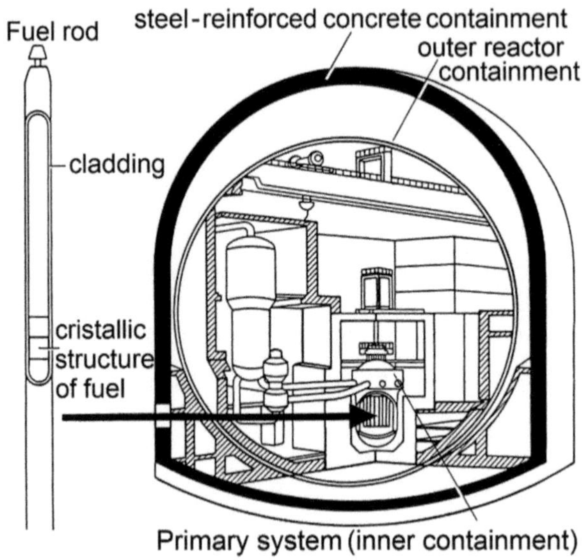

Fig. 5.1 Multiple barrier containment concept against release of radioactivity [4]

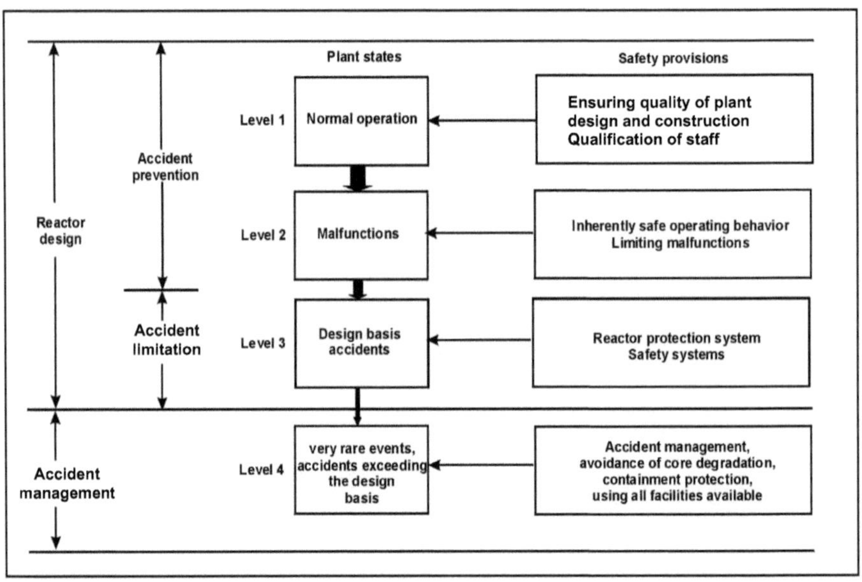

Fig. 5.2 Schematic representation of the multiple level safety principle providing safety in nuclear power plants [4]

the shutdown system as a function of the ranges of temperature and power, as well as the proper setting of the boron or gadolinium poisoning concentrations for the range of fuel burnup envisaged [3].

All nuclear components of the reactor core and the components of the cooling system are designed with high safety margins and must meet stringent requirements with respect to the choice of materials and the quality of manufacture (e.g., basic safety, leak-before-break criterion). In-service inspections and, if necessary, replacement of the components as well as great care in plant operation must ensure a high standard of technical safety quality throughout the entire operating life. This is required to make malfunctions extremely unlikely.

Basic rules in technical safety must be applied such as the failsafe principle, redundant design of cooling systems and safety systems, as well as the principle of diversity to avoid common-mode failures.

More details are given in Sect. 5.6.

5.3.2.2 Second Safety Level: Measures of Accident Prevention

Measurement and detection systems (instrumentation), control, monitoring and limiting systems (e.g. for temperatures, power, pump speeds, pressures, etc.) prevent accidents by early detection of malfunctioning. Limiting and control systems take credit of inherent safety properties to counteract disturbances in an adequate way. After correction of the malfunction, continued operation of the nuclear power plant is easily possible. Even in such cases of malfunction, the release levels for radioactive substances permitted in normal operation must not be exceeded.

More details are discussed in Sect. 5.6.3.

5.3.2.3 Third Safety Level: Design Basis and Measures to Limit Accident Consequences

As the occurrence of an accident cannot be completely excluded, nuclear power plants are equipped with safety systems. These safety related systems include, e.g., the reactor protection and shutdown system, the emergency cooling system and residual heat removal system, and the containment. After having been initiated by the reactor protection system, the safety systems operate largely automatically so as to meet the goals of protection referred to above (Sect. 5.2) and limit the damage arising from an accident. The design takes into account that one redundancy level of the safety system may be under repair and another system may not be available upon request (n + 2 principle) [4, 5].

The plant must accommodate a number of design basis accidents which must be proved by analysis during the licensing procedure of the reactor plant (Sect. 5.4).

More details are discussed in Sect. 5.6.6.

5.3.2.4 Fourth Safety Level: Measures Taken to Reduce Damage if Design Basis Is Exceeded—Accident Management Measures

On the fourth safety level, accident management measures take credit of existing design margins up to failure of in-plant systems and of additionally installed components which will be used in case of failure of the plant protection system.

Accident management measures also take credit of existing in-plant systems which are not classified as safety systems.

Measures are taken to minimize the potential damage caused by plant-internal accidents beyond the design basis and external impacts, e.g. airplane crash, chemical explosion or beyond design earthquakes [4].

5.4 Design Basis Accidents

The reactor plant and the protection systems must be designed and built on the basis of the design basis accident concept. This includes a number of design basis accidents which must be accommodated safely, even if a fault independent of the original cause of accident initiation occurs. The safety must be demonstrated by advance calculation (design basis accident analysis). This analysis must be based on conservative assumptions wherever uncertainties exist.

Selected design basis accidents require proof to be supplied, that certain limits (temperatures of the fuel elements, pressures, stresses and strains in components of the primary cooling systems) are not reached, with the provision that this requires no manual measures to be taken in the first 30 min. The dose levels listed in the Radiation Protection Ordinance for accidental radioactivity releases (Sect. 5.6.6) must not be exceeded.

5.4.1 Events Exceeding the Design Basis

Sequences of events exceeding the design basis and leading to severe accidents—despite measures taken by severe accident management measures—must have a probability of occurrence of less than 10^{-5} to 10^{-6} per reactor year [4–6]. This must be shown by probabilistic safety analysis (Sect. 5.4.2).

5.4.2 Probabilistic Safety Analyses (PSA)

Probabilistic safety analyses are not part of the valid licensing procedures, which use only deterministic criteria. However, they have proved their value as supplements to the deterministic approach. Probabilistic safety analyses begin with the assessment of initiating events (malfunction and defects in plant components) as well as external events like earthquakes etc.. Analyzing the sequence of events requires exact knowledge of all safety systems of a nuclear power plant.

The results of a PSA are frequencies per reactor year for the failure of specific components of the safety systems and for the occurrence of specific accident sequences. In this way, weak spots in technical safety can be identified and the engineered safeguards design of a nuclear power plant can be optimized. When used in risk analyses, these values also serve for relative comparisons, e.g. between various others energy systems.

The results of PSA show that the frequency of events gradually decreases from safety level 1 to safety level 2 to safety level 3 to safety level 4. The calculated occurrence of core meltdown upon failure of the safety systems lies beyond safety level 4 in the range of a target of 10^{-5} to 10^{-6} per reactor year to be reached. The results of probabilistic safety analyses are associated with uncertainties stemming from uncertainties of the data, assumptions, and methods applied in various ways [4, 5, 7–9].

5.5 Atomic Energy Act, Ordinances, Regulations

Most countries operating nuclear reactors or fuel cycle facilities issued an Atomic Energy Act. The "Act on the Peaceful Uses of Nuclear Energy and the Protection against Its Hazards" (Atomic Energy Act) establishes the legal frame for the peaceful utilization of nuclear power.

The provisions of the Atomic Energy Act are supplemented by additional ordinances (in case of Germany as an example), such as

- the Radiation Protection Ordinance,
- the Ordinance on Safety Commission and Reporting Duties under the Atomic Energy Act,
- the Ordinance on Insurance Cover under the Atomic Energy Act.

In addition, there are regulations about technical safety which serve as a basis in the licensing practice of the licensing and supervisory authorities, such as

- safety criteria and guidelines for the design of nuclear power plants,
- guidelines, e.g., about the specialized knowledge required for nuclear power plant personnel.

5.6 Detailed Design Requirements at Safety Level 1

At safety level 1 (Sect. 5.3.2.1), the key role is played by the thermodynamics, neutron physics and mechanical design of the nuclear reactor and the properties of used materials in components, such as the reactor pressure vessel, pumps and pipes. In addition, training of the operating personnel must be ensured.

5.6.1 Thermodynamic Design of LWRs

For achieving higher redundancy, the plant design is split into several identical coolant systems for heat removal from the reactor core. Present pressurized and boiling water reactors have three or four identical cooling circuits with associated coolant pumps, steam generators, feedwater systems, emergency core cooling systems etc. connected to the reactor pressure vessel. Figure 5.3 shows the primary coolant circuit system of a modern PWR as described in Chap. 3. This includes the pressurizer for coolant pressure control and stabilization. In the pressurizer, electric heaters increase pressure through evaporation of the pressurized water. A pressurized-water spray system in the pressurizer condenses the steam, thereby lowering the pressure. When the pressure becomes too high, relief valves above the pressurizer can automatically release steam into an expansion vessel in the reactor containment and thus prevent overpressure failure of the primary cooling system.

The coolant pressure in the cooling circuits and in the pressure vessel of a PWR is chosen such (15.5 MPa) that nominal power of the fuel rods of the reactor core cannot give rise to local or subcooled boiling. In addition, there must be a sufficient margin relative to the critical heat flux. Because of corrosion and embrittlement problems, the temperature of the zircaloy cladding of a fuel rod should not exceed 350 °C [10–12].

Critical heat flux on the surface of a fuel rod would give rise to departure from nucleate boiling (DNB). At this critical level of the heat flux a vapor film is produced on the surface of the fuel rod. This causes the temperature on the surface of the fuel rod to rise so strongly as to cause failure (break) of the zircaloy cladding. The Departure from Nucleate-Boiling Ratio (DNBR) is defined as the ratio between the critical heat flux and the current heat flux on the surface of the fuel rod:

$$DNBR = \frac{q''(critical)}{q''(actual)}$$

where q" is the heat flux [W/cm^2] on the surface of the fuel rod.

In the design of the PWR core, this ratio is chosen as DBNR = 1.80. The critical heat flux is determined on the basis of empirical relations [10, 11].

As a design criterion for the maximum power of a fuel rod, the associated central fuel rod temperature must not reach the melting point of UO_2 fuel (2,865 °C) under

Fig. 5.3 Primary coolant circuit system of a PWR (fourfold redundant primary cooling system) [2]

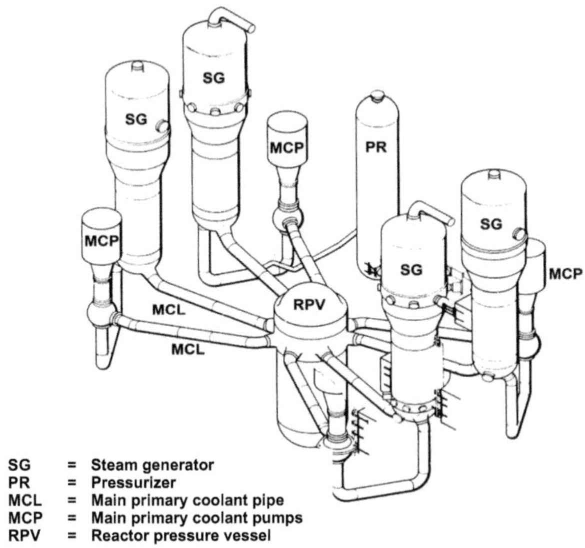

any condition. In a loss-of-coolant accident (LOCA), the maximum cladding temperature of the fuel rod should not exceed 1,200 °C [1, 2, 13].

5.6.2 Neutron Physics Design of LWRs

Preparatory critical experiments must be run to confirm the calculated enrichment by fissile U-235 in the fuel, the number of fuel elements, and criticality, k_{eff}. This implies the slight supercriticality necessary at the beginning, which must be compensated by burnable poisons (boron, gadolinium), boric acid, and poisons in partly inserted control rods. This initial positive excess in k_{eff} compensated by burnable poisons is consumed during reactor operation mainly by the negative contribution arising from the buildup of fission products and also by the depletion of fissile material. This is followed by tests confirming the negative contribution of control/shutdown rods (see also Chap. 2).

The aggregate negative contribution of all shutdown rods to k_{eff} must compensate the positive contribution arising between full load (including a positive contribution assumed in accident analysis) and zero power. At zero power, a relatively low coolant temperature must be assumed.

These are the most important engineered safeguards design parameters of LWRs:

The negative Doppler fuel temperature coefficient and the negative coolant temperature coefficient.

For a 1.3 GWe PWR, these are a

– Doppler coefficient in the range of -2.5×10^{-5} (K^{-1})
– coolant temperature coefficient in the range of -2×10^{-4} (K^{-1})

For a 1.3 GWe BWR, they are a

– Doppler coefficient in the range of -2×10^{-5} (K^{-1})
– coolant void coefficient in the range of -1.3×10^{-3} (per % steam volume increase)

The effective prompt neutron life time l_{eff} is about 2.5×10^{-5} (s) for all LWR cores. The negative safety coefficients, together with the delayed neutrons (see Sect. 2.10), guarantee a safety-oriented feedback and control behavior of LWRs. This will be demonstrated for two examples below.

5.6.2.1 Stable Time Behavior of Power When Absorber (Control) Rods Are Withdrawn in a PWR

Starting from a constant reactor power level, P_0, of a PWR, which is lower than the nominal power, limited withdrawal of the absorber (control) rods by a few cm (Fig. 5.4)—as an example—shall produce a positive ramp type increase in k_{eff} resulting in a $k_{eff} = 1.002$ within an interval of 20 s (Fig. 5.5). Initially, this raises the relative power, $P(t)/P_0$, and the fuel temperature in the reactor core. After a delay of several seconds, radial heat conduction in the fuel rods of the reactor core also increases the cladding tube temperature and, as a consequence, the coolant temperature T_C as well. An increase in fuel temperature by $\Delta T_F(t)$, through the negative fuel Doppler coefficient, practically instantaneously causes a negative Doppler reactivity,

$$\Delta\rho_D = \Delta T_F \cdot \left(-2.5 \times 10^{-5}\right)$$

and, after a short delay, through radial heat conduction in the fuel rods, the coolant temperature increase ΔT_C causes a negative coolant temperature reactivity,

$$\Delta\rho_C = \Delta T_C \cdot \left(-2 \times 10^{-4}\right).$$

Both negative feedback reactivity contributions counteract the positive initial reactivity produced by withdrawal of the absorber (control) rods until the total reactivity becomes zero. This stabilizes the reactor power at a slightly higher level of roughly $P(t)/P_0 = 1.17$ when

Fig. 5.4 Fuel element of a pressurized water reactor with a rod cluster control element [2, 3]

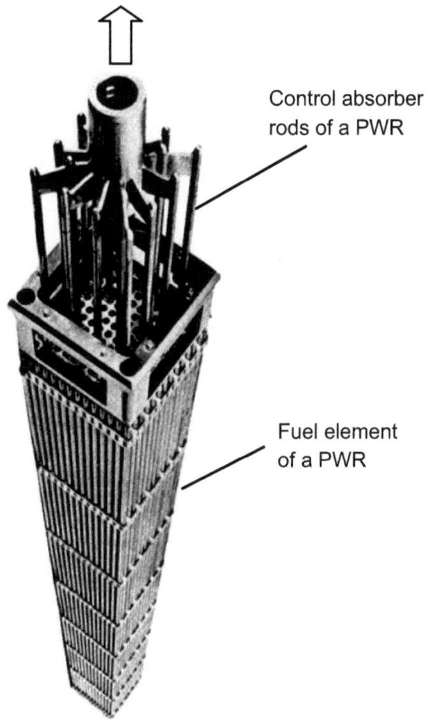

$$\Delta\rho_{CR} + \Delta\rho_{D} + \Delta\rho_{C} = 0$$

is reached.

Figure 5.5 shows the different time-dependent reactivity contributions, ρ_{CR}, ρ_{D}, and ρ_{C}, and the associated time-dependent relative reactor power, $P(t)/P_0$. This indicates the importance of the negative safety coefficients to power stabilization at higher relative reactor power levels. If the coefficients of reactivity (Doppler fuel temperature coefficient and coolant temperature coefficient) were not negative, the relative reactor power will rise uncontrolled (dotted line).

Conversely, inserting the absorber (control) rods would give rise to a negative reactivity ramp. The relative reactor power $P(t)/P_0$ will drop and, as a consequence, also the fuel and coolant temperatures would decrease. These negative changes of temperatures now produce a positive reactivity feedback.

This type of automatic feedback control of a PWR core can be supplemented by minor movements of the control rods. Such support by the control rods allows the change in power to occur faster [2, 3].

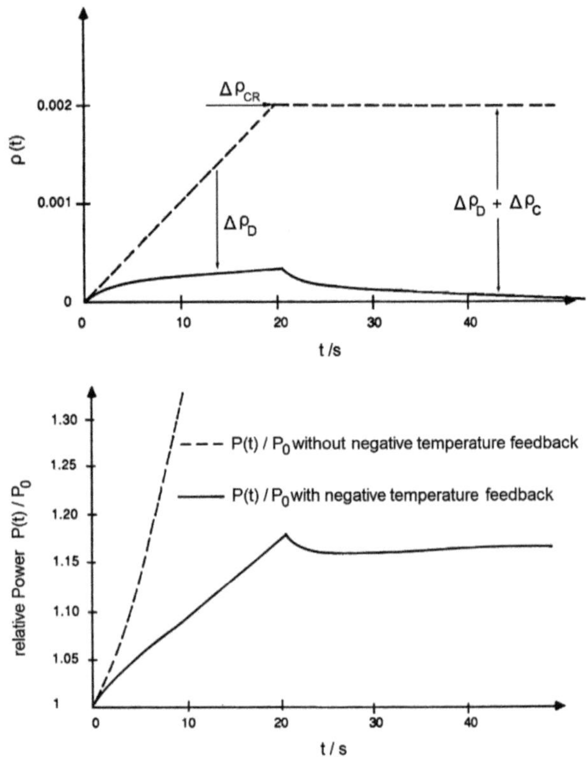

Fig. 5.5 Reactivities and relative power $P(t)/P_0$ as a function of ramp type axial movement of absorber (control) rods (without and with temperature feedbacks of reactivity) [3]

5.6.2.2 Self-Regulation Characteristics of a BWR Under Required Power Changes

BWRs have a negative Doppler coefficient of fuel temperature (Sect. 5.6.2)

$$\text{of approx.} - 2 \times 10^{-5} \text{K}^{-1}$$

and a negative coolant void coefficient

$$\text{of approx.} - 1.3 \times 10^{-3} \text{per\% steam volume increase.}$$

In this way, BWRs can be controlled directly via the speed of the internal coolant recirculation pumps (Fig. 5.6). When more power is needed at the turbine-generator system or more steam is required by the turbine requiring an opening of the steam valve at the turbine (Fig. 3.17), the speed of the recirculation pumps is raised. This increases the cooling water throughput, and rises the axial level of the boiling zone in the reactor core rises. As a result of the decreasing void volume fraction in the reactor core, a positive reactivity increase (positive contribution to k_{eff}) is initiated via the negative coolant void coefficient. The power and the fuel temperatures

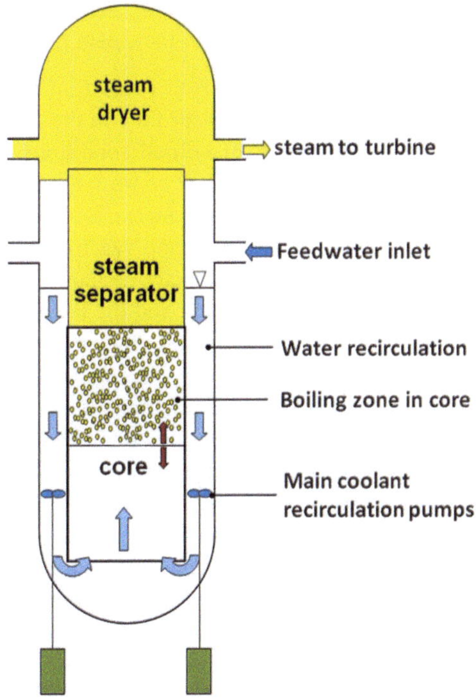

Fig. 5.6 Self-regulation characteristics of reactor power in a BWR through changes in speed of the internal main coolant pumps [3]

increase. The raising fuel temperatures in connection with the negative Doppler coefficient of fuel temperature cause a negative reactivity feedback which stabilizes the process at a higher reactor power level. In this way a BWR can be controlled in a power range above about 60 % of nominal power solely by varying the speed of the axial coolant pumps located in the annulus between the reactor core and the inner wall of the reactor pressure vessel [3].

In lower power ranges control by the recirculation pumps is supported by movements of the absorber (control) rods.

5.6.3 Instrumentation, Control, Reactivity Protection System (Safety Level 2)

Instrumentation implies monitoring important measured data by

- in-core instrumentation, such as the aeroball system, miniature fission chamber detectors, continuously measuring self-powered neutron detectors [6, 14–17];
- out-of-core neutron flux measurements covering the whole range of power from startup to nominal power output; the out-of-core neutron flux instrumentation furnishes important signals to the reactor protection system; it comprises the pulse range at zero power, a medium range, and the power range [6];

– temperatures, pressures, pump speeds, water levels in the reactor pressure vessel, in the pressurizer, and in the steam generators.

The thermal reactor power is determined by measurement of the inlet and outlet temperatures and the coolant flow in the four cooling circuits.

Disturbances and off-normal conditions, respectively, initiate countermeasures by the control systems. The control system is no safety system. Its actions counteract the course of disturbances. In case of power changes, the control system supports self-regulation of the reactor by control rod movement (PWR) or by changing the speed of the main coolant pumps and initiate control element movements (BWR). The pressure and the level of water in the pressurizer are regulated by heating the water or spraying water for condensation of the steam in the pressurizer. On the whole, the control system keeps a number of important safety-related measured parameters stable within preset limits, in this way preventing unnecessary actuation of the reactor protection system.

When limits of normal operating ranges are exceeded, e.g. 112 % of nominal power, the reactor protection system automatically intervenes to support the control system (Fig. 5.7). It can shut down the reactor by dropping the absorber rods (scram) or reduce reactor power specifically by feeding boric acid. In the case of a reactor scram, the reactor protection system at the same time automatically initiates emergency core cooling, emergency power supply, and isolation of the reactor building (containment).

The reactor protection system captures the data necessary for accident detection, e.g., reactor power too high, water levels too low, pump speed too low, etc.. It has triple redundancy and operates in a 2-out-of-3 logic, i.e. when the initiation criteria are exceeded in two out of three redundant lines, the reactor is shut down.

5.6.4 Mechanical Design of a PWR Primary Cooling System

The primary cooling system of a PWR consists of the reactor pressure vessel, steam generator, pumps, a pressurizer, and the piping connected to the reactor pressure vessel (Fig. 5.3). In normal operation, this system is under a coolant pressure of 15.5 MPa. The reactor pressure vessel e.g. is made of high-strength 16 MND5 steel in the EPR (Chap. 3) or 22NiMo37 in a German BWR, with a stainless steel liner on the inside. Also the pipes, pump casings, pressurizer, and parts of the steam generators have stainless steel liners.

5.6.4.1 Reactor Pressure Vessel Design

German reactor pressure vessels made by KWU for the PWR-1300 design—as an example—are made up of forged rings joined by circumferential welds (Fig. 5.8). The pressure vessel is designed to a pressure of 17.6 MPa, a temperature of 230 °C,

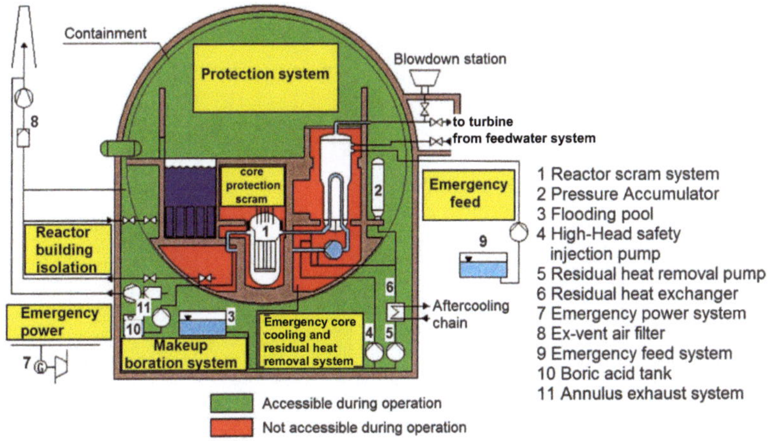

Fig. 5.7 Reactor protection system of a PWR [1, 4]

and a neutron fluence of 10^{19} nvt for neutrons with a kinetic energy >0.1 MeV. The wall of the pressure vessel is 25 cm thick in the cylindrical part [2, 4].

The mechanical stresses in the pressure vessel wall are caused by

- loading as a result of dead weight,
- internal pressure,
- thermal stresses due to temperature gradients.

The mechanical stresses produced must be determined in accordance with the ASME Boiler and Pressure Vessel Code [18] and the RSK Guidelines [13]. A distinction must be made between primary and secondary stresses. Primary stresses are caused by the internal pressure and dead weight and cannot be relieved by plastic deformation. Secondary stresses are relieved by plastic deformation (thermal stresses). One important criterion in the ASME codes [18] and RSK Guidelines [13] is that the primary stresses in the undisturbed part, e.g., the cylindrical wall, must not exceed the value of $0.33 \cdot \sigma_B$ or 0.67 $\sigma_{0.2}$ at operating temperatures ($\sigma_B =$ compressive stress, $\sigma_{0.2}$-stress at 0.2 % strain). This must be demonstrated by stress and fatigue analyses for all load cases occurring.

Compliance with these criteria is able to exclude so-called failure by ductile fracture or brittle fracture (failure by crack growth with limited leakage) in a reactor pressure vessel. This conclusion is supported, inter alia, by the fact that higher internal pressures of approx. 24–26 MPa cause the seal of the top shield to leak as a result of straining of the top shield screws. The leakage area then would correspond to up to a 69 cm^2 leak.

Compliance with the provisions of the ASME code [18] must always be verified by several independent expert consultant organizations. Those provisions were laid down internationally between 1970 and 1980 after the basic principles had been

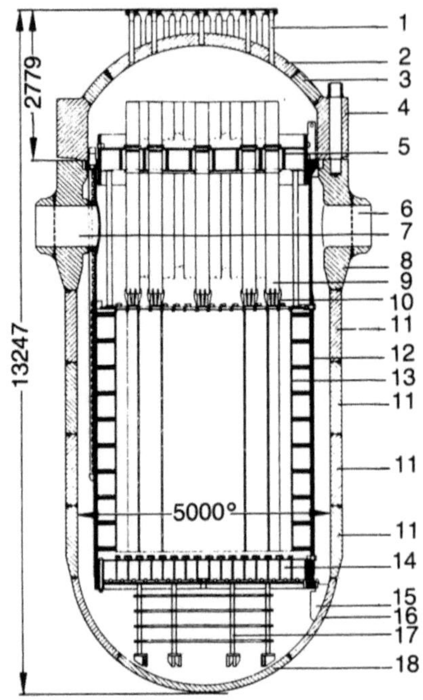

Fig. 5.8 Reactor pressure vessel of a PWR-1300 with internals [2, 4]

1 Control rod nozzle
2 Top hemispherical dome
3 Vessel closure head ring
4 Vessel closure head flange ring
5 Top grid plate
6 Coolant inlet nozzle
7 Coolant outlet nozzle
8 Core shroud flange ring
9 Control rod shroud tube
10 Grid plate
11 Forged ring
12 Core barrel
13 Core baffle
14 Bottom grid plate
15 Core barrel support structure
16 Bottom zone ring
17 Core support structure
18 Bottom hemispherical dome

confirmed in experimental programs (Heavy-Section Steel Technology (HSST) program in the United States [19–27]).

Unlike the four cooling circuits of a PWR, the reactor pressure vessel cannot be built redundant (Fig. 5.3). For this reason, the rules valid today were elaborated with particular care. A special role in this effort was played by brittle fracture behavior and the changes in brittle fracture characteristics as a result of welding processes during manufacture, temperature gradients or materials fatigue due to corrosion and neutron exposure.

The steel of the reactor pressure vessel or the welds could contain minute cracks or slag inclusions. When certain stresses in the material are exceeded, such minor cracks could become unstable (continue to grow larger) and cause the vessel or other components of the cooling circuits to fail. These problems were clarified through the development of fracture mechanics techniques [25–30] and in many notch impact tests. The change in the so-called NDT (Nil Ductility Temperature) for assessing notch impact toughness can be verified by advance specimens in the reactor pressure vessel during reactor operation. In this way, it is also possible to determine in advance the point when the maximum permissible neutron fluence is reached for the wall of the reactor pressure vessel.

The NDT is influenced, above all, by the existence of small fractions of copper, phosphorus, and sulfur in the steel alloy. As a consequence, these elements must be kept below preset concentration levels (Cu <0.1 %, S <0.01 %, P <0.01 %) [2, 30].

Precisely defined notch impact specimens of the base metal of the reactor pressure vessel must exhibit not less than 68 J notched bar impact work at a temperature of NDT + 33 K [4, 30]. The prescribed minimum temperature for the reactor pressure vessel under nominal pressure during reactor operation or under accident conditions is 50 °C.

5.6.4.2 Quality Assurance and In-Service Inspections (Basis Safety)

In addition to the design conditions and rules referred to above, quality assurance throughout the manufacturing process must ensure that all components have the required toughness of the base metal and the welds. For purposes of fracture mechanics the size of any small cracks must be below critical lengths [2, 4, 28–30].

Non-destructive test methods (such as surface crack inspection and ultrasonic inspection) are used to determine crack size and crack distribution. Figure 5.9 shows the manipulators for ultrasonic inspection within the reactor pressure vessel and the manipulators for ultrasonic inspection of the coolant inlet and coolant outlet nozzles in the upper region of the reactor pressure vessel.

5.6.4.3 Hydrostatic Test of the Reactor Pressure Vessel

As an integral test, a hydrostatic water test is carried out at 1.3 times the design pressure (22.8 MPa). In this pressure test, the temperature is not more than 55 °C above the NDT temperature. This is far below the operating temperature. As the fracture toughness decreases with temperature, such a pressure test is close to the design limits of the reactor pressure vessel. After this pressure test, the ultrasonic tests must be repeated. These in-service inspections must be repeated—as an example in Germany—during plant life time at regular intervals of 8 years [2, 4, 28].

5.6.4.4 Leak-Before-Break Criterion

In accordance with the fracture mechanics findings, small through-cracks far below the critical crack length already would give rise to leakages of the reactor pressure vessel (leak-before-break criterion) [4, 29, 30]. These can be detected in reactor operation.

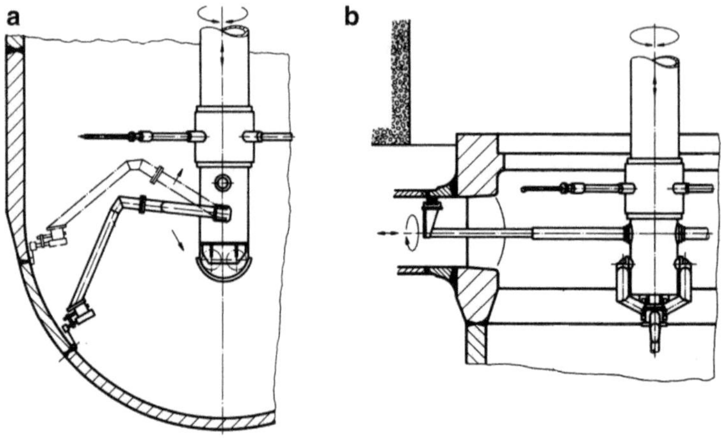

Fig. 5.9 Manipulators for ultrasonic inspection in the hemispherical bottom (**a**) and in the region of the inlet nozzle (**b**) of the reactor pressure vessel [2]

5.6.4.5 Experimental Findings About Pressure Vessel Failure

Within the HSST Program of the Oak Ridge National Laboratory in the United States [2, 19–24], model pressure vessels with large artificial cracks were made to rupture at high overpressure. However, the vessel material was found to be so tough in these experiments that major plastic deformation occurred before break. Such deformations made the cracks applied less sharp-edged, thereby reducing the stress peaks arising from notch action. Ductile failure without any artificial crack faults took at least twice the design pressure level.

5.6.5 Reactor Containment

The cooling systems, which carry the high primary coolant pressure of 15.5 MPa, must be enclosed in an outer containment (Chap. 3) with the following functions and capabilities (see Figs. 5.7 and 5.10):

- In normal operation and under accident conditions, keep releases of radioactivity into the environment within permissible limits;
- accommodate the heat stored in, and released from, the primary cooling system in a loss-of-coolant accident and remove it through active cooling systems together with part of the decay heat (afterheat);
- protect the primary system and steam generators against external impacts.

The design pressure of the containment is determined in terms of its ability to accommodate all the water released and evaporated from the primary system (full-pressure containment). Moreover, the underlying assumption in PWRs is that a

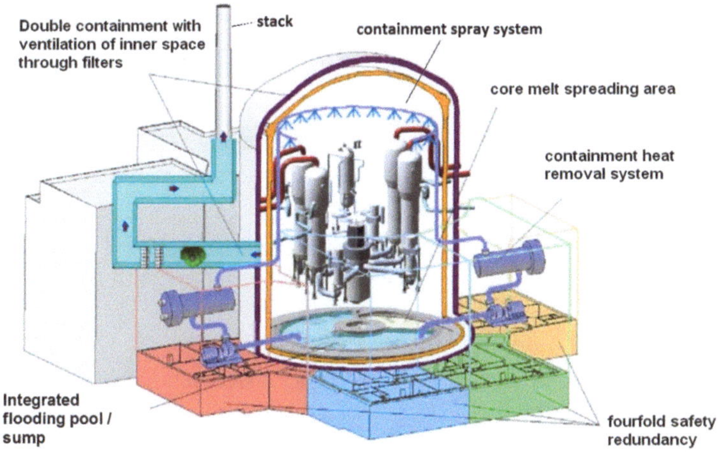

Fig. 5.10 EPR double containment with RPV, cooling systems and molten core spreading area [1]

steam generator fails in addition and its secondary-side water content is taken up by the containment. Finally, also the steam needs to be taken into account which is produced as a result of the emergency cooling water taking part of the energy stored in the secondary water of the steam generators [13]. This leads to design values of approx. 0.6 MPa, depending on the volume of the containment. The design criteria of the outer containment with respect to primary and secondary stresses are similar or identical to those applying to the reactor pressure vessel. The required leak rate of e.g., 0.25 % per day must be verified prior to commissioning and during reactor operation at prescribed intervals. Pressure tests must be conducted at regular intervals before startup and afterwards.

5.6.5.1 Different Designs for Reactor Containments

The containments of PWR and BWR plants have some characteristic differences in design:

- There are containments made of prestressed concrete with an inner steel liner, and containments made entirely out of steel (see Chap. 3).

  For protection against external impacts, the reactor containment is, e.g. additionally surrounded in Germany by a thick steel reinforced prestressed concrete shell which sustains

  - an impulse-type load associated with, e.g. a postulated airplane crash,
  - a pressure wave in connection with, e.g. a postulated chemical explosion.

This design also covers other external impacts, such as tornados, hurricanes, flooding or tsunamis. In more recent plants, the outer concrete shell has a wall thickness of 1.80 m (Kraftwerk Union PWR or BWR) or about 2.6 m (EPR). It also serves as a shield against radiation towards the outside in case the inner containment was radioactively contaminated in an accident.

The space between the containment and the prestressed outer concrete shell is held at a slightly negative pressure relative to the internal pressure and the atmospheric pressure of the external environment by means of a blower. This makes uncontrolled leakages to the outside impossible in normal operation. The air taken in is discharged from the stack through filters.

5.6.5.2 Safety Systems in the Containment

The safety systems in the containment are summarized schematically in the illustration of Fig. 5.10 for the EPR containment.

- The containment spray system, which cools the atmosphere of the containment after a loss-of-coolant accident, condenses the steam released and thus accelerates pressure reduction. The water for spraying is taken from the in-containment refueling water storage tank (IRWST) (see Fig. 5.10 or Sect. 3.1 for PWRs and Sect. 3.2 for BWRs).
- During the recirculation phase the low pressure emergency core cooling system takes in the water from the containment sump for cooling (Kraftwerk Union PWR) or from the IWRST (EPR and AP1000 or US-APWR in Sect. 3.1) or from the water pool or the pressure/suppression chamber of BWRs (Sect. 3.2).
- A containment heat removal system decreases the temperature and pressure in the containment over the medium term (Fig. 5.10).
- Two redundant valves are used for containment isolation, one of which is installed inside, the other one outside the containment. Building isolation is initiated especially after a loss-of-coolant accident or when higher radioactivity levels are detected in the containment.

5.6.6 Analyses of Operating Transients (Safety Level 3, Design Basis Accidents)

The course of various operating transients must be studied for the following accidents with and without failure of the scram system. Transients with failure of the scram system are also referred to as ATWS (Anticipated Transients without Scram). The operating transients listed below—as an example—must be studied as a design basis:

- Failure of the main heat sink, e.g., as a result of closing of the main steam valve with the off-site (auxiliary) power supply functioning.
- Failure of the main heat sink with the off-site (auxiliary) power supply unavailable.

- Faulty opening of the main steam line valves.
- Complete failure of feedwater supply.
- Maximum reactivity increase as a consequence of withdrawal of control elements or groups of control elements at full power.
- Depressurization as a consequence of inadvertent opening of the pressure vessel safety valve.
- Maximum reduction of core inlet temperature due to disturbances on the steam generator secondary side.

In these accidents, the permissible stresses and temperatures of the reactor pressure vessel and the cooling system must not be exceeded. The boration system (secondary shutdown system) and the heat removal systems must be designed so that the reactor core can be shut down safely in these accidents and remains subcritical (Fig. 5.7).

5.6.6.1 Operating Transients of LWRs with the Reactor Shutdown System Functioning (Safety Level 3)

Disturbances of steady-state reactor operation arise from imbalances between heat generation and heat removal. This raises the coolant temperatures and coolant pressure. In all cases, the reactor protection system will shut down the reactor in the shortest possible time when limits of power or coolant temperatures have been exceeded or the pump speed limit of the reactor protection system has been underrun. For instance, the shut down rods of the reactor scram system, operating by the failsafe principle, drop into the reactor core from the top under gravity in about 2 s (PWR) or are pushed into the reactor core under pressure from below (BWR) in a similar time period. There must be another, diverse shutdown system (boric acid system) in case the first shutdown system was to fail (Sects. 3.1 and 3.2).

One example is described below for a PWR. The accident sequence for a BWR is similar.

5.6.6.2 Loss of Off-Site (Auxiliary) Power Supply (Emergency Power Case) with Scram Functioning

Failure of off-site power supply causes the emergency power diesel systems to start up and supply the most important components of the PWR with electricity. The instruments and some smaller electrical components are supplied from batteries. However, the power of the emergency diesel systems is generally not sufficient to supply the large main coolant pumps of the primary system, the main feedwater pumps for the steam generators, and the main cooling water pumps of the turbine condenser. Figure 5.11 shows the key components of the steam circuit of as, e.g. the KWU-PWR.

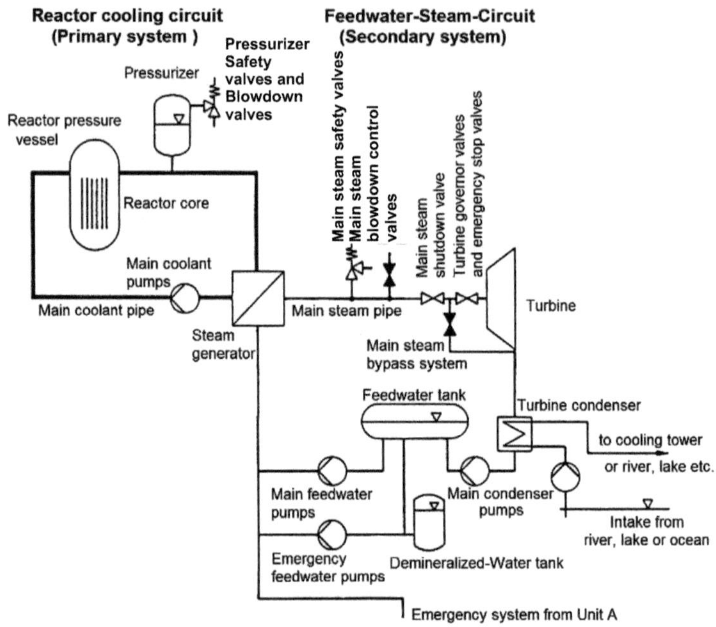

Fig. 5.11 PWR with the key components for steam production [4]

In this accident, the turbine and the generator are separated first from each other. The main coolant pumps of the primary cooling circuits, the main feedwater pumps, and the main cooling water pumps loose speed and coast down. When 93 % of the design speed of the main coolant pumps is underrun, the absorber-shutdown rods drop into the core (reactor scram) and the turbine emergency stop valve closes. Also the valve for the main steam bypass closes, as cooling the turbine condenser fails when the cooling water pumps coast down. Closure of these valves (turbine, condenser) and steam production in the steam generators, which is continued for the time being, cause the main steam pressure to rise. However, this rise of steam pressure and coupled steam temperature on the secondary side of the steam generators can be limited by opening the main steam blowdown valves downstream of the steam generators when 7.0 MPa are exceeded. The escaping steam acts as a temporary heat sink. Initially, heat conduction in the steam generator pipes slightly raises the coolant temperatures and coolant pressure on the primary side. The pressurizer valve opens briefly, limiting the primary pressure. As a result of scram, the cooling water temperature in the reactor core decreases. Also the primary coolant pressure drops at the same time. On a medium term, the primary coolant temperature and the primary coolant pressure rise again slightly because the main coolant pumps coast down to lower speeds. Over the longer term, however, the primary coolant temperatures and the coolant pressure drop because the afterheat heat production decreases and afterheat cooling systems start working (Fig. 5.7).

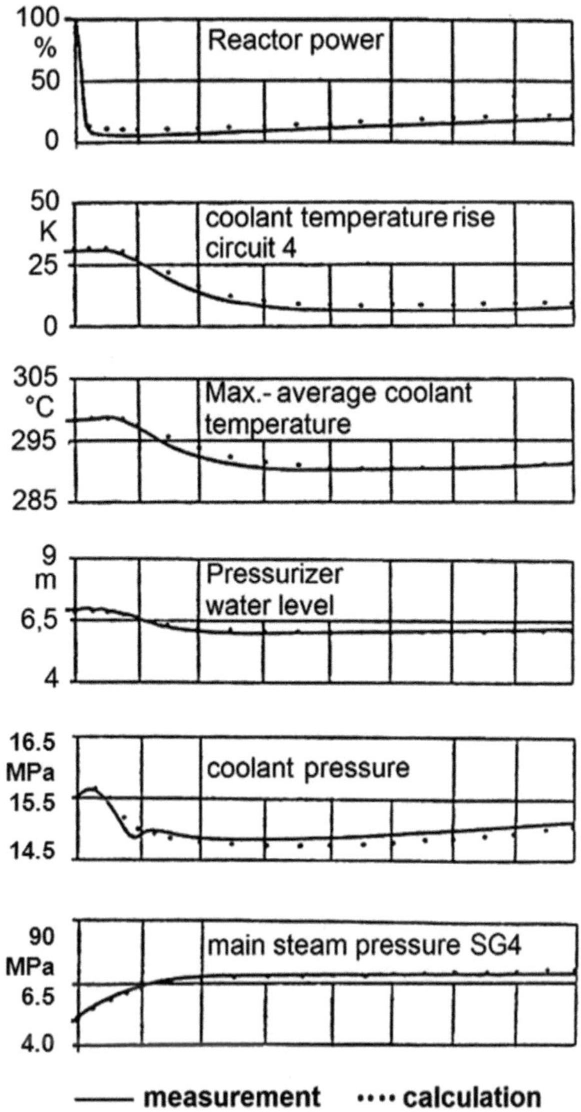

Fig. 5.12 Power, temperatures and pressures in the reactor pressure vessel and cooling circuits in case of loss of off-site (auxiliary) power with scram functioning [2, 3]

This is shown in Fig. 5.12. At the same time, calculated results are compared with measured data for power, primary coolant pressure, and primary coolant temperatures as well as the water level in the pressurizer. The experiments were performed in a German 1.2 GWe PWR (Biblis) [2, 3].

5.6.6.3 Computer Codes for Accident Calculations for PWR and BWR

Computer codes, e.g. RELAP [31] or TRAC [32–35] etc. are available to compute the course of accidents in PWRs and BWRs. The theoretical models underlying these computer codes were verified repeatedly in out-of-pile test rigs.

5.6.7 Transients with Failure of Scram (Safety Level 3)

Safety level 3 of the safety concept also requires accidents including failure of the scram system to be considered. This is why, again by way of example, the emergency power case with failure of the off-site (auxiliary) power supply for the main pumps etc. and subsequent failure of the scram system will be described below. Also in this case, accident behavior is similar for BWRs.

In the previous case with an example for the KWU-PWR (Sect. 5.6.6.2) the reactor scram was supposed to function. As the scram system now is supposed to fail and the nominal power will remain constant in the beginning of the accident, the pressure and coolant temperature in the primary cooling system will rise. When the pressurizer relief valves would not open (lower pressure limit), the primary coolant pressure is limited to 17.6 MPa by opening of the pressurizer safety valves. As the primary coolant temperature (Fig. 5.13a) rises strongly together with the primary pressure (Fig. 5.13b), the negative coolant temperature coefficient takes effect, initially automatically reducing the reactor power to roughly 25 % of nominal power (Fig. 5.13c). The high temperature in the primary cooling system also causes temperatures and pressure on the secondary side in the steam generator to rise. The main steam blowdown valves do limit the pressure to 7 MPa (escaping steam is the temporary heat sink), but the emergency feed pumps are unable to supply enough water as water for heat transfer of about 25 % of nominal power is still needed. Consequently, the secondary steam temperature continues to rise and the steam generators gradually run dry. This causes the primary coolant temperature and the primary pressure to rise again. As a consequence, the reactor is shut down to the afterheat level via the negative coefficient of coolant temperature after about 450 s. Afterwards the reactor must be kept shut down by the boric acid (secondary) shutdown system. In addition the pressure must be further decreased until the low pressure emergency and afterheat cooling systems provide for further cooling after about 800 s.

5.6.8 Loss-of-Coolant Accidents (LOCAs)

Loss-of-coolant accidents can arise from breaks or cracks of pipes or from faulty sticking of valves in the open position. For large pipes, the leak size is assumed in

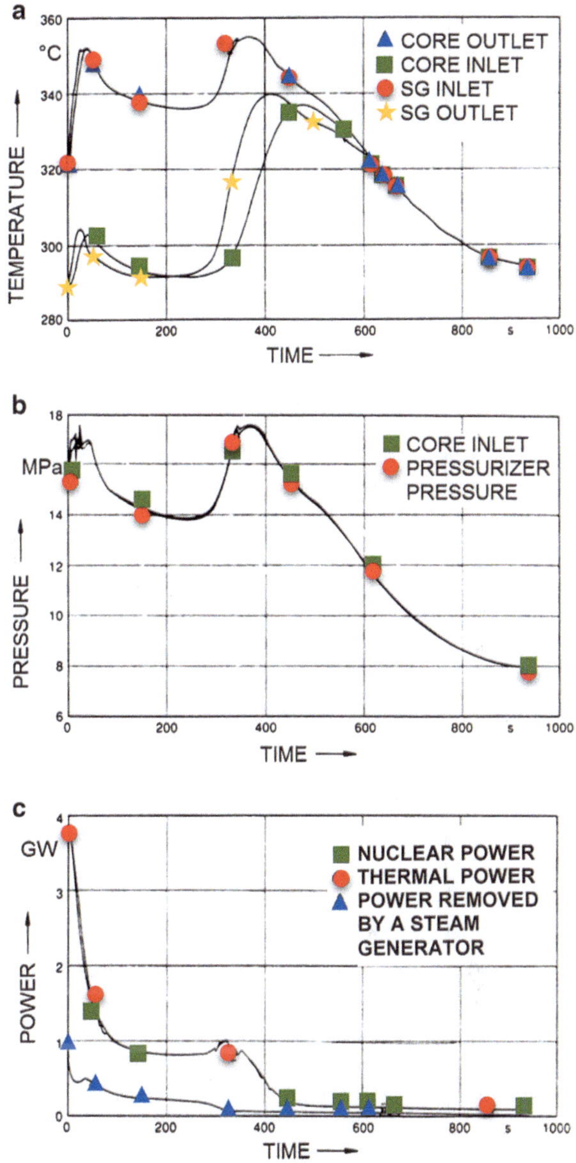

Fig. 5.13 Behavior of temperature, pressure, and power as a function of time (s) in the core and steam generator (SG) after an emergency power case without scram [3] (example of analysis for a KWU-PWR)

accordance with fracture mechanics (Sect. 5.6.4) as 0.1F (F = pipe cross section). However, for the design of the emergency core cooling system, the larger so-called 2F break (guillotine type rupture of pipe) is analyzed conservatively [13]. This highly conservative case will be described here by way of example in Fig. 5.14.

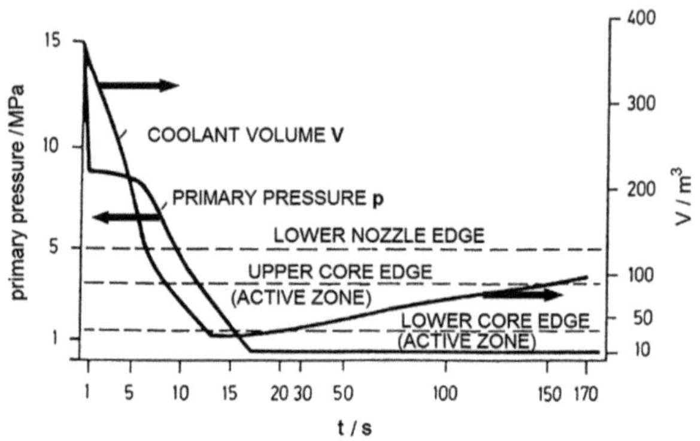

Fig. 5.14 Development as a function of time of coolant pressure and coolant volume in the reactor pressure vessel after a large leak in the cold leg [2, 3]

5.6.8.1 Loss-of-Coolant Accident Due to 2F Break of the Main Coolant Pipe

Large coolant leakages (2F break) do not necessarily require the scram system. Although the scram signal is initiated when the lower limit of primary pressure is underrun, void formation starting in the reactor core shuts down the reactor very quickly automatically via the negative reactivity produced.

When the leak has opened, a pressure relief wave passes through the reactor pressure vessel and piping system. At the leakage points, the critical outlet velocity of a two-phase mixture (steam—water) is established. The pressure of the primary cooling circuit drops to about 0.5 MPa within approx. 17 s. All primary coolant (water) leaves the primary coolant system within approx. 13–15 s (Fig. 5.14). Film boiling starts in the cooling channels of the fuel elements, and cladding temperatures rise very sharply to 750–1,000 °C (Fig. 5.15). Afterwards, automatic shutdown of the reactor power to the level of decay heat power (afterheat level), and simultaneous cooling by steam, cause a temporary decrease of temperature at the fuel cladding. However, this temperature again rises slightly until the borated cooling water of the pressure accumulators (Fig. 5.16) takes over core cooling after some 50–90 s. These pressure accumulators start feeding borated water when the primary pressure falls below 2.5 MPa (Phase 2, Fig. 5.16). The water level in the reactor pressure vessel rises up between 20 s and 170 s (Fig. 5.14) to the upper core edge again covering the core. It then fills the pressure vessel up to the lower inlet and outlet nozzle edges. (Height of the rising center level in the pressure vessel (Fig. 5.8) and the volume on the right scale of Fig. 5.14 are correlated).

This borated water cools the reactor core and keeps the reactor subcritical. After further decrease of the primary pressure to <1.0 MPa the low pressure emergency water injection system takes over (Phase 3, Fig. 5.16). When the reservoir of borated water has been depleted, the water originating from the loss of coolant and collecting in the reactor building sump is taken in, cooled in the residual heat exchangers, and returned to the primary system (Phase 4, Fig. 5.16). In this way cooling by borated water from the pressure accumulators is supported by the

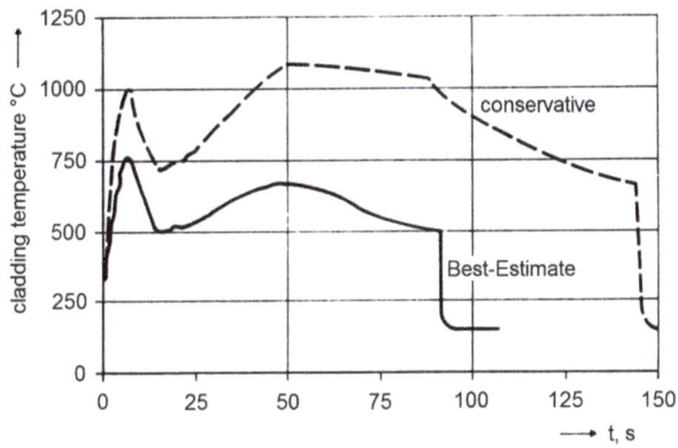

Fig. 5.15 Cladding temperature of fuel rod during 2F-Break accident (average fuel rod and fuel rod with highest conservative temperatures) [2–4]

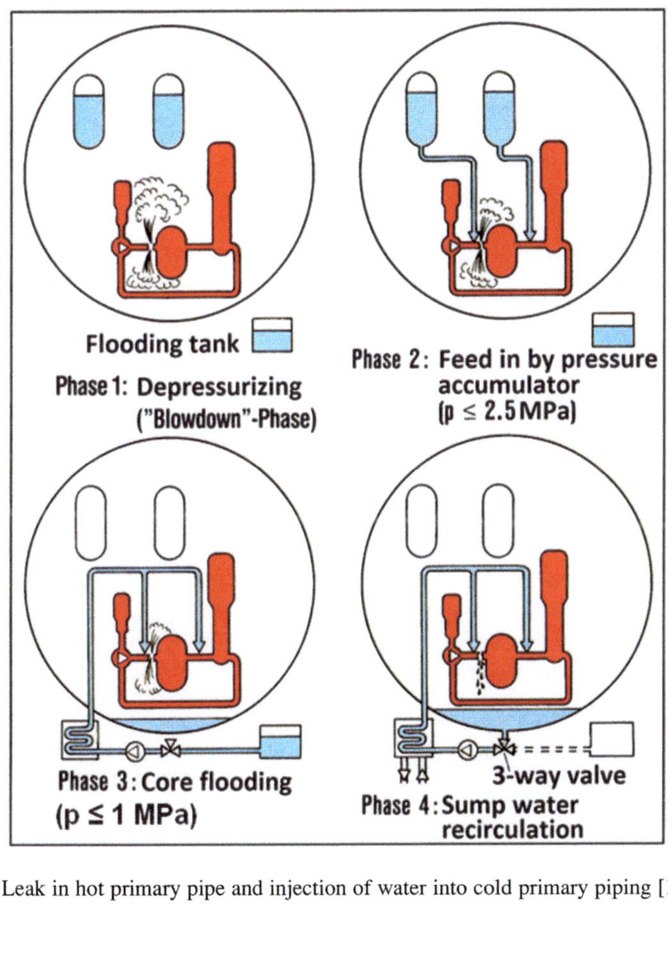

Fig. 5.16 Leak in hot primary pipe and injection of water into cold primary piping [1]

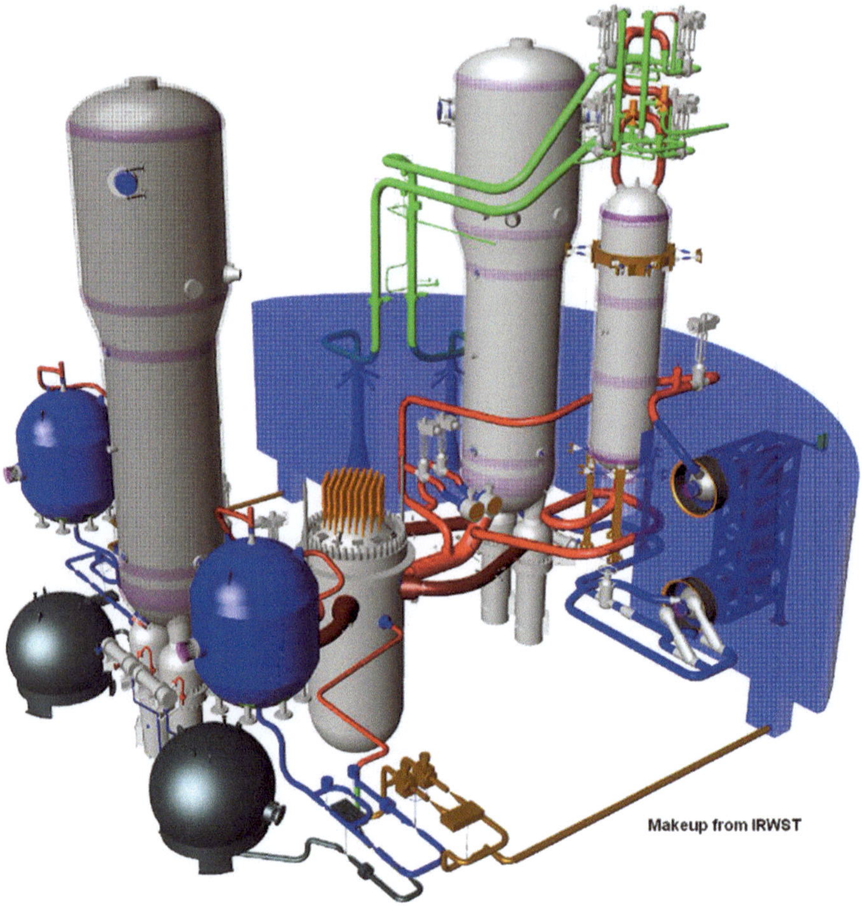

Fig. 5.17 Emergency core cooling systems of AP1000 [36]

low-pressure emergency core cooling systems (Kraftwerk Union PWR) provided their power supply is available.

As described already in Sect. 3.1 the emergency cooling systems of other modern PWRs (AP1000, US-APWR, EPR) are slightly different. The collected sump water in KWU-PWRs is replaced by the large water volume of the IWRST in AP1000, US-APWR and EPR. This is shown in Fig. 5.17 for the case of AP1000. The high pressure core flooding system and the accumulators are also installed in the AP1000. The automatic depressurization system (ADS) allows depressurizing the primary coolant system directly into the IWRST in AP1000, US-APWR and EPR (In the older KWU-PWRs this is achieved by accident management measures).

5.6.8.2 Loss-of-Coolant Accident Due to Minor Coolant Leakages

Regarding the possible consequences the large 2F break is not necessarily the most extreme or most severe accident involving leakages. Minor leakage accidents rather follow a different sequence of events with similar consequences. In a small leak, the primary coolant pressure will drop, and the filling level in the pressure vessel will decrease. This initiates scram. When a primary pressure of 11 MPa (Kraftwerk Union PWR) or 9.2 MPa (EPR) has been reached, the high-pressure safety pumps feed borated water from the flooding tanks into the primary system. In case of AP1000 high pressure borated water is fed from the core make-up tanks.

When the water supply in the flooding tanks has been depleted, pressure in the primary cooling system must be reduced further. This is done by opening the main-steam-blowdown valves (secondary system depressurization). This decreases pressure and temperature on the secondary side. When the pressure on the primary side drops below 1 MPa, the low-pressure emergency core cooling systems take over further cooling. In case of EPR, AP1000 and US-APWR high capacity relief valves are actuated to depressurize the primary coolant system to <1 MPa within a short time period.

Depending on the size of the coolant leak, at least (in case of the KWU-PWR as example)

- one or two out of the four high-pressure feed systems,
- one out of the two pressure accumulator feeds,
- one or two out of the four low-pressure emergency core cooling systems

must be available for feeding or for recirculation operation as a minimum requirement for accident control. In that case, the PWR in the long run can be transferred into the safely coolable mode.

However, serious damage to the reactor core can develop when three or all four systems of the emergency core cooling and residual heat removal systems or the emergency power supply fail. In that case, severe core damage will arise and the core will melt down.

References

1. Kessler G (2012) Sustainable and safe nuclear fission energy. Springer, Heidelberg
2. Smidt D (1979) Reaktorsicherheitstechnik, Sicherheitssysteme und Störfallanalyse für Leichtwasserreaktoren und Schnelle Brüter. Springer, Berlin
3. Emendörfer D et al (1993) Theorie der Kernreaktoren, Band 2: Der instationäre Reaktor. BI Wissenschaftsverlag, Mannheim
4. Deutsche Risikostudie Kernkraftwerke Phase B (1990) Verlag TÜV Rheinland, Köln
5. Kersting E et al (1993) Safety analysis for boiling water reactors. A summary, GRS-98. Gesellschaft für Anlagen- und Reaktorsicherheit, Garching
6. Boland JF (1970) Nuclear reactor instrumentation (in-core). Gordon and Breach Science, New York, NY

7. Deutsche Risikostudie Kernkraftwerke Phase A (1980) Gesellschaft für Reaktorsicherheit (GRS). Verlag TÜV Rheinland
8. (1975) Reactor safety study: an assessment of accidents risks in US commercial nuclear power plants. In: Rasmussen NC (ed) US Nuclear Regulatory Commission, WASH-1400 (NUREG-75/014), Washington
9. Lewis EE (1977) Nuclear power reactor safety. Wiley, New York, NY
10. Märkl M (1976) Core engineering and performance of pressurized water reactors. Kraftwerk Union AG, Erlangen
11. Tong LS et al (1979) Thermal analysis of pressurized water reactors. American Nuclear Society, LaGrange Park, IL
12. Lahey RT et al (1977) The thermal hydraulics of a boiling water nuclear reactor. American Nuclear Society, LaGrange Park, IL
13. RSK-Leitlinien für Druckwasserreaktoren (1996) Fassung 11.96, BAnz Nr. 214 vom 05.11.1996
14. Bachmann G et al (1971) Leittechnik des Kernkraftwerks Stade. Atomwirtschaft 16:600–602
15. Aleite W et al (1971) Regeleinrichtungen des Kernkraftwerks Stade. Atomwirtschaft 16:597–599
16. Aleite W et al (1987) Leistungsregeleinrichtungen und Begrenzungen von Druck- und Siedewasserreaktoren. Atomwirtschaft 32:129–134
17. Aleite W et al (1987) Leittechnik in Kernkraftwerken. Atomwirtschaft 32:122–128
18. ASME Boiler and Pressure Vessel Code (2007) Section III-Div. 1, Rules for construction of nuclear power plant components. The American Society of Mechanical Engineers, New York, NY
19. Laufs P (2006) Die Entwicklung der Sicherheitstechnik für Kernkraftwerke im politischen und technischen Umfeld der Bundesrepublik Deutschland seit 1955. Dissertation, Historisches Institut, Abteilung Geschichte der Naturwissenschaften, Universität Stuttgart, Germany
20. Merkle JG et al (1975) An evaluation of the HSST program intermediate pressure vessel tests in terms of light water reactor pressure vessel safety. ORNL-TM-5090
21. Derby RW et al (1974) Test of 6-inch-thick pressure vessel, Series 1: Intermediate test vessels V-1 and V-2. ORNL-4895
22. Bryan RH et al (1975) Test of 6-inch-thick pressure vessels, Series 2: Intermediate test vessels V-3, V-4, and V-6. ORNL-5059
23. Merkle JG et al (1976) Test of 6-in.-thick pressure vessels, Series 3: Intermediate test vessel V-7. USNRC Report ORNL/NUREG-1
24. Whitman GD (1976) Heavy section steel technology program quarterly progress report for January thru March 1976. ORNL-TM-28
25. Griffith AA (1921) The phenomena of rupture and flow in solids. Phil Trans R Soc Lond A221:163–198
26. Irwin GR (1958) Fracture. In: Handbuch der Physik, Bd. VI, Elastizität und Plastizität. Springer, Berlin
27. Williams PT et al (2012) Fracture analysis of vessels – Oak Ridge FAVOR, v 12.1, computer code: theory and implementation of algorithms, methods, and correlations. ORNL/TM-2012/567
28. Kußmaul K et al (1978) Maßnahmen und Prüfkonzept zur weiteren Verbesserung der Qualität von Reaktordruckbehältern für Leichtwasserreaktoren, 2. VGB-Kraftwerkstechnik 58 (6):439–448
29. Hoffmann H et al (2007) Das Integritätskonzept für Rohrleitungen sowie Leck- und Bruchpostulate in deutschen Kernkraftwerken. VGB Power Tech 7:78–91
30. Kußmaul K (1984) German basis safety concept rules out possibility of catastrophic failure. Nucl Eng Int 12:41–46
31. Bruch CG (1976) RELAP4/Mod.5, a computer program for transient thermal-hydraulic analysis of nuclear reactors and related systems, user's manual, vol 1: RELAP4/MOD5 description, SRD-113-76. Aerojet Nuclear Company

32. Vigil JC et al (1981) Accident simulation with TRAC. Los Alamos Sci 3:36–52
33. Liles DR et al (1978) TRAC-P1, an advanced best-estimate computer program for PWR LOCA analysis, vol 1, methods, models, user information and programming details, LA-7279-MS, vol 1. NUREG/CR-0063
34. Liles DR et al (1979) TRAC-P1A, an advanced best-estimate computer program for PWR LOCA analysis. Report LA-7777-MS, NUREG/CR-0665. Los Alamos Scientific Laboratory
35. Liles DR et al (1981) TRAC-PD2, an advanced best-estimate computer program for pressurized water reactors loss-of-coolant accident analysis. Report LA-8709-MS, NUREG/CR-2054. Los Alamos Scientific Laboratory
36. Foulke L (2011) Safety systems for pressurized water reactors. IAEA Workshop 2011. http://www.iaea.org/NuclearPower/Downloads/Technology/meetings/2011-Oct-Simulators-WS/Foulke.3-Passive.Safety-1.pdf
37. US Nuclear Regulatory Commission (1999) Regulatory Guide 3.54, spent fuel heat generation in an independent spent fuel storage installation (Revision 1, Jan. 1999). http://www.nrc.gov/NRC/RG/03/03-054rl.html
38. Xu Z et al (2005) Impact of high burnup on PWR spent fuel characteristics. Nucl Eng Des 151:261–273

Chapter 6
Probabilistic Analyses and Risk Studies

Abstract The first comprehensive study to determine the risk of LWRs by probabilistic methods was the US Reactor Safety Study WASH-1400 in 1975. Similar studies in other countries, e.g. Germany, followed. The methodology starts with the event tree analysis followed by the probabilistic analysis. This is continued by an analysis of the radioactivity release for the different accident sequences. Subsequently meteorological data and models for atmospheric diffusion and aerosol deposition are used to determine the radioactivity concentration and radiation dose to individuals in the areas around the plant. Countermeasures can be taken, e.g. evacuation or relocation, to lower the radioactive exposure of the population. Finally the results of event tree and fault tree analysis for different PWRs and BWRs (presently operating and more recent (future) designs) are presented. In addition, the results of reactor risk studies in the USA (WASH-1400) and in Germany are reported and discussed.

Modern Light Water Reactors (LWRs) have up to four redundant cooling systems, four redundant emergency cooling systems, four redundant emergency power supplies etc. as well as diverse emergency power supplies (diesel engines, steam turbines, gas turbines etc.). Severe core damage can occur when, e.g. three out of four of all four emergency cooling systems fail.

Failure of all of these redundant safety systems ultimately will cause a core meltdown accident. In probabilistic risk studies, such a failure of all systems is assumed conservatively, without accounting for the fact that the operating team could e.g. repair the failed components in due time.

Experience has shown that any component of specific safety systems can fail within a given period of time. In that case, event tree diagrams (fault tree analyses) and assignment of individual probabilities of component failure can be employed to compute the failure of major single systems and the overall probability of an accident sequence (probabilistic safety analysis, PSA).

G. Kessler et al., *The Risks of Nuclear Energy Technology*, Science Policy Reports,
DOI 10.1007/978-3-642-55116-1_6,

6.1 General Procedure of a Probabilistic Risk Analysis

The first comprehensive study performed to determine the risk of LWRs by probabilistic methods, the US Reactor Safety Study WASH-1400 [1], was published in 1975. Similar studies were performed later also in other countries, such as Germany [2].

The risk, $R_{m,i}$ attributed to a type i accident initiated by a type m event (e.g., leak in a primary coolant pipe) in a reactor plant can be described in a simplified way by this relationship:

$$R_{m,i} = F_{m,i} \cdot D(C_{m,i})$$

where

- $F_{m,i}$ is the annual frequency of occurrence of a type i reactor accident initiated by a type m event,
- $C_{m,i}$ is the amount of radioactive material, expressed in Bq, released into the environment from the reactor outer containment during a type i accident initiated by a type m event,
- D is the damage to the population resulting from the release $C_{m,i}$ of radioactivity. D depends on a number of other environmental parameters, such as atmospheric conditions, population distribution, etc. (Figs. 6.1 and 6.2).

The annual frequency, $F_{m,i}$, of occurrence of an accident is determined in detailed probabilistic analyses applying event tree and fault tree methods [1–7]. In those studies, the failure probability of all relevant components of a safety system is taken into account (Fig. 6.2). In determining the radioactivity release, $C_{m,i}$, the sequence of accident events must be assessed as a function of time in the reactor core, the pressure vessel, and the surrounding containment. This then results in the radioactivity, $C_{m,i}$, (fission products, activation products and actinides), released into the environment from the containment after containment failure. Subsequently, meteorological data and models of atmospheric diffusion and aerosol deposition are used to determine the radioactivity concentration and the radiation dose to which individuals in the environment of the plant are exposed, countermeasures being taken into account. Finally, health physics data (Chap. 4) are used to determine the probability of disability or death as a result of the exposure dose.

6.2 Event Tree Method

An accident sequence is started by an initiating event, e.g., a leak in a pipe in the primary coolant system. The safety system of the reactor reacts to this initiating event, and the consequences of the sequence of accident events are controlled, provided that the safety system functions sufficiently well.

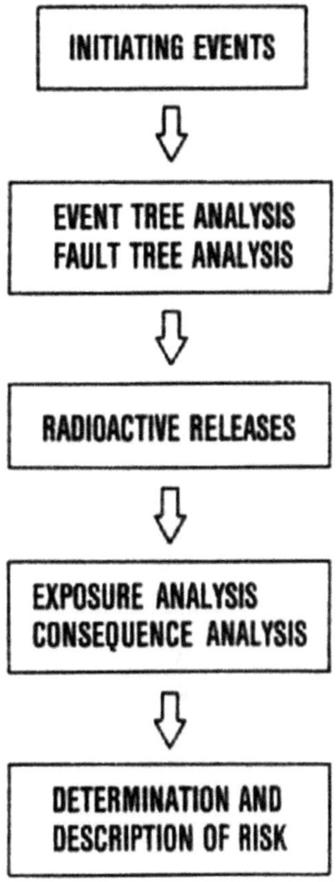

Fig. 6.1 Major steps in a reactor risk study [3]

Table 6.1 shows some data for the frequency of initiating events used in the first risk studies for a KWU-PWR in Germany [8].

Only if components of the safety system fail on a major scale there will be a release of radioactivity. Figure 6.2 shows a simplified event tree for a loss-of-coolant accident in a PWR. In this case, the accident is initiated by the break of a pipe. This pipe rupture is assumed to occur with a frequency of f_m per reactor year.

The further development of this accident is then mainly determined by the availability of the electricity power supply. Failure of the electric power supply to operate the emergency core cooling system (ECCS) is assigned the probability of p_1. Since electricity is either available or not, the probability of power being available and the ECCS functioning properly is $(1-p_1)$ If there is no electricity available the ECCS will not work and the core, after having lost its coolant, will melt down partially or entirely for lack of cooling. In that case, there may well be major releases of radioactivity into the environment as a result of a failure of the containment.

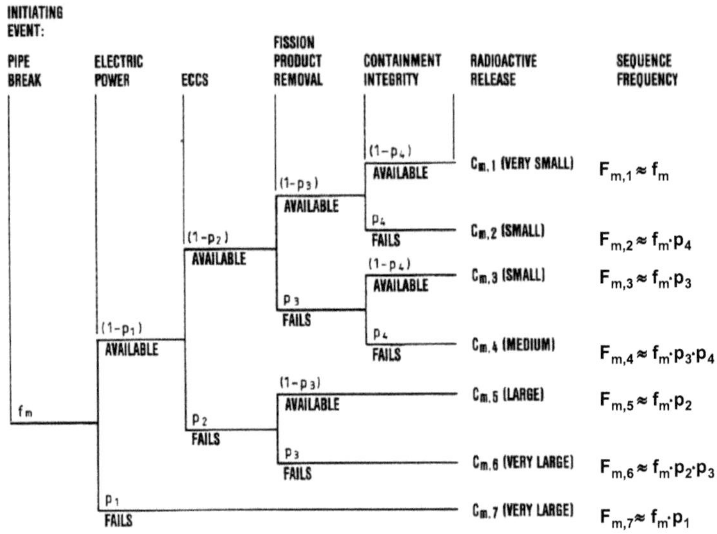

Fig. 6.2 Simplified event tree for a loss-of coolant accident in a water cooled reactor [3]

Table 6.1 Frequencies per year of accident initiating events for a 1.3 GW(e) KWU-PWR at Neckarwestheim (Germany) [8]

Initiating event	Frequency of initiating event per year (f_m)
Loss of electrical auxiliary power supply	2.5×10^{-2}
Loss of main heat sink and loss of main feed water supply	3.8×10^{-2}
Small leak (25–80 cm^2) in main primary coolant pipe	1.5×10^{-4}
Small leak at pressure vessel (1–6 cm^2)	2.5×10^{-3}

If power is available, the next possible event will be a potential failure of the ECCS, which must be assigned the probability of p_2. The availability of the ECCS is again characterized by $(1-p_2)$.

If fission products are released in the course of an accident, the fission product removal system mitigates the radioactivity release into the containment. The failure probability of this system is characterized by p_3, its availability $(1-p_3)$.

The final barrier against the release of radioactivity is the leak tightness (integrity) of the outer containment. The probability of this containment function failing is called p_4 the availability of that function, $(1-p_4)$. If the containment integrity is preserved, releases of radioactivity can only be slight, but if the containment leaks, radioactivity can escape into the environment, depending on the size of the leak.

From the results of this simplified event tree of Fig. 6.2 it can be seen that radioactivity releases can vary between very small and very large releases, depending on the level at which the safety systems fail.

Since the individual components of the safety system are characterized by their high availabilities and, consequently, very low failure probabilities (p_1, p_2, p_3, $p_4 \ll 1$), the probability of the availability of $(1-p_1)$ etc. of these safety components can each be assumed to be approximately equal to 1. Consequently, the sequence frequency at the upper end of the branching of the event tree of Fig. 6.2

$$F_{m,1} = f_m\,(1-p_1)\,(1-p_2)\,(1-p_3)\,(1-p_4) \approx f_m$$

The radioactive release caused in this case, $C_{m,1}$ is negligible. On the other hand, the radioactive releases, $C_{m,6}$ or $C_{m,7}$ as a consequence of a failure of the electricity supply followed by a failure of the ECCS and of the integrity of the outer containment would be very large, because the core would melt and a large fraction of the radioactive inventory would be released from the outer containment. The frequency of occurrence, however, of this maximum accident is extremely low, amounting to $F_{m,6} = f_m \cdot (1-p_1) \cdot p_2 \cdot p_3 \approx f_m \cdot p_2 \cdot p_3$ and $F_{m,7} = f_m \cdot p_1$.

In a detailed event tree analysis, many more details must be considered, such as the individual functions of the ECCS, etc. Interdependencies of the different events may lead to systematic consequential failures and to the elimination of branches in an event tree.

6.3 Fault Tree Analysis

The fault tree analysis approach is used for numerical assessment of the failure probabilities of larger units of the safety system. It breaks these larger systems down into single components, concluding about the failure probability of a larger unit from the failure probabilities of such individual components by taking into account the way in which the logical functions of the single components are interrelated. If common mode failures are possible they must be accounted for. Often, fault trees must be developed to such detail that available data on single equipment components or human error can be applied from experience. Uncertainties in reliability data are taken into account by entering not only single values, but distribution functions for the failure probabilities of single components. For other components, such as emergency power diesel systems, statistical data directly available from experience are applied. When determining the failure rate of pipings under pressure, methods of probabilistic fracture mechanics must be used in addition.

6.4 Releases of Fission Products from a Reactor Building Following a Core Meltdown Accident

6.4.1 *Initiating Events*

Initiating events controlled by the safety system will not contribute to risk. Accordingly, major contributions to the overall risk must be expected to arise only with large scale failure of fuel rod claddings and from those events in which the reactor core will melt down partially or completely because of an extensive failure of the safety systems. Event tree studies show that the occurrence of a major leak in a main coolant pipe followed by a failure of the respective safety systems (emergency cooling systems and afterheat removal systems) will cause the reactor core to melt down; within a few hours the molten core can even penetrate the reactor pressure vessel. In an early superheated phase of the reactor core, hydrogen will be generated in a reaction between water and the zirconium in the fuel claddings.

After having penetrated through the reactor pressure vessel, the hot core material will contact the concrete foundation slab of the reactor building and gradually melt into the concrete. This will cause water bound in the concrete to be released and react with the melt, which will generate hydrogen. Depending on the type of concrete used, also CO, may be released. For the further sequence of accident events it was assumed in a pessimistic estimate in the US and German risk studies [1, 2, 7, 9] that the molten core contacts and evaporates the sump water, thus increasing the pressure in the containment.

6.4.2 *Failure of the Containment*

A number of penetrations through the containment building for locks, pipes and cables may develop leaks with a certain failure rate. In this case, radioactivity could escape to the outside. If, on the other hand, the containment is assumed to remain tight, i.e., preserve its integrity, the core meltdown accident described above would generate vapor, H_2, CO and CO_2, and raise the pressure in the containment so that the permissible design pressure of the outer containment could be exceeded. After failure of the outer containment integrity, radioactivity could be released into the environment.

In the US Reactor Safety Study, WASH-1400 [1], and the German Risk Study [2, 9], also the case of large scale hydrogen detonation and of a potential steam explosion resulting from a contact between molten hot core material and water was discussed. This was assumed to occur with a certain probability in the bottom part of the reactor pressure vessel.

6.4.3 Releases of Radioactivity

Radioactivity may be released from the reactor core in the following events [10–13]:

- In cladding tube failures the gaseous and highly volatile fission products are released.
- As the fuel is heated to melting temperature and melts, fission products as well as chemical compounds with lower melting points will be released as aerosols.
- During interaction of the melt with the concrete, aerosols are generated.

In a pipe leak of the primary circuit, or if the reactor pressure vessel has been penetrated by molten core material, the gaseous and volatile fission products will enter the containment. They can be retained there by active removal (e.g. spray) systems and by diffusion, coagulation, condensation, sedimentation and thermophoretic processes of the aerosols. Radioactive decay makes the retention by the containment more effective, the longer its integrity can be ensured. Studies (German Risk Study, Phase A [2]) showed that the time after which the maximum pressure would be exceeded and the containment fails would be like 5–12 days (depending on the concrete composition). Within such a time period of about 5 days or more, the concentration of airborne aerosols decreases already by orders of magnitude [11]. This is shown by Fig. 6.3.

6.4.4 Distribution of the Spread of Radioactivity After a Reactor Accident in the Environment

In reactor risk studies, releases of radioactivity are determined for various accident categories and all kinds of meteorological conditions at all the different reactor sites. This is used to determine a mean value for the consequences of radioactive exposure (early deaths, late health consequences, soil contamination). However, an accident at a specific reactor site is determined only by the weather conditions prevailing at that time.

Computer codes, such as COSYMA [14] and RODOS [15], were written to describe this situation on the basis of the release of radioactive gases and aerosols during the accident and further spreading of this radioactivity in the atmosphere as a function of weather conditions and wind direction (Fig. 6.4). For each point at a certain distance from the reactor site it is possible in this way to determine the radioactivity in the atmosphere and contamination of the soil. The radioactive exposure of the population and the environment is determined on this basis.

The radioactive cloud causes external and internal radiation exposures of persons. The external radiation exposure is the result of

- cloud-borne radiation (radioactive nuclides),

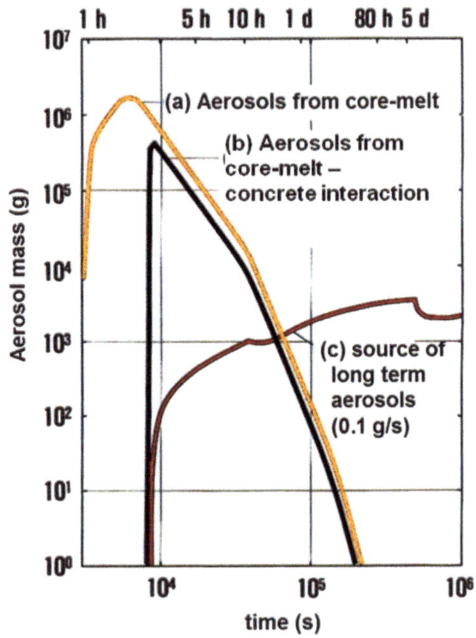

Fig. 6.3 Concentration of aerosols in the outer containment atmosphere as a function of time after release during a core melt accident [9, 12, 13]

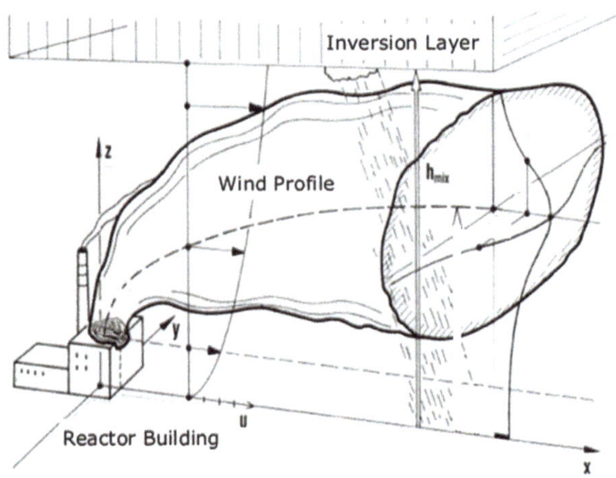

Fig. 6.4 Spread of radioactivity after a core melt accident with radioactivity release to the environment [16]

– ground-borne irradiation after surface contamination following precipitation of the radionuclides.

Internal exposure to radiation is the consequence of inhalation and ingestion of contaminated food items.

In Fig. 6.4, the cloud emitted, which is loaded with radioactive gases and aerosols, extends to the inversion layer h_{mix} and then spreads horizontally.

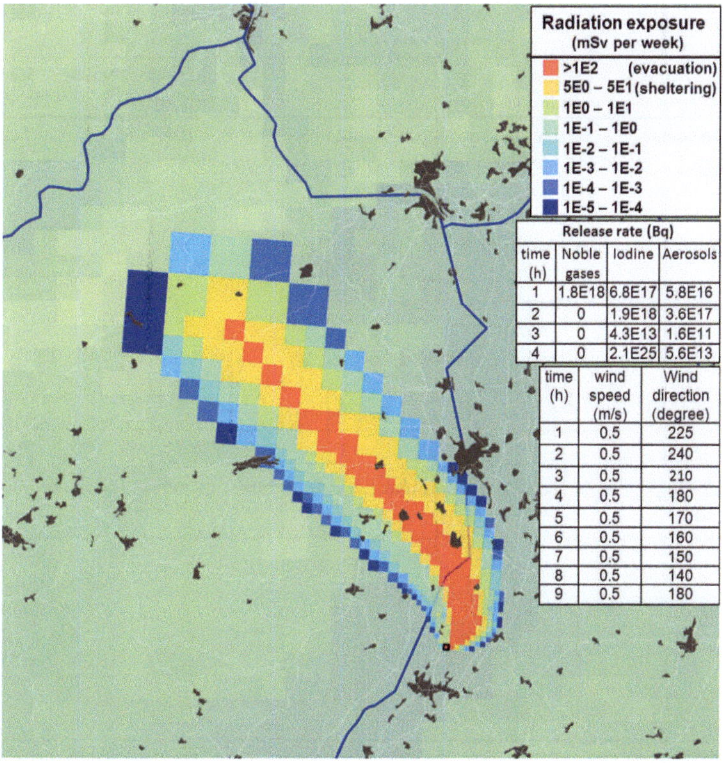

Fig. 6.5 Example for radioactivity release after a core melt accident with major release of radioactivity. The different *colors* indicate areas where the population would have to be evacuated or remain in shelters [17]

Figure 6.5 shows by way of example the spread of radioactivity released after a core melt accident into the environment. The release of radioactivity as a function of time (noble gases, iodine, aerosols), the wind speed and wind directions are assumed for this example. The different colors indicate areas where evacuation or sheltering of the population would be required.

See also [18] for more details in this book.

6.5 Protection and Countermeasures

Radioactive exposure of the population can be affected by these protective measures and countermeasures:

- sheltering in buildings and protective rooms,
- evacuation of the population,
- distribution of iodine tablets,

Table 6.2 Reference exposure dose values for initiation of protection and countermeasures [16, 19, 20]

Action	Reference dose	Reference value [mSv]	Integration time
Remain in house sheltering	Effective dose through inhalation and external radiation	10	7 days
Taking of iodine tablets	Children <18 years	50	7 days
	Persons 18–45 years	250	7 days
Evacuation	Effective dose through inhalation and external radiation	100	7 days
Temporary relocation	External radiation deposition of radioactivity	30	1 month
Longterm relocation	External radiation deposition of radioactivity	100	1 year
Food ban	Effective equivalent dose by ingestion	5	1 year

Table 6.3 Upper limits for adults in Europe for concentration of radioactive materials in food [16, 19, 20]

Radioactive nuclide	Limits of radioactivity concentration nuclide [Bq/kg] or [Bq/l]	
	Milk products	Other food
Strontium isotopes especially Sr-90	125	750
Iodine isotopes especially I-131	500	2,000
α-Emitters especially Pu-239, Am-241	20	80
Other nuclides with half-lives more than 10 days especially Cs-134, Cs-137	1,000	1,250

- ban on consuming contaminated food and water,
- relocation and blocking of areas,
- decontamination of urban areas and agricultural land.

In Europe, decisions by the authorities (Table 6.2) are based on these guiding values for various protective measures and countermeasures [16, 19, 20]. These are lower and upper limits—the population would receive during the first week—for which measures like sheltering, evacuation or relocation (1 month or 1 year) must be initiated. In addition there are limits for the different isotopes released during a reactor accident which do collect in the food like milk, vegetables, meat etc.. These limits are given in Table 6.3 in case these limits are exceeded the food must be banned.

Table 6.4 Expected frequencies for core melt down for a 1.3 GW(e) KWU-PWR at Neckarwestheim (Germany) [8]

	Initiating event	Frequency per year
Loss of coolant	Medium to large leak (80–200 cm^2) in a primary coolant pipe	9.0×10^{-8}
	Small leak in primary coolant pipe (25–80 cm^2)	1.6×10^{-7}
	Very small leak in primary coolant pipe (2–25 cm^2)	1.3×10^{-6}
	Leaks at pressurizer (40 cm^2)	4.0×10^{-7}
	Very small leak at reactor pressure vessel (1–6 cm^2)	1.8×10^{-7}
Operational transients	Loss of auxiliary power supply	1.4×10^{-6}
	Loss of main feedwater supply without loss of main heat sink	2.2×10^{-6}
	Loss of main heat sink without loss of main feedwater supply	1.8×10^{-6}
	Loss of main feedwater supply and loss of main heat sink	3.0×10^{-7}
	Break of steam line out of outer containment	1.3×10^{-7}
	Break of feedwater line out of outer containment	2.9×10^{-7}
Sum total		8.2×10^{-6}

6.6 Results of Reactor Safety Studies

6.6.1 Results of Event Tree and Fault Tree Analyses

The methods and analyses described above were applied to both PWRs and BWRs, e.g. in the US Reactor Safety Study [1] and in the German Risk Studies [2, 8, 9] for PWRs and for BWRs [7]. The results of recent risk studies for modern PWR- and BWR-designs built by Kraftwerk Union, as described in Sects. 3.1 (PWRs) and 3.2 (BWRs) will be reported as examples below.

Tables 6.1, 6.4 and 6.5 list the main findings of the event tree and fault tree analyses in a German risk study for a PWR operating at Neckarwestheim [8]. In analyzing all possible initiating events, mainly four groups of accident sequences were identified to result in core damage for the PWR:

- loss-of-coolant accidents initiated by a leak or break in the reactor coolant system,
- operational transients leading to an imbalance between the heat generated in the core and the heat removed from the core.
- reactor plant internal initiating events like fire
- reactor plant external events e.g. earthquakes, airplane crash and flooding.

The data in Table 6.4 as well as Table 6.5 result from detailed studies of the above accident sequences.

The sum total of the frequency per year of all conceivable loss of coolant accidents and operational transients is 8.2×10^{-6} per year.

Table 6.5 Internal and external initiating events [9]

	Initiating event	Frequency per year
Plant internal	Fire	1.7×10^{-7}
	External flooding	$<10^{-7}$
Plant external	Earthquake with loss of main heat sink and main feedwater supply	3×10^{-6}
	Airplane crash in reactor containment	$<10^{-7}$

The frequency for core melt due to internal fire and flooding is 1.7×10^{-7} per year. The frequency for core melt due to plant external events (earthquake and airplane crash) was analyzed in an earlier study [9] to be about 3×10^{-6} per year.

6.6.2 *Severe Accident Management Measures (Safety Level 4)*

The detailed analyses of accident sequences in the German Risk Study Phase B [9] showed that it is possible, even after the failure of safety systems, to control accident sequences by introducing so-called severe accident management measures. These can then prevent core meltdown.

In a first step the following accident management measures were investigated in the German Risk Study Phase B [9]

- depressurization of the secondary coolant system and feed in of water by mobile pumps
- depressurization of the primary coolant system by opening the pressurizer valves to the pressure level such that the high pressure emergency and other emergency coolant systems, e.g. mobile pumps can feed in water.

A number of accident sequences initiated by loss of coolant or loss of off-site (auxiliary) power supply are seen to have as their main cause in the failure of cooling water supply to the steam generators. This failure of steam generator supply would ultimately lead to core meltdown.

Accident management measures go beyond the automatic safety measures provided for by the reactor protection system [21, 22] allowing interventions by the operating personnel. Thus, for instance, the steam generators, in the late accident phase of running dry can be fed with water from the feed water tank or, by means of mobile pumps, from water reservoirs inside and outside the reactor. The decay heat (afterheat) can be removed by blowing off the steam through the main steam blowdown station (secondary feed and bleed procedure) (Fig. 5.11).

The opening of the pressurizer valves as an accident management measure in older PWR designs, e.g. KWU-PWR (see above) lowers the pressure in the primary system within about half an hour. This allows

- the high-pressure safety feed pumps to be activated at 11 MPa (KWU-DWR),
- the pressure accumulators to start feeding at 2.5 MPa (KWU-DWR),
- and the residual heat removal systems to be activated at about 1.0 MPa.

In this way, for instance, core meltdown under high pressure can be avoided, or this accident sequence can be changed into core meltdown at low pressure.

Accident management measures include the actuation of pressure relief valves or the actuation of smaller pumps of auxiliary circuits by means of batteries. For this reason the internal battery capacity is increased and reinforced accordingly.

German LWRs, in addition, were equipped after these risk studies [2, 8, 9] with a so-called sheltered control room. It is possible to shut down the reactor to decay heat (afterheat) level from this completely independent control room in case the main control room can no longer be used, e.g. because of fire or radioactive gases. Additional emergency instrumentation, e.g. filling level probes in the reactor pressure vessel or measurements of radioactivity at various points of the cooling system, allow more precise knowledge to be obtained about the operating status of the reactor [9]. Emergency electric power can be supplied by underground cables from more distant power plants. Cooling water can be supplied by deep wells situated on plant site [22].

The sum total of the core melt frequency per year applying these severe accident management measures could thereby be lowered by one order of magnitude [9].

6.6.3 Core Melt Frequencies per Reactor Year for KWU-PWR-1300, AP1000 and EPR

The core melt frequencies and radioactivity release frequencies for different PWR designs, e.g. KWU-PWR-1300, AP1000 and EPR are given in Table 6.6. The more recent PWR-designs have by an order of magnitude lower overall core melt frequencies per reactor year and so-called large radioactivity release frequencies per reactor year than the older design of the KWU-PWR-1300. EPR is designed such that even in case of a core melt accident large radioactivity releases are excluded (see Chaps. 3 and 10).

6.7 Results of Event Tree and Fault Tree Analyses for BWRs

Probabilistic risk analyses were carried out also for BWRs. Table 6.7 lists some of the most important initiating events of accident sequences in the German BWR-1,300 at Gundremmingen (Sect. 3.2.1) [7].

Table 6.6 Core melt frequencies and large release frequencies per reactor year for KWU-PWR-1,300, AP1000 and EPR [8, 9, 23, 24]

		KWU-PWR-1,300 [8, 9]	AP1000 [23]	EPR 1,600 [24]
Core melt frequency per reactor year	Internal events	8.2×10^{-6a}	4.2×10^{-7}	8.4×10^{-8}
	External events	3×10^{-6}	2.4×10^{-7}	6.4×10^{-7}
Large radioactivity release frequency per reactor year	Core melt down followed by containment failure	About 10^{-7}	1.8×10^{-8}	No evacuation No re-housing No sheltering

[a]Can be decreased by about one order of magnitude by application of severe accident management measures

Table 6.7 Frequency of initiating events (BWR-1,300) [7]

Initiating event	Frequency of initiating event per year
Loss of electrical power supply from grid	0.04
Loss of main heat sink and loss of main feed-water supply	0.3
Inadvertent opening of turbine valve or by-pass valve	0.2
Failure to close a safety and relief valve	0.1
ATWS with loss of main heat sink	$<10^{-7}$

Table 6.8 Expected frequencies for core melt down (German Reactor Risk Study [7]) for the BWR-1,300 (Gundremmingen) described in Sect. 3.2.1

	Initiating event	Frequency per year
Loss of coolant	Small leak (5–150 cm^2) in main steam line (outside containment)	2.0×10^{-7}
	Small leak (5–150 cm^2) in main steam line (inside containment)	4.0×10^{-7}
	Small leak (5–150 cm^2) in feed water line	3.0×10^{-7}
ATWS	Loss of main heat sink, no scram	10^{-6}
Operational transients	Loss of electrical power from grid	3.2×10^{-6}
	Loss of main feedwater supply	5.5×10^{-6}
	Loss of main heat sink	2.0×10^{-5}
	Loss of main heat sink and loss of main feedwater	1.5×10^{-5}
	Failure to close safety and relief valve	4.1×10^{-6}
Sum total		5×10^{-5}

Table 6.8 lists the frequencies of core melt per year for the most important accident sequences which can result in core meltdown, such as

- loss-of-coolant accident,
- anticipated transients without scram (ATWS),
- operational transients.

Table 6.9 Core melt frequencies per reactor year for KWU-BWR-1,300, ABWR, ABWR-II and SWR-1000 (KERENA) [8, 9, 23, 24]

		KWU-BWR-1,300 [7]	ABWR [25]	ABWR-II [25]	KERENA [26, 27]
Core melt frequencies [year]	Internal events	5×10^{-5}*[a]	1.6×10^{-7}	4.5×10^{-8}	$<10^{-7}$
	External events	10^{-6}	Not given in literature		

[a]Can be decreased to 4.4×10^{-6} per year by application of accident management measures

While the frequencies of occurrence of loss-of-coolant and ATWS accident sequences are in the range of 10^{-6} per reactor year, the frequencies of occurrence—initiated by operational transients—are in the range of 10^{-5} per reactor year. The frequencies of occurrence of core melts initiated by fire or earthquake are of a similar order of magnitude as in a German PWR, i.e. in the range of 10^{-6} per reactor year. This leads to a total frequency per reactor year of 5×10^{-5} [7].

Accident management measures can reduce the frequency of occurrence per reactor year by roughly one order of magnitude to 4.4×10^{-6} [7].

Such additional accident management measures [21, 22] are e.g.

- primary pressure relief by opening of the pressure relief valves and steam blowdown to the water pools (condensation pool) of the pressure/suppression chamber
- additional emergency electrical power and water supply (see Sect. 6.6.2)

6.7.1 Core Melt Frequencies for KWU-BWR-1300, ABWR, ABWR-II and SWR-1000 (KERENA)

The core melt frequencies per reactor year for other BWRs, e.g. ABWR, ABWR-II and KERENA are listed in Table 6.9. Again, as in case of more recent PWR-designs, the more recent BWR-designs, as e.g. ABWR-II and SWR-1000 (KERENA) have lower core melt frequencies per reactor year.

6.8 Release of Radioactivity as a Consequence of Core Melt Down

The results in Tables 6.4, 6.5 and 6.7 can be combined with results obtained from accident consequence analyses for assumed core meltdown and subsequent containment failure. This analysis can be simplified by grouping the results into release categories for accident sequences with the same containment failure mode. In addition to the frequencies of occurrence, also the schedule of radioactivity release from the outer containment after the onset of an accident and the fractions of fission products released can be determined. These fission product fractions refer to the total radioactive inventory of the PWR or BWR core.

In the US Reactor Risk Study [1] and in the German Risk Study Phase A [2, 9] a very conservative approach was applied, e.g. it was assumed that a core meltdown accident is followed by a steam explosion with a certain probability of occurrence. For such a steam explosion it must be assumed that the molten core is mixing with water and fragmenting into very small particles which transfer heat very quickly to the water, thus rapidly producing a large amount of steam. This release category includes the highest radioactivity release which would occur at about 1 h after the initiating event with subsequent core meltdown. More recent studies described in Sect. 10.1, however show that after a large scale steam explosions a subsequent failure of the pressure vessel and of the outer containment can be considered to be impossible.

Other release categories of e.g. the German Study Phase A [1, 2, 10] did comprise core meltdown accidents with a subsequent large scale hydrogen-air detonation and failure of the outer containment (Sect. 10.2) or so-called containment bypass accidents (Sects. 10.2.2 and 10.2.3). In these cases, openings in the outer containment of 25–300 mm equivalent diameter were assumed. Lower release categories of the Risk Studies [1, 2, 10] represented core meltdown with late containment overpressure failure. The lowest release categories cover loss-of-coolant accidents controlled by the emergency cooling systems. Since, in these cases, the reactor core is cooled sufficiently by the emergency core cooling system, the fuel element claddings will be damaged only partially and the reactor core will not melt.

6.9 Accident Consequences in Reactor Risk Studies

The US Reactor Safety Study, WASH-1400 [1], was performed for 100 reactor plants (PWRs and BWRs) in the United States. It was published in 1975, thus preceding the similar German Risk Study [2, 10] which was performed in 1979 for 25 German reactor plants. Compared with the results of the German Risk Study, and aside from slightly different safety designs of German LWRs, it was mainly the meteorological data, the population density, the purely linear dose-risk relationship as well as protective measures and countermeasures, which differed in the two studies.

It must also be emphasized that in these reactor risk studies the results are averaged over 68 different sites (WASH-1400 [1]) or 19 different sites (German Reactor Risk Study, Phase A [2]) with several hundred different weather conditions.

The frequency of core meltdown accidents was determined to be 5×10^{-5} per reactor year in the US Reactor Safety Study [1]. The largest number of early fatalities was approx. 3,300, with a probability of occurrence of 10^{-7} per reactor year. The cases of early illness are 45,000, with a frequency of occurrence of 10^{-9} per reactor year.

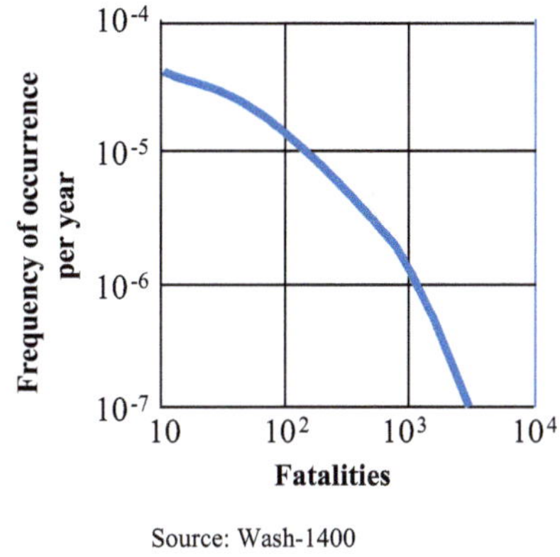

Fig. 6.6 Results of the US-Risk Study WASH-1400 [1] for 100 nuclear reactor plants

Figure 6.6 shows the complementary, cumulative frequency distribution of early fatalities, determined by the WASH-1400 Risk Study [1], which could be caused by radiation exposure after accidental radioactivity release. Results of the German Risk Study, Phase A [2, 10] are similar.

Of course, a large number of early fatalities will occur if major releases are encountered on sites with high population densities, the wind blows in the direction of the sector with the highest population density, and rain falls in the immediate vicinity, thus creating high radioactivity concentration levels on the ground.

In the US and the German Risk Studies [1, 2, 10] late fatalities at all dose levels are reported since a linear dose rate (Sect. 10.5.1) without threshold value was assumed according to ICRP (1977). The occurrence of late fatalities is therefore not restricted to only the immediate vicinity of the reactor plant, as in case of early fatalities. Also a considerable fraction of the late fatalities determined in the Risk Studies [1, 2] was due to low radiation exposures of <50 mSv (The average natural background radiation on earth is about 2.4 mSv per year or 170 mSv over 70 years life time (Sect. 4.3). Therefore, these results are still discussed controversially.

6.9.1 Use of Results of Reactor Risk Studies

The value of Reactor Risk Studies should not be exaggerated or misinterpreted or even used for forecasts in which time periods core melt accidents could occur. As their methodology is based on probabilistic considerations their results can only be used for

- comparison with the results of risk studies for other energy production systems (coal, oil, gas etc.) which are based on the same probabilistic methodology
- for the optimization of the design of the different safety systems to reach a well balanced overall safety concept for that nuclear plant for which the probabilistic safety analysis was performed.

6.9.2 Safety Improvements Implemented in Reactor Plants After the Risk Studies

As mentioned in the previous sections, rather simplified and pessimistic assumptions were made in the German Risk Study Phase A and the US Reactor Safety Study [1, 2] about the sequence of accident events in which the molten core penetrates the bottom of the reactor pressure vessel leading to subsequent accident conditions which can lead to a relatively early loss of containment integrity. This leads to an overestimation of the accident consequences and risks. More recent theoretical studies and preliminary experimental results indicate that (Chap. 10)

- the assumed large steam explosions leading to early containment failure can be considered to be impossible
- the assumed early containment failure after core melt through under high primary coolant pressure can be counteracted by primary coolant depressurization or strengthening of the anchorage of the reactor pressure vessel
- the penetration of the molten core into the concrete does not necessarily lead to contacts with the sump water. However, if it contacts sump water, a loss of containment integrity would occur only after time periods of approximately 5–12 days. Water spray systems can lower the internal pressure in the containment. Exventing filters introduced after the publication of the reactor risk studies (WASH-1400 and German Risk Study Phase B) can avoid overpressurization of the outer containment. However, the danger of a hydrogen detonation still remains. Hydrogen recombiners can decrease the amount of hydrogen released. The containment can be proven to even withstand to large severe hydrogen detonations. For reasons of aerosol physics [11], a considerable percentage of radioactive aerosols will have settled within the containment under fog or rainlike conditions and shortlived radioisotopes will have decayed away by the time containment failure was assumed to occur (Fig. 6.3). The release of radioactivity into the environment then decreases by several orders of magnitude.

These results were confirmed during the decades after the reactor risk studies appeared by large scale experiments (Chap. 10).

References

1. (1975) Reactor safety study: an assessment of accidents risks in US Commercial Nuclear Power Plants. In: Rasmussen NC (ed). US Nuclear Regulatory Commission, WASH-1400 (NUREG-75/014), Washington
2. Deutsche Risikostudie Kernkraftwerke Phase A (1980) Gesellschaft für Reaktorsicherheit (GRS). Verlag TÜV Rheinland
3. Kessler G (2012) Sustainable and safe nuclear fission energy. Springer, Heidelberg
4. Lewis EE (1977) Nuclear power reactor safety. Wiley, New York, NY

5. Farmer FR (1967) Reactor safety and siting: a proposed risk criterion. Nucl Saf 8:539–548
6. Dunster HJ (1980) The approach of a regulatory authority to the concept of risk. IAEA Bull 22 (5/6):123–128
7. Kersting E et al (1993) Safety analysis for boiling water reactors. A summary, GRS-98. Gesellschaft für Anlagen- und Reaktorsicherheit, Garching
8. Risikostudie Neckarwestheim, Bewertung des Unfallrisikos fortschrittlicher Druckwasserreaktoren in Deutschland (2001) Gesellschaft für Anlagen- und Reaktorsicherheit, GRS 175, München
9. Deutsche Risikostudie Kernkraftwerke Phase B (1990) Verlag TÜV Rheinland, Köln
10. Bayer A, Heuser FW (1981) Basic aspects and results of the German Risk Study. Nucl Saf 22:695–709
11. Bunz H et al (1981) The role of aerosol behavior in light water reactor core melt accidents. Nucl Technol 53:141–146
12. Hassmann K et al (1987) Spaltproduktfreisetzung bei Kernschmelzen. Verlag TÜV Rheinland GmbH, Köln
13. Kuczera B (1990) Aktueller Stand der Reaktorsicherheitsforschung dargestellt anhand von Ergebnissen aus der deutschen Risikostudie Kernkraftwerke – Phase B, Radioaktivität – Risiko – Sicherheit, Herausgeber Kernforschungszentrum Karlsruhe (2. veränderte und aktualisierte Auflage 1991)
14. COSYMA (1991) A new program package for accident consequence assessment, EUR-13028. European Commission, Brussels
15. Ehrhardt J et al (2000) RODOS: decision support system for off-site nuclear emergency management in Europe. Report EUR 19144. European Commission, Brussels
16. Ehrhardt J (1991) Probabilistische Unfallfolgenabschätzungen. In: Radio-aktivität/Risiko – Sicherheit. Kernforschungszentrum Karlsruhe, p 51–61
17. Hasemann I (2011) Personal communication, Karlsruhe Institute of Technology (KIT)
18. Landman C et al (2014) Relevant radiological phenomena, fundamentals of radiological emergency management, modeling of radiological situation. In: Kessler G et al (eds) The risks of nuclear energy technology: safety concepts of light water reactors. Springer, Berlin
19. Rahmenempfehlungen für den Katastrophenschutz in der Umgebung kerntechnischer Anlagen (2008) GMBI Nr. 62/63 vom 19. Dezember 2008. http://www.bmu.de/files/pdfs/allgemein/application/pdf/rahmenempfehlung_katastrophenschutz.pdf
20. Leitfaden für den Fachberater Strahlenschutz der Katastrophenschutzleitung bei kerntechnischen Notfällen (1989) Veröffentlichungen der Strahlenschutzkommission, Band 13, Gustav Fischer Verlag, Stuttgart. Also: Radioaktivität und Strahlung Grenzwerte und Richtwerte. http://osiris22.pi-consult.de/userdata/l_20/P-105/Library/data/grenzwerte-und-richtwerte-04-03-internetversion.pdf
21. Birkhofer A (1989) Anlageninterner Notfallschutz, Achtes deutsches Atomrechts-symposium, München, 1.–3. März 1989. Carl Heymann Verlag KG, Köln
22. Schenk H (1990) Maßnahmen zum anlageninternen Notfallschutz. Atomwirtschaft November:514–520
23. Cummings WE et al (2013) Westinghouse AP1000 advanced passive plant. In: Proceedings of ICAPP 2003, Paper 3235, Cordoba, Spain
24. IAEA generic review for UK HSE of new reactor designs against IAEA safety standards, EPR. http://hse.gov.uk/newreactors/reports/eprriaea.pdf
25. IAEA status report 98 – advanced boiling water reactor II [ABWR-II]. http://www.iaea.org/NuclearPower/Downloadable/aris/2013/3.ABWR-II.pdf
26. IAEA status report 98 – KERENA™ [KERENA™]. http://iaea.org/NuclearPower/downloadable/aris/2013/24.KERENA.pdf
27. Stosic ZV et al (2008) Boiling water reactor with innovative safety concept: the generation III+ SWR-1000. Nucl Eng Des 238:1863–1901

Chapter 7
Light Water Reactor Design Against External Events

Abstract LWRs must be designed against earthquakes, air plane crashes, chemical explosions, flooding, tsunamis and tornados. The design of LWRs against earthquakes must meet certain guidelines required by regulatory authorities. These distinguish between the design basis earthquake and the safe shut down earthquake. The design basis earthquake is the highest intensity earthquake which can occur according to scientific findings at the site of the nuclear power plant. In a safe shut down earthquake the fundamental safety functions of the LWR must remain fulfilled. The mechanical loads and stresses acting on nuclear power plants in an earthquake are determined by horizontal and vertical displacements and accelerations as well as the associated frequencies of vibration and the duration of the earthquake. Besides the rules recommended by regulatory authorities also two- and three-dimensional finite-element codes are employed on the mechanical analysis of the plant. Where horizontal or vertical displacements and the resultant stresses are too high, pipings and components may be supported by means of damping elements. Also the entire nuclear plant may be built on thousands of damping elements located in the foundation bottom concrete slab of the reactor building. LWR plants are designed against air plane (military or commercial) crashes into the plant. Impulse models and experiments form the basis for a shock load versus time curve which has to be applied for the design of the plant.

LWRs must also be designed against a given pressure wave resulting from chemical explosions in the vicinity of the plant.

The risk of flooding by a maximum-level flood must be taken into account on the basis of scientific findings about floods for the past 10,000 years. Similar requirements exist for tsunamis and for tornados.

Nuclear power plants are designed against the impacts of external events, such as earthquakes, airplane crashes, chemical explosions, flooding, or tsunamis and tornados.

G. Kessler et al., *The Risks of Nuclear Energy Technology*, Science Policy Reports,
DOI 10.1007/978-3-642-55116-1_7,

7.1 Earthquakes

Nuclear power plants, e.g. in Germany and in other countries must be designed against vibrations caused by earthquakes. The designs must meet specific design guidelines [1–10]. Internationally, a difference is made between [2, 11, 12]:

- Design basis earthquake:
  The nuclear plant must withstand an earthquake undamaged in such a way that it can be restarted and continue operation after automatic scram. In Germany, for instance, the probability of this design basis earthquake being exceeded must be as low as 10^{-5} per year (statistically, an earthquake exceeding that level would have to be expected once in 100,000 years) [6–9].
- Safe shutdown earthquake:
  In a safe shutdown earthquake, the fundamental safety functions of the nuclear power plant must remain fulfilled, i.e. the nuclear power plant must be shut down safely, and the residual heat must be removed safely. The radioactive materials must remain safely contained by barrier systems (containment) of the nuclear power plant.

In Germany, as an example, the guidelines contained in KTA 2201 [6, 7] and the RSK-LL [10] apply. The design basis conditions for earthquakes are defined by seismologists on the basis of historical data and the tectonic conditions in the perimeter of the nuclear power plant.

7.1.1 Definition of the Design Basis Earthquake According to KTA 2201

The design basis earthquake to be assumed is the earthquake of the highest intensity on the respective site which can occur according to scientific findings, taking into account a wider environment of the site (up to approx. 200 km around the site).

Since 1935, earthquake intensities according to Richter have been referred to as magnitudes, M (deflection of the seismograph) at approx. 100 km distance from the epicenter of the earthquake as determined by several earthquake measuring stations and by calculation. This is the applicable formula:

$$M = \log_{10}(A/A_0)$$

with A as the deflection of the seismograph and A_0 as the reference amplitude. An increase in M by one unit means a tenfold increase in earthquake intensity. However, the magnitude, M, is not a direct measure of the horizontal and vertical accelerations and vibration frequencies shaking the nuclear power plant and subjecting it to mechanical loads and stresses. Moreover, such measurements

Table 7.1 Intensities (between VI and XII) of the MSK/EMS-98 seismic intensity scale [13, 14]

Intensity	Description
VI	Minor damage to buildings (fine cracks)
VII	People leave houses; cracks and crevices develop in walls and stacks
VIII	Walls, monuments, and stacks crumble
IX	Buildings are shifted from their foundations, walls and roofs collapse, underground pipes break, cracks develop in the ground
X	Many buildings collapse, gaps up to 1 m wide develop in the ground
XI	Many clefts are produced in the ground; landslides occur in the mountains
XII	Pronounced changes in the surface of the earth

have been recorded only since 1935. Consequently, following Medvedev-Sponheuer-Kárnik [1, 5] (MSK-64), an intensity scale of I to XII has been introduced (today referred to as the European Macroseismic Scale 1998 (EMS-98)). The intensity of an earthquake is a measure of its impact on persons, buildings and their underground, etc. Intensities between VI and XII in Table 7.1 are listed here as an example of this intensity scale (EMS-98) [13, 14].

There are semi-empirical formulae showing the correlation between the magnitude, M, and the different displacements and accelerations from a certain distance [1, 15]. On the other hand, relations have also been quoted between the intensity scale and local accelerations [16, 17]. In this way, it is ultimately possible also to establish correlations between the magnitude and the intensity scale [1, 5]. However, because of the rough classification of the intensity scale, these are only rough approximations.

The earthquake with the highest intensity reported in Central Europe occurred in Basel in 1356, causing churches, many houses, sections of the walls around the city, and even castles in the environment of Basel to collapse. That earthquake in Basel was assigned the intensity of IX to X [18–20] (Table 7.2). All later earthquakes in Central Europe are supposed not to have exceeded intensity VIII [18, 19].

The mechanical loads and stresses acting on nuclear power plants in an earthquake are determined by the horizontal and vertical displacements and accelerations, the associated frequency scales, and the duration of the earthquake. Moreover, the qualities of the ground underneath the nuclear power plant play a major role (Fig. 7.1).

Taking into account many measured seismic data and theoretical considerations, the technical safety codes define criteria for seismic design in a conservative way.

The U.S. Regulatory Guide in the United States contains definitions of so-called response spectra for horizontal and vertical displacements and accelerations [2–4, 49]. In Germany, KTA rule 2201 applies [6–8]. In France similar Guidelines exist [5].

The German, e.g. rule contains these requirements:

- "The design basis earthquake must be defined on the basis of information about expected maximum accelerations, duration of vibrations, response spectra, etc. taken from seismological opinions also considering local geological conditions."

Table 7.2 Intensities (EMS-98) and magnitudes (Richter scale) of various earthquakes according to Mohrbach [20]

Earthquake	Intensity, EMS	Magnitude, M
Tohoku [2011] (Fukushima)	≈XI	9.0
Lisbon [1755]	≈XI	9.0
Basel [1356]	IX–X	6.9
	VIII–IX	6.25[a]
Düren [1756]	VIII	5.9
Roermond [1992]	VII	5.3

[a]Ahorner [17]

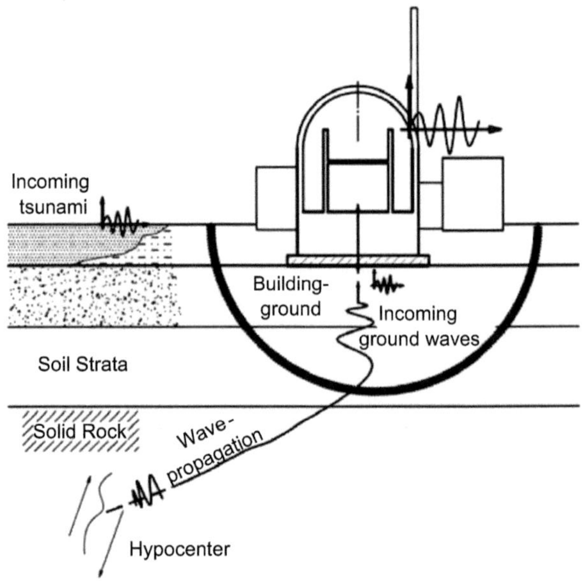

Fig. 7.1 Changes in the response spectra of a nuclear power plant in an earthquake [21]

- "Horizontal and vertical accelerations must be assumed to act simultaneously. The maximum vertical acceleration must be assumed to be 50 % of the maximum horizontal acceleration."

The seismic waves change amplitude and frequency-vs.-time characteristics when impacting local ground strata around the building site of a nuclear power plant. Figure 7.2 shows such a typical ground response spectrum for horizontal accelerations as a function of frequency and damping levels, D, of the reactor building vibrating in the soil strata (the frequency, period, and the damping levels, respectively, correspond to the theoretical description of a damped spring mass system vibrating under an external force [1]).

Besides the rules proposed (KTA 2201.1 [6, 7]), also two-dimensional and three-dimensional finite-element models can be employed in mechanical analysis. Materials in these cases can be modeled either linearly or non-linearly. Various damping models (viscous or viscous with hysteresis) have been developed. The results obtained as a function of height in the reactor building are the maximum horizontal

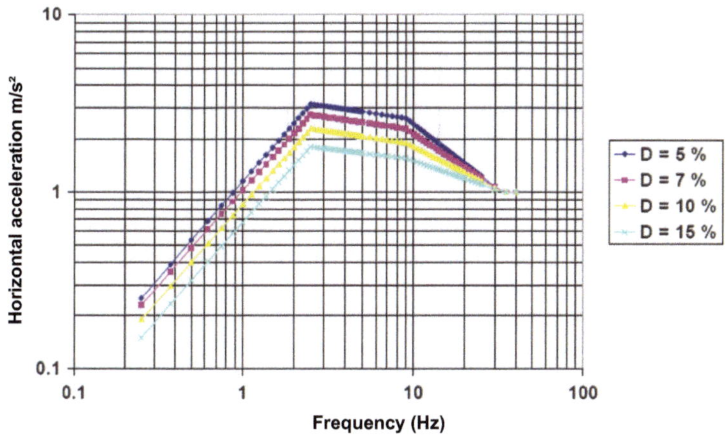

Fig. 7.2 Typical ground response spectra for a reactor building (damping, D, as a parameter) [22]

displacements and accelerations as well as the seismic forces (loads) and frequencies for the different floors of a reactor building (floor response spectra).

Figure 7.3 shows the example of three-dimensional results (magnified scale) for the displacements of and in the reactor building caused by horizontal seismic accelerations [23, 24].

The vibration characteristics and the failure, if any, of components, pipes, switchgear, and cable ducts (Fig. 7.4) can be derived from this solution (KTA 2201.4 [8]). In addition to the calculations performed, internal components of reactors also were tested on vibrating tables (in Japan, components weighing up to 1,000 ton) [21].

As an example German nuclear power plants (pre-konvoy PWR, konvoy-PWR and BWR-72) on the average have been designed to design basis earthquakes of intensity levels VI to VIII (Table 7.3) [25]. However, they can withstand at least an earthquake one intensity level higher, and can be shut down safely under those conditions [25].

According to KTA 2201.5 [6, 7], nuclear power plants must not only have sensors for scramming the plant but also seismic instruments recording accelerations in an earthquake and, after the seismic event, allowing a comparison to be made with the assumptions in the design basis earthquake [9].

7.1.2 Seismic Loads Acting on Components in Nuclear Power Plants

After calculating and defining the seismic forces acting on the components of a nuclear power plant (steam generators, pumps, pipes etc.), the mechanical loads in these components due to earthquakes must be determined. Mechanical stresses

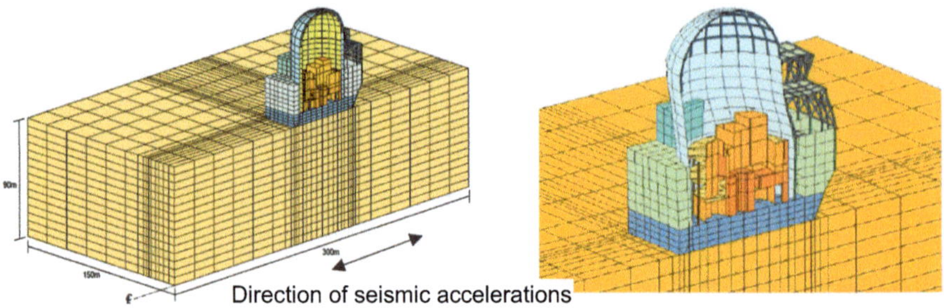

Fig. 7.3 Example of a numerical solution (magnified scale) for displacements of and in the reactor building [23]

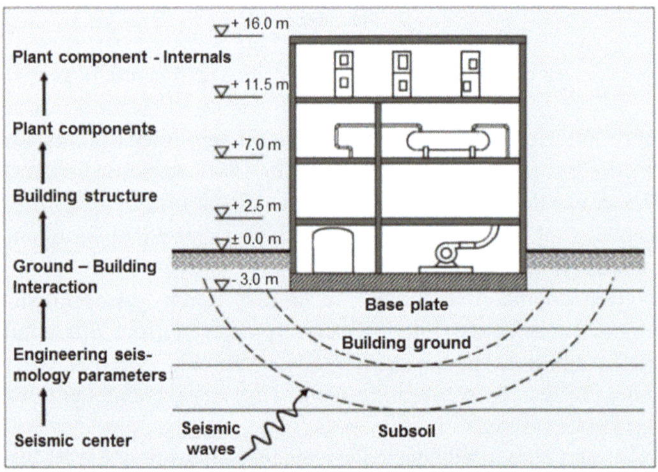

Fig. 7.4 Various floors with components in a reactor building [22]

must not exceed the upper limits prescribed (RSK-LL [10]). In other countries similar requirements are valid [2–4].

Where horizontal or vertical displacements and the resultant stresses are too high, the components may be supported and protected by means of damping elements.

For nuclear power plants in seismic zones, the entire plant may be built on thousands of damping elements located in the foundation bottom slab of the reactor building (Fig. 7.5 [26, 27]).

This type of construction nowadays is considered a mature technology and is employed in large non-nuclear buildings in seismic areas (Tokyo, Los Angeles etc.). There are four nuclear power plants at Cruas, France, and two nuclear power plants at Koeberg, South Africa, as well as the spent fuel pool at the reprocessing plant La Hague, France, featuring this design [26, 27].

Table 7.3 Design basis earthquakes of German nuclear power plants as an example [25]

Reactor	Design basis intensity	Subsoil class	Probability of exceeding	Max. horizontal acceleration (m/s^2)	Max. duration of earthquake (s)
Isar II konvoy PWR	VI 1/4	Loose sediment	$1.1 \times 10^{-5}/a$	0.75	3.5
KGG B+C BWR-72 Gundremmingen	VII	Loose sediment/ sediment	$10^{-5}/a$	1.0 0.5 vertical	10
KKP-II Philippsburg pre-konvoy PWR	VII 1/2	Thick sediment	$10^{-5}/a$	2.1	9
GKN-2 konvoy	VIII	Rock	$10^{-5}/a$	1.7	8
KKE Emsland konvoy-PWR	VIII		$10^{-5}/a$	1.3	2.6
Brokdorf pre-konvoy PWR	VI	Silt, sand, peat	$7.3 \times 10^{-6}/a$	0.5	4

Fig. 7.5 Damping elements in the foundation of a nuclear power plant [26]

In addition, pipes and other components inside the reactor building can be supported by restraining and damping measures to prevent excessive oscillatory movements [28] (Fig. 7.6).

The dampers of rubber or neoprene in the bottom foundation of the reactor building attenuate both vertical and horizontal accelerations in the reactor building during an earthquake. Peak accelerations are shifted into a higher frequency range, which mitigates the risk of damage [26, 27, 29].

These restraining and damping measures applied to pipes and components, and the use of rubber and neoprene dampers in the bottom foundation of reactor buildings, allow new nuclear power plants as well as commercial high risers in big cities to be designed safely against higher seismic intensities of the kind which may occur in seismic areas.

Fig. 7.6 Example of a pipe section equipped with restraints and damping units protecting against dislocations caused by seismic events [28]

7.1.3 Comparison Between Seismic Design and Seismic Damage in Existing Nuclear Power Plants

Nuclear power plants are equipped with acceleration sensors which trigger a reactor scram in earthquakes of sufficiently high intensity levels (below the design basis earthquake), and then automatically change into the residual heat removal mode. Nuclear power plants are inspected after an earthquake to compare any damage with the design features laid down for the design basis earthquake.

In Japan, Armenia, and the United States of America, the reactor scram system (RESA) of nuclear power plants has always functioned successfully in earthquakes. The highest seismic loads so far have occurred in the Kashiwazaki-Kariwa nuclear power plant caused by the Niigata-Chuetsu-Oki earthquake (November 15, 2007) with a magnitude of $M = 6.2$. The epicenter was only 16 km away from the nuclear power plant. The Kashiwazaki-Kariwa nuclear power plant comprised seven boiling water reactors designed for a safe shutdown earthquake with maximum seismic accelerations of 1.67–2.73 m/s^2. In all reactors, scram and transition to the residual heat removal mode worked according to the rules. The maximum accelerations measured during the earthquake in the bottom slabs of the seven BWR nuclear generating units, however, were 3.22–6.80 m/s^2, which is roughly twice the level of the design basis earthquake. Nevertheless, all safety-related components remained functional. Because of other damage to the building structures, these nuclear power plants were down for repair for 3 years after the seismic event [21].

7.2 Design Against Airplane Crash

Since the 1970s, the protection of nuclear power plants against airplane crashes has been a topic of discussion in Germany. At the same time, also underground designs were analyzed and discussed [30, 31]. Finally, however, preference was given to construction above ground with an external concrete containment [31].

The main point of interest was the crash of a high-speed military aircraft and, implicitly, the demand to protect against intentional impacts by third parties (such as terrorist attacks) [1]. The probability of a commercial aircraft crashing into a nuclear power plant was much lower, however. After lengthy debates, the decision was taken to design the outer steel-reinforced concrete containment (outer containment) of a nuclear power plant in such a way that it withstood the crash of a Phantom II military aircraft of the time (in the 1970s, this was the fastest and heaviest military aircraft) weighing approximately 20 ton and reaching a crashing speed of approx. 215 m/s or 774 km/h. This at the same time covered crashes of a smaller aircraft [1, 32–35].

On the basis of an impulse model by Riera [36, 37], and after evaluation of penetration and shock experiments with bullets fired at steel-reinforced concrete walls, the shock load-vs.-time curve shown in Fig. 7.7 was defined and laid down around 1977 for new nuclear power plants to be built [10].

The penetration tests at the time had shown that full protection, i.e. no penetration of such a military aircraft through the concrete wall, and no spallation of concrete on the back of the steel-reinforced concrete wall between approx. 105 cm and approx. 145 cm thick, would be ensured (depending on the stability of steel reinforced concrete) [32, 35]. Present-day pre-konvoy PWR, konvoy-PWR, and the SWR-72 BWR line have concrete walls 180 cm thick (outer containment).

The area of impact of the Phantom II aircraft is assumed to be 7 m^2 in size (circular). The angle of impact is assumed normal to the tangential plane of the spherical or cylindrical reactor building. The vibrations caused by the impact of the aircraft must be taken into account in the design of the nuclear power plant (building and components) [10].

The shock load-vs.-time curve defined in the RSK-guidelines [10] as shown in Fig. 7.7 was based on computer models by Riera [36, 37] and Drittler [32–34]. In 1980, Riera showed a comparison of these models and findings. That comparison is reflected in Fig. 7.8. It is evident that the peak shock loads occurring were smoothed for the shock load-vs.-time curve in Fig. 7.7 to be assumed in accordance with RSK guidelines [10]. It is also seen that an aircraft not hitting the concrete wall in a perpendicular direction (angle of impact 30° off the vertical direction (Fig. 7.8, right)) causes the shock load to be broken down into a smaller vertical and a still smaller horizontal shock load.

The shock load-vs.-time curve laid down in Germany for the crash of a Phantom II aircraft was tested by Sandia National Laboratories in Albuquerque, New Mexico, USA in April 1988. The test was initiated by a Japanese research institute which had been required to protect the buildings of the Japanese reprocessing plant of

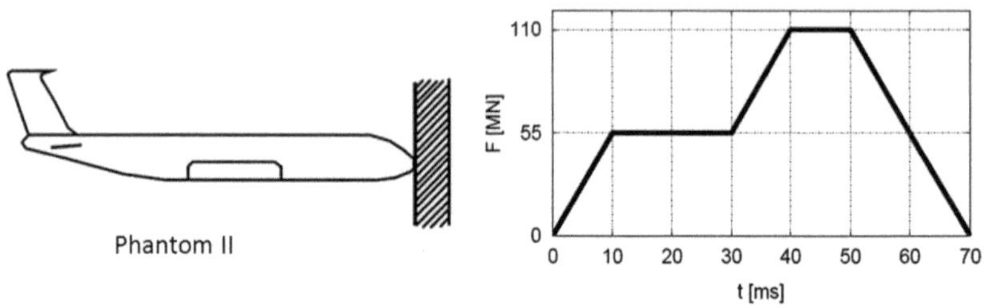

Fig. 7.7 Shock load-vs.-time curve for the impact of a Phantom II aircraft on the steel-reinforced concrete containment of a reactor building [10, 32]

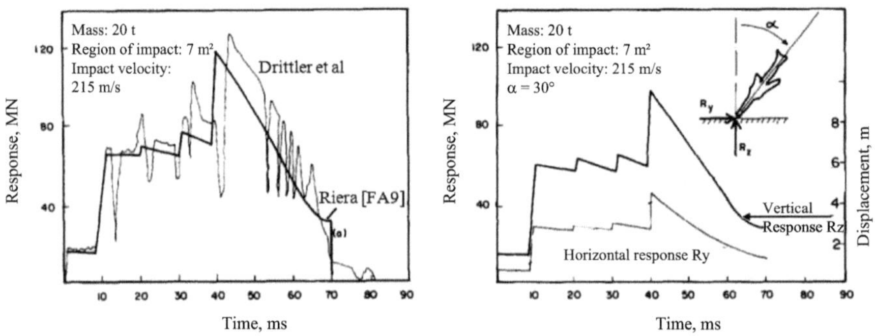

Fig. 7.8 Shock load-vs.-time-diagrams for a Phantom II aircraft according to models by Riera [36, 37] and Drittler et al. [32–34]

Rokkasho-mura against crashes of Phantom military aircraft stationed at a U.S. airfield nearby. The experiment was run to determine the shock load-vs.-time curve produced by the impact on a rigid concrete block. In addition, the burst characteristics of the aircraft and the dispersion of its fuel were to be determined (replaced by water in the test).

The Phantom II, with an impact weight of roughly 20 ton, was accelerated to 215 m/s on a straight path by a two-stage rocket propulsion system. It hit a block of steel-reinforced concrete (7 × 7 × 3.66 m) which, being supported on air bearings, was moveable nearly without any friction. Acceleration sensors installed along the body of the Phantom allowed the speed reduction of the aircraft during the impact to be determined. Accelerations and displacements were measured on the concrete block. The behavior of the engine was studied by high-speed photography. Figure 7.9 shows the Phantom II hitting the target; small parts of the wings were sheared off while the other parts of the aircraft were destroyed completely (the body disintegrated). The water simulating fuel was distributed over only a small area. The surface of the concrete block was damaged very little. Roughly 15 cm of concrete were broken out of the surface of the concrete block as a result of the impact of the engine shaft [38–40].

Fig. 7.9 F-4 Phantom II crash test (Sandia, USA) [38]

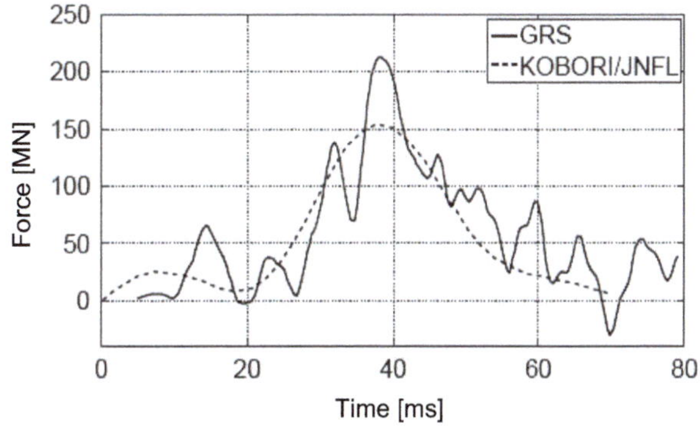

Fig. 7.10 Force-vs.-time curves determined in the Sandia experiments, from [39]

The shock load-vs.-time curve determined in the experiment is shown in Fig. 7.10. It shows the force-vs.-time curves measured (In the dashed smoothed Japanese evaluation, more peaks were filtered out of the data determined experimentally than in the GRS curve) [39].

Given the fact that the brief shock load peaks occurring in the Sandia experiment were covered by the slightly longer curve of the smoothed RSK-LL shock load-vs.-time relation, relatively good agreement was found between theory, experiment, and the given shock load-vs.-time diagrams.

The demand to include load assumptions made for airplane crashes in the construction of nuclear power plants in accordance with the RSK-guidelines [10] in this strict form so far had existed only in Germany (one exception being the Japanese reprocessing plant of Rokkasho-Mura). For the EPR, German-French guidelines were elaborated which roughly correspond to the German RSK-guidelines [10]. Consequently, the EPR has a wall thickness of the concrete containment of two times 1.3 m (Chap. 3).

Figures 7.11 and 7.12 show the containment wall thicknesses of the French N4 PWR and those of the Russian VVER reactors compared with the German

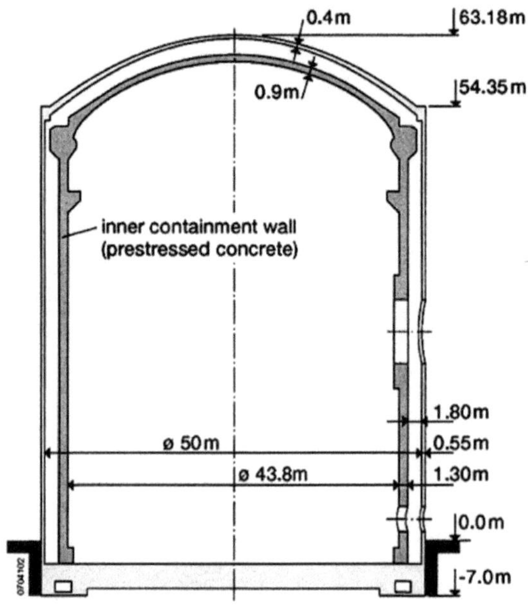

Double wall PWR-containment of a French NPP, type „N4"

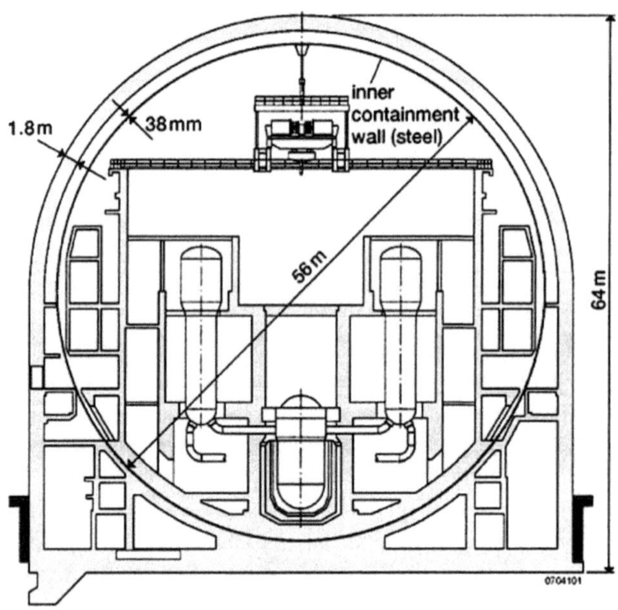

Double wall PWR-containment of a German NPP, type „KONVOI"

Fig. 7.11 Containment of a French ("N4" type) and a German ("konvoy" type) nuclear power plant

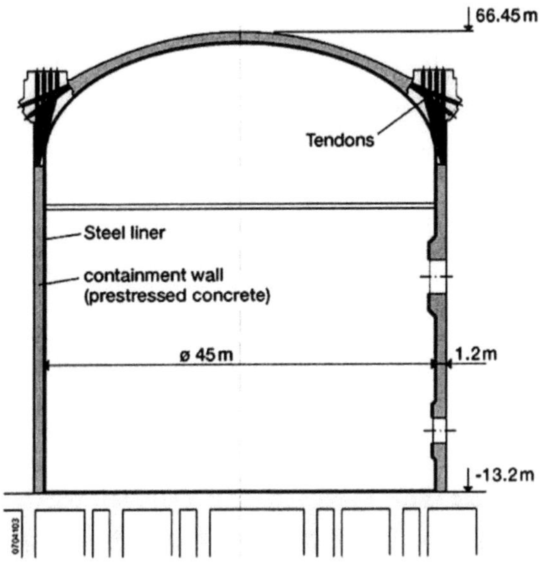

Single wall PWR-containment of a Russian NPP, type „WWER 1000"

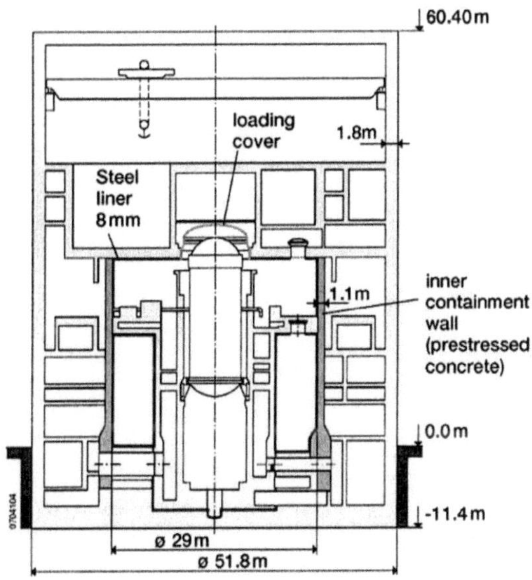

Double wall PWR-containment of a German BWR, type „SWR 72"

Fig. 7.12 Containments of a Russian PWR ("VVER 1000" type) and a German BWR ("SWR-72" type)

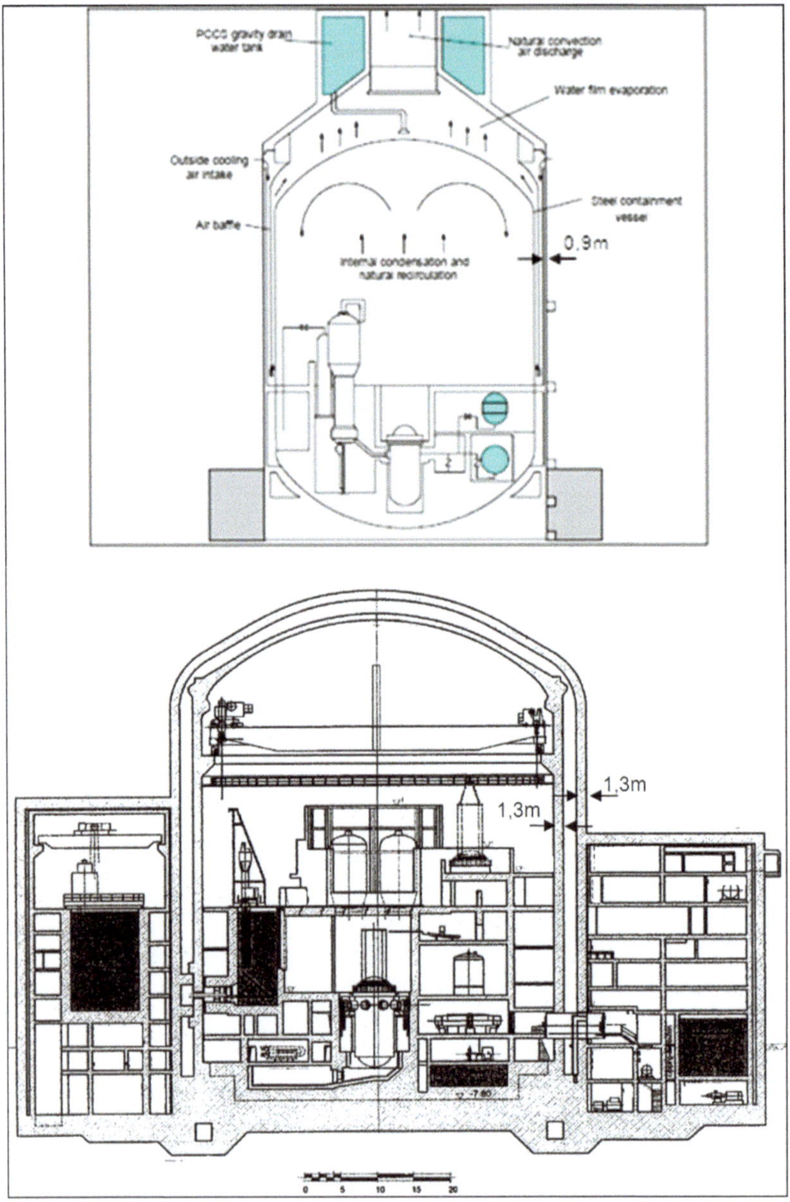

Fig. 7.13 Wall thickness of AP1000 and of EPR as described in Chap. 3 [41]

konvoy-PWR and the SWR-72 BWR lines currently in operation. Figure 7.13 shows the containment wall thickness of AP1000 and of EPR as described in Chap. 3.

Crashes of large commercial airliners were described and treated theoretically in [42–45]. After thorough analyses, Eibl SMP Engineering Consultants, Karlsruhe [46], found that the impact of a Boeing 747 jumbo jet at an approach velocity of 175 m/s hitting a convoy-type PWR with 180 cm of concrete of the outer concrete shell could give rise to cracks in the concrete containment of convoy-type PWR plants. However, kerosene would not penetrate (no fire). This finding applies to the 180 cm thick concrete containment of the Neckarwestheim-II nuclear power plant but can be transferred to all German convoy-type PWR, pre-convoy PWR, and Gundremmingen B + C, each with outer steel-reinforced concrete containments of 180 cm thickness. The European Pressurized Water Reactor, EPR, has a total wall thickness of the inner and outer containment of 2.6 m (Fig. 7.13).

7.3 Chemical Explosions

Chemical explosion hazards can arise in transports of explosible materials by barges on rivers and ocean-going vessels or nearby railway lines and roads. In Germany, nuclear power plants must be designed against a given pressure wave resulting from such hazards. In this pressure wave, the pressure is supposed to rise to 0.045 MPa (0.45 bar) overpressure within 0.1 s and then drop to 0.03 MPa (0.3 bar) within 2 s and remain at that level for another 1 s (Fig. 7.14) [10].

The buildings adjacent to the reactor building should be arranged in such a way that any focusing arrangements and structures promoting turbulences are avoided.

All these requirements are met in all pre-convoy PWR, convoy-type PWR, and the Gundremmingen B and C BWR [25].

Toxic gases which could enter the ventilation system can be detected. The ventilation system can be shut off against the entry of toxic gases.

7.4 Flooding

In German nuclear power plants, the risk of flooding by a maximum-level flood must be taken into account on the basis of scientific findings about maximum floods for the past 10,000 years [10, 47]. Flooding by tsunamis must be considered for nuclear power plants located at the coasts of zones endangered by seismic events [20, 48]. Table 7.4 shows a list of historically documented tsunamis initiated by earthquakes along Japanese coasts.

Table 7.4 shows that very high tsunami waves can be produced even by earthquakes of relatively low magnitudes (Richter scale).

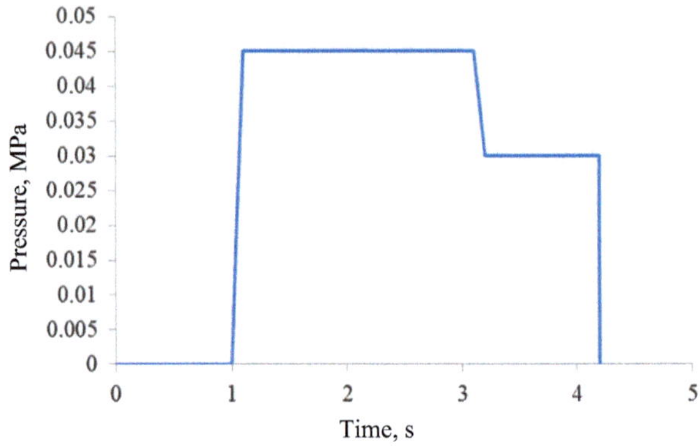

Fig. 7.14 Pressure profile of chemical explosions in the vicinity of German nuclear power plants

Table 7.4 List of historically documented tsunamis initiated by earthquakes at Japanese coasts to the Fukushima accident [48]

Year	Region	Magnitude	Max. level [m]	Casualties
1498	Enshunada Sea	8.3	10.0	31,000
1605	Nankaido	7.9	10.0	5,000
1611	Sanriku	8.1	25.0	5,000
1703	Off the Boso peninsula	8.2	10.5	5,233
1707	Enshunada	8.4	11.0	2,000
1771	Ryukuyu Islands	7.4	85.4	13,486
1854	Nankaido	8.3	28.0	3,000
1896	Sanriku	7.6	38.2	27,122
1923	Sagami Bay	7.9	13.0	2,144
1933	Sanriku	8.4	29.0	3,022
1944	Off the south-eastern coast of the Kii peninsula	8.1	10.0	1,223
1983	Noshiro	7.8	14.5	100
1993	Japanese Sea	7.7	54.0	208
2011	Northeastern Honshu	9.0	23.0	>15,000

References

1. Smidt D (1979) Reaktorsicherheitstechnik, Sicherheitssysteme und Störfallanalyse für Leichtwasserreaktoren und Schnelle Brüter. Springer, Berlin
2. Regulatory Guide 1.60 (1973) Design response spectra for seismic design of nuclear power plants (Revision 1). http://pbadupws.nrc.gov/docs/ML0037/ML003740207.pdf
3. McGuire RK et al (2001) New seismic design spectra for nuclear power plants. Nucl Eng Des 203:243–257
4. Regulatory Guide 1.61 (2007), Damping Values for Seismic Design of Nuclear Power Plants (Revision 1). http://www.nrc.gov/reading-rm/doc-collections/reg-guides/power-reactors/rg/01-061/01-061.pdf
5. Résistance au séisme des installations nucléaires en France. http://fr.wikipedia.org/wiki/R%C3%A9sistance_au_s%C3%A9isme_des_installations_nucl%C3%A9aires_en_France

6. KTA 2201.1 (1990) Auslegung von Kernkraftwerken gegen seismische Einwirkungen, Teil 1. Grundsätze, Kerntechnischer Ausschuß (KTA). http://www.kta-gs.de/d/regeln/2200/2201_1.pdf
7. KTA 2201.1 (1990) Auslegung von Kernkraftwerken gegen seismische Einwirkungen; Teil 1: Grundsätze. http://www.kta-gs.de/e/standards/2200/2201_1e_2011_11.pdf
8. KTA 2201.4 (1990) Auslegung von Kernkraftwerken gegen seismische Einwirkungen – Teil 4: Anforderungen an Verfahren zum Nachweis der Erdbebensicherheit für maschinen- und elektrotechnische Anlagenteile, Fassung 6/90. http://www.kta-gs.de/d/regeln/2200/2201_4_2012_11.pdf
9. KTA 2201.5 (1996) Auslegung von Kernkraftwerken gegen seismische Einwirkungen – Teil 5: Seismische Instrumentierung, Fassung 06/96. http://www.kta-gs.de/d/regeln/2200/2201_5.pdf
10. RSK-Leitlinien für Druckwasserreaktoren (1996) Fassung 11.96, BAnz Nr. 214 vom 05.11.1996. http://www.rskonline.de/downloads/8110dwr.pdf
11. IAEA Safety Guide NS-G-3.3 (2002) Evaluation of seismic hazards for nuclear power plants. International Atomic Energy Agency, Vienna
12. IAEA Safety Guide NS-G-1.6 (2003) Seismic design and qualification for nuclear power plants. International Atomic Energy Agency, Vienna
13. DIN 4149 (2005) Bauten in deutschen Erdbebengebieten – Lastannahmen, Bemessung und Ausführung üblicher Hochbauten.
14. Eurocode 8 (2004) Design of structures for earthquake resistance. http://shop.bsigroup.com/Browse-By-Subject/Eurocodes/Descriptions-of-Eurocodes/Eurocode-8/
15. Kanai K (1961) An empirical formula for the spectrum of strong earthquake motions. Bull Earthq Res Inst 39:85–95
16. Werner SD (1976) Engineering characteristics of earthquake ground motions. Nucl Eng Des 36:367–395
17. Werner SD (1976) Procedures for developing vibratory ground motion criteria at nuclear plant sites. Nucl Eng Des 36:411–441
18. Leydecker G (1986) Erdbebenkatalog für die Bundesrepublik Deutschland mit Randgebieten für die Jahre 1000–1981. Geolog. Jg., Hannover E36, p 3–81
19. Ahorner L, Rosenhauer W (1978) Seismic risk evaluation for the Upper Rhine Graben and its vicinity. J Geophys 44:481–497
20. Mohrbach L (2011) Unterschiede im gestaffelten Sicherheitskonzept: Vergleich Fukushima Daiichi mit deutschen Anlagen. http://www.kernenergie.de/kernenergie/service/fachzeitschrift-atw/hefte-themen/2011/apr-mai/02_unterschiede-fukushima-deutsche-anlagen.php
21. Sadegh-Azar H et al (2009) Bautechnische Auslegung von Kernkraftwerken für Erdbeben. Atomwirtschaft 54(12):753–759
22. Henkel F-O et al (2009) Auslegung der Anlagenteile von Kernkraftwerken gegen Erdbeben – Stand und Tendenzen. Atomwirtschaft 54(12):760–766
23. Nakamura N et al (2008) Analyses of reactor building by 3D nonlinear FEM models considering basemat uplift for simultaneous horizontal and vertical ground motions. Nucl Eng Des 238:3551–3560
24. JEAG (1991) Technical guidelines for aseismic design of nuclear power plants. Japan Electric Association Guideline, Ministry of International Trade and Industry, Tokyo
25. Stellungnahme der RSK (2011) Anlagenspezifische Sicherheitsüberprüfung RSK(SÜ) deutscher Kernkraftwerke unter Berücksichtigung der Ereignisse in Fukushima (16.05.2011)
26. Forni M (2011) Seismic isolation of nuclear power plants, state of the art and future applications. Seminary at Karlsruhe Institute of Technology IKET, January 2011
27. Tang Y et al (2011) Seismic isolation for advanced fast reactors. Nucl Technol 173:135–152
28. Gundremmingen. http://www.kkwgun.de/
29. Okamura S et al (2009) Seismic isolation design for JSFR –, FR09. Paper No. IAEA-CN-176-08-28P, Kyoto, Japan, 7–11 December 2009

30. Kröger W (1978) Unterirdische Bauweise von Kernkraftwerken. Informationstagung: Die Sicherheit des Leichtwasserreaktors, Deutsches Atomforum, Mainz
31. Müller WD (1981) Kernkraftwerke unter die Erde (Leitartikel). Atomwirtschaft 5(81):289
32. Drittler K et al (1973) Zur Auslegung kerntechnischer Anlagen gegen Einwirkungen von außen, Teilaspekt: Flugzeugabsturz, Bericht IRS-W-7, Dez. 1973. Institut für Reaktorsicherheit der technischen Überwachungs-Vereine e.V. in Köln
33. Drittler K et al (1976) Calculation of the total force acting upon a rigid wall by projectiles. Nucl Eng Des 37:231–244
34. Drittler K et al (1976) The force from impact of fast-flying military aircraft upon a rigid wall. Nucl Eng Des 37:245–248
35. Nachtsheim W et al (1982) Interpretation of results of Meppen slab tests – comparison with parametric investigations. Nucl Eng Des 75:283–290
36. Riera JD (1968) On the stress analysis of structures subjected to aircraft impact forces. Nucl Eng Des 8:415–426
37. Riera JD (1980) A critical reappraisal of nuclear power plant safety against accidental aircraft impact. Nucl Eng Des 57:193–206
38. Ruch D (2010) Bestimmung der Last-Zeit-Funktion beim Aufprall flüssigkeitsgefüllter Stoßkörper. Dissertation, Karlsruhe Institut für Technologie (KIT), Fakultät Bauingenieur-, Geo- und Umweltwissenschaften
39. Bachmann P et al (1994) Bewertung der Ergebnisse der bei SANDIA durchgeführten japanischen Stoßversuche hinsichtlich der Sicherheit von Kernkraftwerken bei Flugzeugabsturz, Technischer Bericht, Gesellschaft für Anlagen und Reaktorsicherheit (GRS) mbH, 1994
40. von Riesemann WA et al (1979) Full-scale aircraft impact test for evaluation of impact forces. Part 1: Test plan, test method and test results. In: Transactions of 10th international conference on structural mechanics in reactor technology, p 285–294
41. IAEA-TECDOC-1391 (2004) Status of advanced light water reactor designs 2004. International Atomic Energy Agency, Vienna
42. Rebora B et al (1976) Dynamic rupture analyses of reinforced concrete shells. Nucl Eng Des 37:269–297
43. Wolf JP et al (1978) Response of equipment to aircraft impact. Nucl Eng Des 47:169–193
44. Stepan J (2009) Validation of aircraft FE model for impact analyses. In: 20th International conference on structural mechanics in reactor technology (SMIRT 20), Espoo, Finland, 9–14 August 2009. SMIRT 20-Division 4, Paper 1929
45. Stepan J (2007) Large commercial aircraft crash into the light-weight nuclear facility building. In: Transactions, SMIRT 19, Toronto, August 2007
46. Schlüter F-H (2014) Safety of German light-water reactors in the event of a postulated aircraft impact. In: Kessler G et al (eds) The risks of nuclear energy technology: safety concepts of light water reactors. Springer, Berlin
47. KTA 2207 (2004) Schutz von Kernkraftwerken gegen Hochwasser, Sicherheitstechnische Regel des KTA (Kerntechnischer Ausschuß), Fassung 11/04. http://www.kta-gs.de/d/regeln/2200/2207n.pdf
48. Tsunami. http://de.wikipedia.org/wiki/Tsunami
49. Newmark NM et al (1973) Seismic design spectra for nuclear power plants. J Power Div Am Soc Civ Eng 99:2

Chapter 8
Risk of LWRs

Abstract Chapter 8 compares the risk of LWRs as determined in the US-risk study WASH-1400 with the risk of other technical systems, e.g. air plane crashes, fires, dam failures, explosions and chlorine releases in chemical industry. It is shown that the risk, e.g. of 100 nuclear power plants in the USA is smaller than the risk of these other technical systems. Natural disasters, e.g. hurricanes, floods including tsunamis, earthquakes, avalanches and landslides or volcanic eruptions have a higher frequency of occurrence per year and a much higher number of casualties i.e. their risk—the product of frequencies and casualties—is much higher than that of technical systems including nuclear power plants.

8.1 Comparison of the Risk of LWRs with the Risks of Other Technical Systems

The findings about the risk of LWRs as determined in the Rasmussen Risk Study (WASH-1400) [1] and the German Risk Study Phase A [2] were compared to the risks associated with other technical systems. Data from the experience of insurance companies for major accidents in coal mining, on oil platforms, tanker accidents in oil and gas transport, large fires in refineries, gas explosions, and risk studies for large chemical plants in England (Canvey Island) and France were used [3–9].

Figure 8.1 shows the number of deaths fatalities in major accidents as a function of the frequency of occurrence for dam failures, chlorine releases (chemical industry), explosions and fires as well as airplane crashes. The frequency of occurrence of major accidents in power generation by nuclear plants (early fatalities and late fatalities for 100 nuclear power plants in the USA) is seen to be below or in the vicinity of the figures for other systems of energy generation and for the chemical industry, respectively [1].

G. Kessler et al., *The Risks of Nuclear Energy Technology*, Science Policy Reports,
DOI 10.1007/978-3-642-55116-1_8,

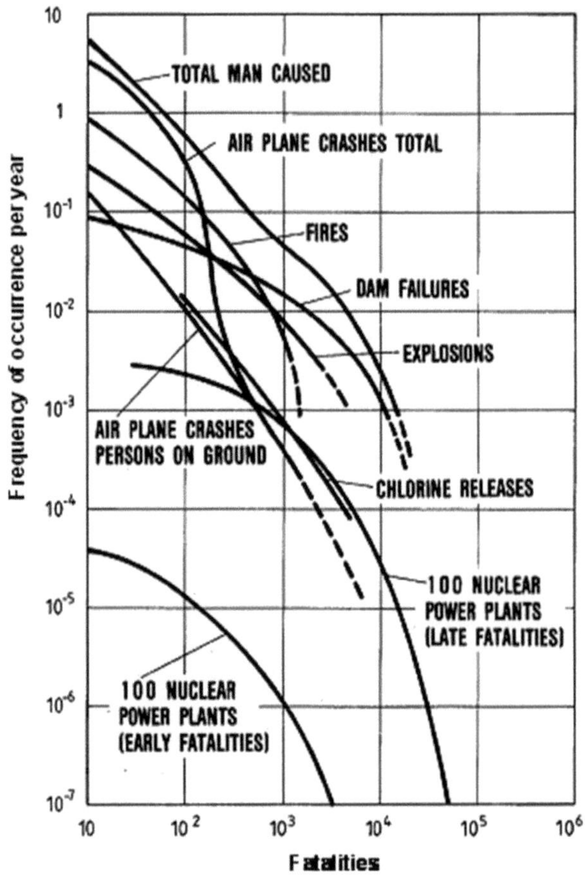

Fig. 8.1 Comparison of man caused fatalities of different technical systems with the results of WASH-1400 for 100 nuclear power plants in the USA [1, 9]

8.2 Major Accidents in the Power Industry

The worst accident in the coal mining industry occurred 1931 in Fushun (Manchuria, China) killing 3,000 persons. Explosions, e.g. during fertilizer production in Oppau (Germany) killed 561 people in 1921 and 207 people in 1948. In the oil industry, the worst accident with 3,000 casualties occurred in 1987 when an oil tanker collided with a passenger ferry boat in the Philippines. In the natural gas industry, the severest accidents were recorded in Ufa, now Azerbaijan, in 1989, with 600 persons killed, and in Ixhuapetek, Mexico, with 500 casualties [4, 6, 10].

When it comes to large dams for hydroelectric plants, the record shows 459 people killed in Fréjus, France, in 1959. In Vajont, Italy, 1914 people were killed in 1967, and 2,500 people were killed at the Indian Machhu II dam in 1979.

The biggest accident in the chemical industry so far occurred in Bhopal, India, killing 3,400–8,000 persons [4, 8].

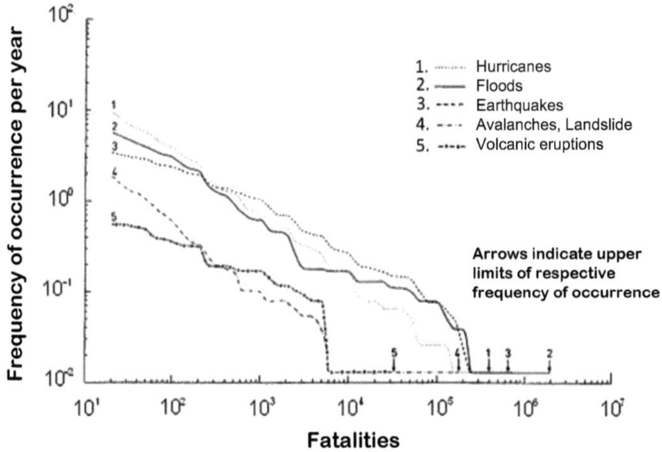

Fig. 8.2 Fatalities (deaths) occurring in natural disasters

In the nuclear industry various accidents occurred with radioactivity releases, causing either no or—e.g. in case of Chernobyl—a high number of casualties (see Sect. 9.2). These are the accidents of Windscale England), Tomsk, Lake Karachay, Chelyabinsk (Russia) and Three-Mile-Island (USA). These reactor accidents are overshadowed by the severe nuclear accidents of Chernobyl (Ukraine) and Fukushima (Japan) described in Chap. 9.

8.3 Natural Disasters

It is interesting in this connection to have a look also at the consequences of natural disasters mankind must suffer as inescapable (Fig. 8.2) [4]. These are

- natural disasters,
- hurricanes,
- floods, including tsunamis
- earthquakes,
- avalanches and landslides,
- volcanic eruptions.

The frequency of occurrence of such natural disasters is far higher than, e.g., that of major accidents in power generation systems. They can cause up to 10^5 deaths [4, 6, 11].

References

1. (1975) Reactor safety study: an assessment of accidents risks in US commercial nuclear power plants. In: Rasmussen NC (ed) US Nuclear Regulatory Commission, WASH-1400 (NUREG-75/014), Washington
2. Deutsche Risikostudie Kernkraftwerke Phase A (1980) Gesellschaft für Reaktorsicherheit (GRS). Verlag TÜV Rheinland
3. Fritsche AF (1988) Gesundheitsrisiken von Energieversorgungssystemen. Verlag TÜV Rheinland, Köln
4. Frischknecht S et al (1998) Project GaBE, comprehensive assessment of energy systems. Severe accidents in the energy sector. PSI-Bericht Nr. 98-16, ISS 1019-0643
5. Hubert P et al (1981) Les impacts sanitaires et écologiques de la production de l'électricité: Le cas français, CEPN-R-3.4.3 (2ème version), Fontenay-aux-Roses, France
6. Rogner HH (1998) Comparing energy options. IAEA Bulletin 40/1/1998. Sustainable development, Nuclear Power, IAEA
7. Canvey-Island Study, Canvey (1978) Summary of an investigation on potential hazards from operations in the Canvey Island/Thurrock area. Health and Safety Executive, London
8. Le programme international sur la securité chimique. http://www.who.int/inf-fs/Fo/am87.htm
9. Kessler G (2012) Fission reactors and their fuel cycles, transmutation, incineration, safety. Springer, Heidelberg
10. Farmer R (1980) What is risk. Atom 282:108
11. Environmental disasters. http://www.umweltbundesamt.de/uba-inde-daten-c/

Chapter 9
The Severe Reactor Accidents of Three Mile Island, Chernobyl, and Fukushima

Abstract Three major severe accidents with core meltdown/core disruption occurred at Three Mile Island (USA) in 1979, Chernobyl (Ukraine) in 1986 and Fukushima (Japan) in 2011.

The LWR of Three Mile Island was a two loop PWR with 880 MW(e) output. The accident started with technical problem in the feedwater loop for the steam generators. As the steam generators were not able to remove the heat, the pressure in the primary system increased and the safety valve of the pressurizer opened thereby releasing steam. The reactor was shut down because of too high pressure. When the pressure in the primary system dropped the safety valve did not shut again and remained open. The operators were given the opposite information by the instruments in the control room. The high pressure emergency core cooling systems started to feed water in the reactor pressure vessel. But the water in the pressure vessel rose too high and the operators throttled the emergency cooling systems. As the primary pumps started to vibrate the operators also shut down both primary pumps. As a consequence the cooling water in the pressure vessel started to boil. The zirconium claddings started to chemically react with water: hydrogen was formed. The reactor core began to melt down. The silver-indium-cadmium control rods did melt. Part of the molten core collected at the bottom of the pressure vessel. A hydrogen explosion occurred in the reactor building. Only the radioactive noble gases and a small part of the fission products iodine and cesium were able to penetrate the filters of the reactor building. The radioactive exposure of the population was therefore very small. Cost for decontamination of the plant and disposal of the destroyed core were very high. The Three Mile Island accident was classified a level 5 accident on the International Nuclear Event Scale (INES).

The Chernobyl accident occurred in one of four RBMK1000 reactors at the Chernobyl site 100 miles north of Kiev. The operators were preparing an experiment in which the energy of rotation of the turbine during shut down should produce emergency electrical power for the support of the diesel generators. Unexpectedly the experiment had to be interrupted for some time to comply with electricity supply which led to the buildup of the fission product Xe-135 (neutron poison). When the experiment could be continued the power level dropped to about

G. Kessler et al., *The Risks of Nuclear Energy Technology*, Science Policy Reports, DOI 10.1007/978-3-642-55116-1_9, © Springer-Verlag Berlin Heidelberg 2014

30 MW(th) because of operator error. This led to additional buildup of Xe-135 (neutron poison). As a consequence the operators had to withdraw the control rods manually to their upper limits after they had shut off the automatic control system. The RBMK1000 was known to have a positive coolant temperature coefficient. This gave rise to instabilities in power production, coolant flow and temperatures in the low power range.

Then the experiment began at the power level of 200 MW(th). Steam to the turbine was shut off. The diesel generators started and picked up loads. The primary coolant pumps also run down. However this led to increased steam formation as the coolant temperature was close to its boiling temperature. With its positive coolant temperature coefficient the RBMK1000 reactor now was on its way to power runaway. When the SCRAM button was pushed the control elements started to run down into the reactor core. However, due to a wrong design of the lower part of the control elements (graphite sections) the displacement of the water by graphite led to an increase of criticality. A steep power increase occurred, the core overheated causing the fuel rods to burst, leading to a large scale steam explosion and hydrogen formation. The reactor core was destroyed and the top shield cover and the fuel refueling machine were lifted up. Fuel elements and graphite blocks were dispersed outside the reactor core. The reactor core was now open to the atmosphere. Fission products and fuel aerosols were distributed over the Ukraine, Belarus, Russia and Europe. Very high radiation doses were received by fire fighters, operators, helicopter pilots and members of the emergency team. Approximately 800,000 military people were involved in rescue teams receiving various levels of high radiation doses. About 135,000 people were evacuated rather late. In total about 3,000 km^2 of land were contaminated with more than 1,500 Bq/m^2, roughly 7,200 km^2 with 600–1,500 Bq/m^2 and about 103,000 km^2 with 40–200 Bq/m^2 of Cs-137. The Chernobyl accident was classified a level 7 accident on the International Nuclear Event Scale (INES).

The severe reactor accidents at Fukushima occurred in 2011 after a severe earthquake with intensity 9 (Richter scale) close to the northeastern coast of Japan. The earthquake was followed by a tsunami wave which hit the six BWRs of the Fukushima-Daiichi plant with a water level up to 14 m. Unfortunately the Fukushima-Daiichi plant was only protected up to a tsunami wave level of 5.7 m. Only three BWRs of the six BWRs of the Fukushima-Daiichi plant were in operation when the earthquake and the tsunami wave hit the reactor site. All BWRs were duly shut down by the seismic instrumentation and changed into the residual heat removal mode. However, the tsunami wave flooded the two diesel generators of each of the three reactor units 1–3, located in the lowest part of the turbine building. The diesel generators and the battery systems failed. The external grid power and heat exchangers transferring afterheat to the ocean water had already been destroyed by the earthquake. In unit 1 due to the lack of electrical power the high pressure coolant injection system did not work. The steam driven isolation condenser system worked only partly in time and failed. The primary coolant system could not be depressurized due to lack of electrical power and pressurized nitrogen. Low pressure emergency pumps, therefore, could not feed water in the primary coolant system. The primary coolant system heated up and exceeded its design pressure. The core became uncovered, the zirconium claddings of the coolant system chemically reacted with water and formed hydrogen. The core melted down. The pressure in the pressure vessel was relieved into the primary containment because core melt penetrated the lower bottom wall. The pressure in the primary containment led to release of hydrogen and fission product gases into the upper reactor building, where a hydrogen explosion occurred destroying the upper building structures.

In units 2 and 3 the accident developed in a similar pattern, though with a larger shift in time. However a hydrogen explosion only occurred in unit 3 (BWR) not in unit 2 (BWR). However, a hydrogen explosion also occurred in unit 4 (BWR) due to a backflow through the common gas treating system. The hydrogen explosion destroyed the upper structures of the reactor building. The spent fuel pools of unit 1, 3 and 4 had to be cooled part time by concrete pumping trucks, water cannons or helicopters dropping water, but no damage occurred to the fuel in the spent fuel pools. After detailed measurements of the radioactivity released into the environment the Japanese government evacuated about 200,000 people. Four persons of the operating crew were killed by the earthquake and the tsunami wave. Some 20 staff members were injured by the hydrogen explosions. Out of the about 23,000 emergency workers 12 received effective radiation doses up to 700 mSv and 75 workers received <200 mSv. The radiation dose of all others was <10 mSv. The contamination of land was measured. About 2,200 persons would not be allowed to return to a no-entry zone because of too high radiation exposure. The Fukushima severe reactor accident was classified level 7 on the International Nuclear Event Scale (INES).

Three major reactor accidents with core meltdown/core disruption occurred at Three Mile Island (TMI), USA, on March 28, 1979; at Chernobyl, Ukraine, on April 26, 1986, and at Fukushima, Japan, on March 11, 2011. The three reactor accidents at Three Miles Island, Chernobyl and Fukushima will be discussed and described briefly below. Prior to the most severe accident at Chernobyl, Ukraine, the reactor accident at Windscale (United Kingdom) in a plutonium production reactor for military purposes in 1957 had been considered the worst nuclear accident with radioactivity release to the environment. It will not be discussed here.

9.1 The Accident at Three Mile Island

On March 28, 1979, a sequence of accidents occurred in unit 2 of the Three Mile Island reactor facility in the United States of America which ultimately resulted in partial meltdown of the reactor core. The pressurized water reactor had been built by Babcock & Wilcox (Fig. 9.1) [1, 2].

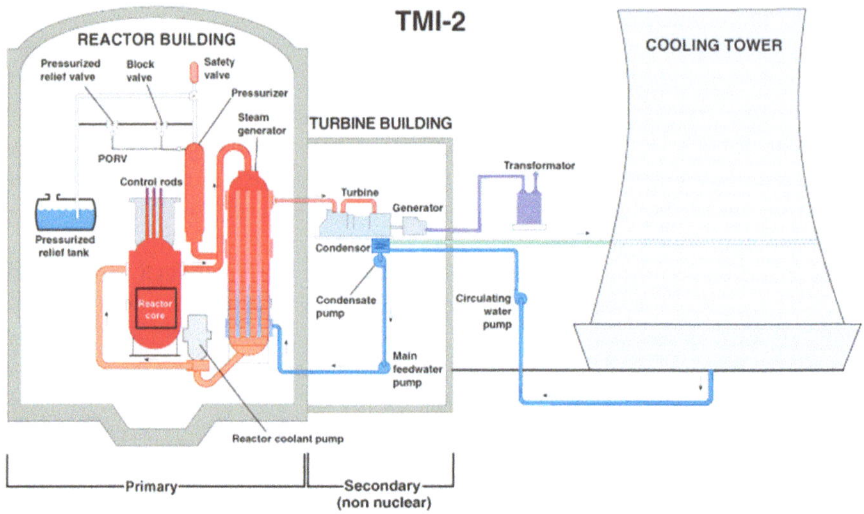

Fig. 9.1 Simplified schematic diagram of unit 2 [1]

It had a reactor power of 880 MW(e) and only two cooling circuits, A and B. The sequence of accidents began with technical problems in the feed water loop for the steam generator. This caused a turbine trip and triggered the startup of the emergency feed water systems. However, the valves in the emergency feed water system were closed by mistake. As the steam generators did not get enough feed water and, for this reason, were not able to remove enough heat, also the temperatures and pressures in the primary cooling circuit started to rise. When an excessively high primary pressure had been reached, the pilot-operated relief valve (PORV) of the pressurizer opened, and the steam was released into the pressure relief tank in the containment. The "primary pressure too high" signal caused the reactor to be shut down. The reactor power dropped to the residual heat level. As a consequence, also the primary pressure dropped. Apart from the mistake mentioned above the events vent on in accordance with the measures planned for that incident condition.

However, the PORV did not shut again at the lower pressure, but remained open. The operators, though, were given the opposite information by the instruments in the control room, namely the indication that the PORV had closed again. The open PORV continued to discharge more steam, and the primary pressure continued to drop. When the primary pressure reached the level at which the high-pressure emergency cooling pumps start to feed, these pumps were activated automatically in order to compensate for the loss of primary coolant. The pressure relief tank was overfilled by the steam released, the water spilled over and collected in the sump of the reactor building. From here it was automatically pumped into storage tanks for radioactive water in the auxiliary systems building (not shown in Fig. 9.1).

The operating crew, who had become confused by the wrong readings of the instruments and did not know precisely the status of the plant, now throttled the high-pressure safety feed system because the water level in the pressure vessel became too high, and opened the valve to a dump pipe. As a consequence of these steps, the water level in the reactor pressure vessel dropped so far that the water coolant began to boil. Although the operating team now succeeded in opening the valves in the feed water line, which had been shut at the beginning of the sequence of accidents, this changed nothing in the further course of the accident.

Roughly 15–30 min after the start of the accident, also the storage tanks filled with radioactive water in the auxiliary systems building started to spill over. Radioactive gases and aerosols entered the atmosphere of the auxiliary systems building. The filters of the auxiliary systems building were able to retain some 99.9 % of the aerosols. But the radioactive noble gases, escaped through to filters to the environment.

As the primary pumps began to vibrate under the impact of steam in the cooling water, the operators first shut down the primary pump of primary cooling circuit B and, slightly later, also that of cooling circuit A. As a consequence, the cooling water in the core began to boil even more violently. The fuel elements, in particular the fuel rod claddings, heated up. At temperatures of the fuel rod claddings above 1,200 °C, steam began to react chemically with the zirconium of the zircaloy cladding, and hydrogen was produced.

$$Zr + 2\,H_2O \rightarrow ZrO_2 + 2\,H_2.$$

This situation changed only gradually after the operators had created a feed-and-bleed procedure by again feeding water through the high-pressure safety feed systems and bleeding the steam through the open pressure relief system. As late as 15 h after the start of the accident it became possible to restart a primary pump and transfer the reactor into a stable residual heat cooling mode.

The hydrogen produced in the reactor core during the accident entered the reactor building together with the steam, initiating an explosion as a result of self-ignition. As iodine and cesium combined chemically to produce, e.g., CsI, and occurred as aerosols, they were largely retained by the filters in the auxiliary systems building.

The small amounts of aerosols, the shortlived radioactive noble gases, and the gaseous I-131 (halflife 8 days) gave rise to only a relatively low mean radioactive exposure of the population of 0.015 mSv [2] (The world wide annual effective dose caused by natural radiation is about 2.4 mSv/year with a typical range of 1–10 mSv/year in various regions of the world (Chap. 4)). The "Kemeny Committee" appointed by the U.S. President to investigate the Three Mile Island accident arrived at this finding: "The Three Mile Island accident would cause so few cases of cancer, if any, that they would not be detectable statistically" [3].

Analysis of the accident (Fig. 9.2) indicated that roughly one third of the zircaloy fuel rods had reacted with steam and produced hydrogen. When the water of the emergency cooling feed systems contacted the hot fuel rods, this resulted in fragmentation of the zircaloy fuel rod claddings and the UO_2 pellets. The silver-indium-cadmium control rods had molten almost completely. Nearly the entire

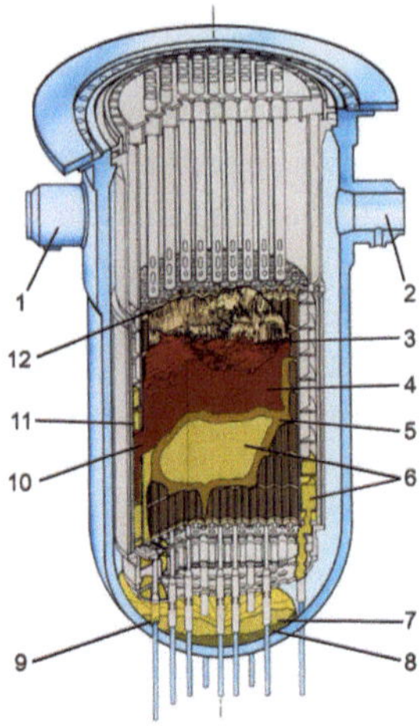

1 connection primary line
2 connection primary line
3 cavity built by melt down
4 loose fragments of the core
5 crust of the melt
6 molten material
7 fragments at bottom of spherical calotte
8 uranium depleted region
9 destroyed duct
10 perforated shield
11 layer of molten material on the surfaces of the bypass channels
12 damages at the upper grid

Fig. 9.2 Molten reactor core of the Three Mile Island accident [1, 2]

fission gas plena of the fuel rods had been destroyed, and gaseous fission products had been released. Parts of the reactor core had molten through at the edge of the grid plate and collected at the bottom of the reactor pressure vessel. However, the hemispherical bottom of the pressure vessel did not melt through.

The reactor plant was decontaminated in many years of work, and the reactor top lid was opened. The partly molten reactor core had to be disposed of. Costs amounted to approximately $1 billion [2].

Lessons Learned The non-availability of the emergency feed system, the erroneous signals produced by the instrumentation with regard to the "open" valves of the pressure vessel, and the lack of knowledge of the real status of the plant as well as the shutdown of the primary pumps led to a partial core meltdown.

Present-day pressurized water reactors have better instruments in the reactor pressure vessel and displays in the control room. In addition they have fourfold redundancy of primary cooling and emergency cooling systems, and the possibility to reduce primary and secondary pressures with subsequent possibilities to feed water.

9.2 The Chernobyl Accident

The Chernobyl accident occurred on April 26, 1986 in unit 4 of the reactor plant of four units of 1,000 W(e) or 3,200 MW(th) power each. The four reactor plants had been build some 130 km north of Kiev and comprised four Russian graphite-moderated, boiling-water cooled so-called RBMK1000 reactors. The UO_2 fuel was enriched with 2 % U-235. The core of these RBMK1000 reactors is about 7 m high and about 12 m in diameter. The RBMK1000 has two coolant loops with four circulation pumps each. One pump is always on standby. The reactor core is controlled by raising and lowering 211 control rods. This combination of low-enriched fuel, graphite as the moderator, and boiling water as the coolant resulted in a positive coefficient of coolant temperature (see also Chap. 2) [4–7].

The operators were preparing an experiment in which—after shut down of the reactor- the energy of rotation of the turbine during coastdown was to be used to produce emergency electrical power. This was considered necessary in case of reactor shut down with subsequent failure of the external electrical grid (station black out). The three emergency diesel generators needed about 1 min after their start up to reach full speed and power to feed one primary coolant pump required for cooling of the afterheat generated in the core. This existing lack of emergency power during roughly 1 min was to be provided by the energy of rotation of the turbine during its cast down. Three such experiments had already been carried out 4, 2 and 1 year before the accident, but they had been unsuccessful. They had shown that the excitation voltage of the turbine- generator system was too low. This had been modified in the meantime and the new experiment was to test the new voltage regulation system The experiment was set to begin at a power level at a power level of about 700 MW(th).

The experimental procedures began with a power reduction from full reactor power. However, this had to be interrupted at 1,000 MW(th) because the electrical grid coordinator (load dispatcher) suddenly requested power again. At this point in time, the emergency core cooling systems had already been shut down in preparation of the test. When the load dispatcher permitted again a further drop in power, the envisaged power level was not reached by the operators. During the period of power production at reduced power level the production of the fission products Xe-135 (neutron absorber) had began. It decreased the effective neutron multiplication factor keff and caused the power level to drop to about 500 MW(th). A following operator error (control rods were inserted too far) led to a further drop of power level to about 30 MW(th) [8]. As a consequence, there was additional buildup of the xenon-135 fission product (neutron absorber) with the associated decrease of the effective multiplication factor, k_{eff}. (The amount of Xenon poisoning and its influence on the effective multiplication factor k_{eff} was not known to the operators at that point in time.)

At this very low power level of 30 MW(th) the operators made the decision to restore power by shutting off the automatic control system and to extract the majority of the control rods by manual control to their upper limits. The power started to rise and could be increased to about 200 MW(th), a value smaller than the planned 700 MW(th).

The positive coefficient of coolant temperature of the RBMK1000 reactors was known to cause instabilities in this low power range. When the eight primary coolant pumps were activated during startup, this gave rise to instabilities in power production, coolant flow and temperatures. Various alarms started going off at this time. The operators received emergency signals regarding the levels in the steam/water drums and large variations in the feed water flow as well as from the neutron flux or power monitors, respectively (Fig. 9.3).

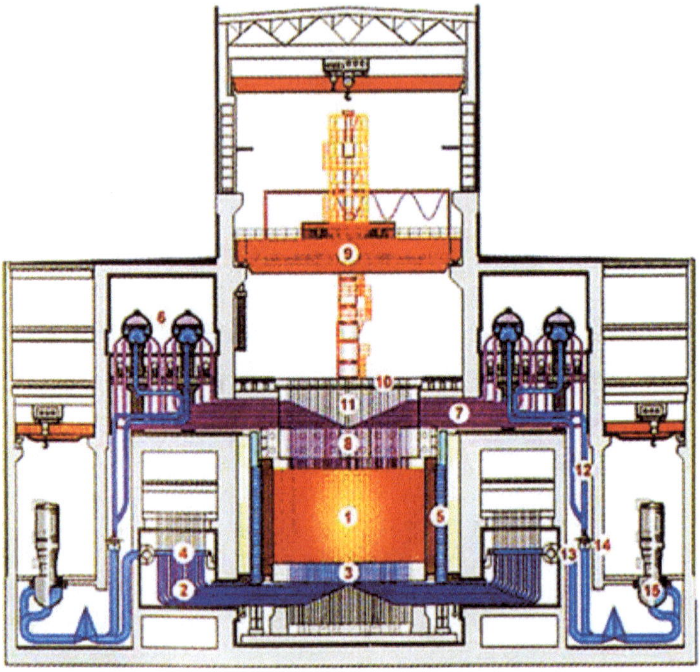

1 Reactor core	9 Fueling machine
2 Core inlet pipe	10 Platform for fueling
3 Bottom radiation shield	11 Vertical channels for fuel elements
4 Collector – distributor	12 Downcomer pipes
5 Lateral radiation shield	13 High-pressure collector
6 Steam separator	14 Low-pressure collector
7 Steam-water pipes	15 Main coolant pipes
8 Top radiation shield	

Fig. 9.3 Schematic design of the Chernobyl reactor [4, 5]

After the more or less stable power level at 200 MW(th) had been reached, preparations of the experiment continued by activating extra water pumps. This led again to coolant temperature variations and the operators responded by turning off two of the circulating pumps. All these actions led to an extremely unstable reactor state before the experiment began.

Almost all control rods had been removed manually to their upper limit. The automatic control system together with other automated safety features had been disabled. **The reactor coolant temperature was close to its boiling temperature.**

The reactor was already outside of the safe operating envelope established by the designers.

The experiment began by shutting off the steam to the turbine. The turbine generators began to run down. The diesel generators started and picked up loads.

Within the time period of about 1 min until the diesel generators reached full power, the running down turbine generators systems was to support the diesel generators. As the coolant flow rate decreased (the four coolant pumps were also running down; only one coolant pump was to obtain its power supply from the diesel generator and the turbine generator) this lead to increased steam formation (coolant near boiling temperature) in the core. With its positive coolant temperature coefficient and increasing steam formation the reactor theoretically, was now on its way to power runaway.

At this point in time the emergency shut down (SCRAM) bottom was pressed manually. All control rods started immediately to be fully inserted. The control rods moved with a speed of 0.4 m/s into the 7 m high core. The control rods contained a graphite section in their bottom parts followed by absorber sections with boron carbide. During insertion into the upper core neutron-absorbing water was displaced by non-neutron absorbing graphite. This led to an additional increase of the effective neutron multiplication factor k_{eff}. A steep power increase occurred causing the core to overheat. The fuel rods ruptured under the pressure of overheated fuel and fission product gases. The finely dispersed fuel abruptly mixed with the cooling water causing of steam explosion (Sect. 10.2.1). According to theoretical analysis the reactor power jumped to about 30,000 MW(th), ten times the normal operational output. The last reading on the control panel showed 33,000 MW(th). The steam generated caused the destruction of the steam boiler and of core structures and lifted the 2,000 tons top shield together with the refueling machine upwards. Fuel elements and red glowing (not burning) graphite parts were ejected from the core. The reactor building was heavily damaged [8]. A second explosion occurred some seconds later terminating the nuclear reaction and destroying the reactor core and building structures even more, dispersing damaged fuel elements and red glowing graphite parts. There are hypotheses that this second explosion was a second steam explosion or a hydrogen explosion (hydrogen generated from a chemical reaction between the zirconium fuel rod cladding [8]). As bitumen had been used for the construction of the reactor building floor and the turbine hall ejected material ignited fires. The remains of the overheated reactor core were now open to the atmosphere. The fission product gases as well as fission product aerosols and fuel aerosols released were driven by the heat release to an altitude of roughly 2,000 m, in some cases even 10,000 m. Strong winds at these altitude distributed the aerosols over the Ukraine, Belarus, western parts of Russia and Europe. The damaged RBMK1000 (unit 4 of the Chernobyl reactor plant) is shown by Fig. 9.4.

Shortly after the accident firemen arrived to extinguish the fires. Many firemen received very high doses of radiation. The fire was finally extinguished by a combined effort with helicopters dropping 5,000 tons of sand, lead, clay and boron carbide onto the burning reactor. However none of the neutron absorbing boron carbide reached the core [8]. Remotely controlled cranes and bulldozers were used to push back the radioactive material into the reactor. Radioactive debris was shoveled by liquidators wearing heavy protective gears. These workers could only spend a maximum of 40 s working because of the high radioactive doses [9]. There

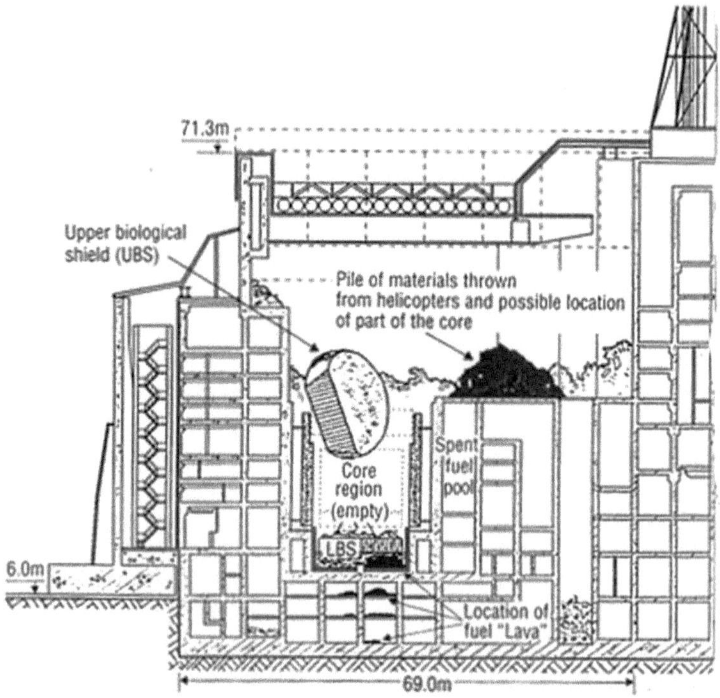

Fig. 9.4 The destroyed Chernobyl reactor after the accident [8]

was fear that the core could melt through the concrete structures. But finally the core mixed with sand, lead etc. and remained within the reactor building.

9.2.1 Radiation Exposure of the Operators, Rescue Personnel, and the Population

Very high, lethal radiation doses and burns in the first phase of the accident were suffered by the firemen, some members of the operating crew, and helicopter pilots dropping, among other things, sand, lead, and boron carbide (31 casualties). Some 1,400 members of the operating crew, scientists, and members of the emergency team (some 200,000 liquidators) were exposed to varying high radiation doses resulting in radiation sickness and radiation injuries.

Different figures are quoted in the literature of up to several thousand additional deaths to be expected in future [6, 7, 9]. According to IAEA data 2011 [6, 7], another 20 persons died afterwards from excessive radiation doses. This number includes roughly five children who died from cancer of the thyroid.

The inhabitants of the nearby cities of Prypjat and Chernobyl were evacuated as late as 30 h after the accident. Their radiation exposure was estimated to be 0.25–

0.5 Sv. On the whole, 135,000 persons were evacuated over the first few days [5]. As a result of varying weather conditions, radioactivity was carried as far as Germany (Chap. 4) and other countries in Western Europe, including Scandinavia.

9.2.2 Chernobyl Accident Management

Hundreds of specialists and approximately 800,000 military men were involved in pushing back the fuel elements into the reactor in "shortest-time activities," piling up sand, lead, boron carbide, and concrete, building a provisional concrete shield, and decontaminating plant compartments. The maximum radiation exposure of these members of emergency teams had been set at 350 mSv [5–9] (see also Chap. 4).

Approximately from 2012 on, the destroyed reactor unit 4 is being enclosed in a new arched sarcophagus [10, 11].

9.2.3 Contaminated Land

As a result of the prevailing weather conditions with precipitation, a number of areas in Ukraine, Belarus and in the western part of Russia were very highly contaminated over the first 10 days. While iodine-131, with a halflife of 8 days, had decayed already after roughly 1 month, cesium-134 (halflife 2 years) for the first 10 years or so and, above all, cesium-137 (halflife roughly 30 years) over approximately 100 years will determine the radiation exposure of the population due to ground-borne exposure and food ingestion [4, 5]. The regions with the highest exposure levels are shown in Figs. 9.5 and 9.6.

Depositions in excess of 40 kBq/m^2 cover large areas in northern Ukraine and southern Belarus. The most highly contaminated zone in Ukraine is the 30-km zone around the RBMK-1000 reactor of Chernobyl with more than 1,500 kBq/m^2.

In the region of Bryansk, Belarus, the highest soil contamination in some villages was measured to be up to 5,000 kBq/m^2. In the Kaluga-Tula-Orel region, contamination of 600 kBq/m^2 was found.

In summary, 3,000 km^2 were contaminated with **more than** 1,500 kBq/m^2 of Cs-137,[1] roughly 7,200 km^2 with 600–1,500 kBq/m^2 of Cs-137, and roughly 103,000 km^2 with 40–200 kBq/m^2 of Cs-137.

Dose exposures for people living in these areas can be estimated from similar dates given in Sect. 9.3.4 for the Fukushima reactor accident.

[1] No upper limit or higher level ranges as in case of Fukushima (Sect. 9.3.4) were published for Chernobyl.

Fig. 9.5 Cs-137 contamination caused by the Chernobyl accident in various regions of Ukraine, Belarus, and Russia [4, 5]

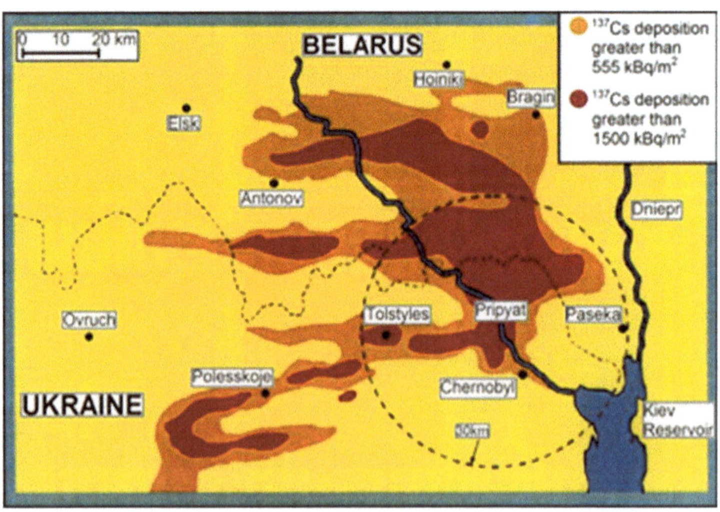

Fig. 9.6 Cs-137 contamination caused by the Chernobyl accident in Ukraine [4, 5]

Lessons Learned The main reasons for the Chernobyl reactor disaster were the wrong design of the RBMK reactors with a positive coefficient of coolant temperature, and also of the control/shutdown systems containing zones of graphite. Moreover, shutting down the automatic protection system and other safety devices by the operators in the course of the test was inadmissible and banned, respectively.

Western light water reactors may be built and operated only with a sufficiently high negative coefficient of coolant temperature.

9.3 The Reactor Accident of Fukushima, Japan

On March 11, 2011, an earthquake of the intensity M = 9 (Richter scale) hit the northeastern coast of Japan east of the city of Sendai (Fig. 9.7). This intensity corresponds to intensity XI on the European MSK/EMS-98 seismic intensity scale [12–21].

Roughly 1 h later, a tsunami wave hit the coast, flooding the Fukushima nuclear power plant up to a water level of 14 m. The earthquake was number four on the list of the severest earthquakes so far registered worldwide; its intensity had not been foreseen by Japanese seismologists when the reactor was designed [15]. Tsunami waves of much greater heights (up to 38 m) had impacted the Japanese coast in the past in the course of earthquakes of lower intensity (Richter scale) (Table 7.14, Sect. 7.4). However, the Fukushima-Daiichi nuclear power plant had been designed only against tsunami waves up to 5.7 m high (Fig. 9.8) [12].

Four nuclear power plants with an aggregate 14 BWRs in the environment of the city of Sendai were hit by the earthquake (with a number of subsequent seismic events) and the tsunami wave on March 11, 2011. These were (Fig. 9.7) the nuclear power plants of Onagawa with three BWRs, Fukushima-Daiichi with six BWRs, Fukushima-Daini with four BWRs, and the Tokai Research Center with one BWR and one research reactor. At that time, only three BWRs were in operation in the Fukushima-Daiichi plant, while the fourth BWR was down (with all fuel elements removed and located in the spent fuel pool), and BWRs 5 and 6 had been shut down for inspection and repair. The eleven BWRs in operation **were duly shut down automatically by the earthquake instrumentation and changed into the residual heat removal mode.**

The seismic waves hitting the Fukushima-Daiichi nuclear power plant from the epicenter caused a horizontal acceleration of 507 cm/s^2 of the most highly loaded reactor of the three units (unit 1). This plant had been designed to 449 cm/s^2. Nevertheless, the BWR was duly shut down. The reactor cooling system and the reactor core were not damaged. In units 2 and 3, horizontal acceleration had not exceeded the design basis levels (Fig. 9.9) [12, 13, 16, 18].

The operating crew immediately started accident management measures in each of the three reactors of the Fukushima plant. First, core cooling was maintained in each of the three reactors by means of the battery power available and a small steam turbine pump system fed by steam from the reactor pressure vessel.

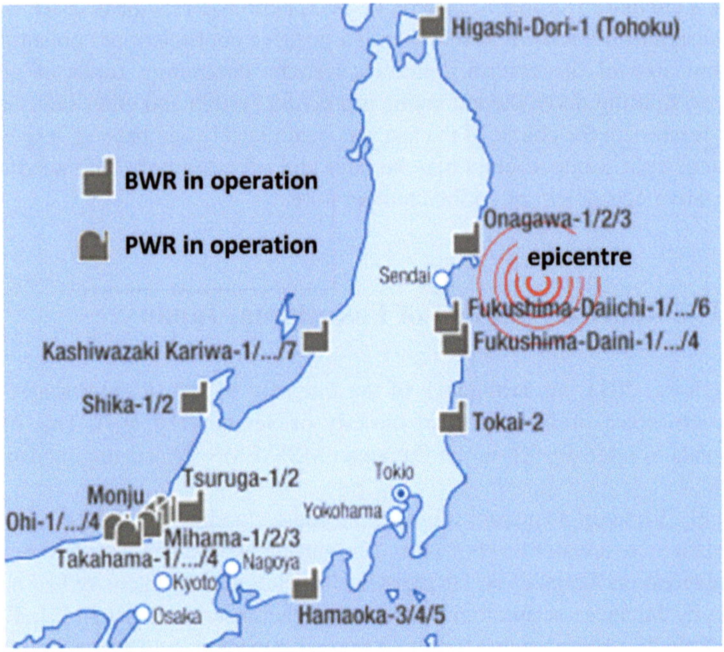

Fig. 9.7 Nuclear power plants operated in Japan (Honshu Island) when the earthquake and tsunami hit the coast [12, 13]

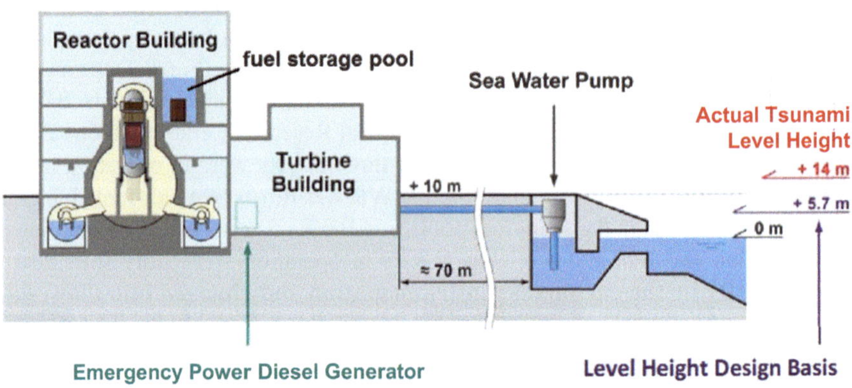

Fig. 9.8 Fukushima-Daiichi reactor plant with countermeasures installed against tsunami waves [12, 13]

However, when the tsunami wave flooded the plants 56 min later, the two diesel generators per reactor of the Fukushima-Daiichi plant units 1–3, installed in the lowest part of the turbine building (Fig. 9.8), were submerged and failed. Also the fuel supply to the diesel generators was partly torn off by the wave. All external grid

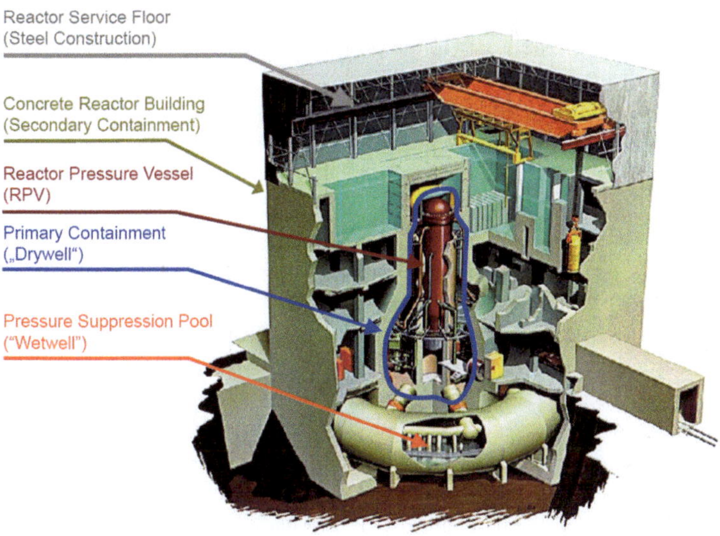

Fig. 9.9 Schematic design of the General Electric BWR-3 boiling water reactor of Fukushima-Daiichi [12, 13]

power supplies to the nuclear power plant had been destroyed already by the earthquake. Also the heat exchangers transferring residual heat to the ocean water outside failed. In addition the direct current batteries which were also located in the basement of the plant were flooded and lost.

In unit 1 due to the loss of all direct current batteries (after flooding) all instrumentation required to control the accident became unavailable. The operator crew had to work in the dark. The high pressure coolant injection system did not work, because of loss of power from the submerged diesel power generators (emergency power supply). The isolation condenser system (steam driven pumps) worked only partially. The primary coolant system with the reactor pressure vessel was not depressurized, because of lack of electrical power or pressurized nitrogen and lack of knowledge about the actual state in the various vessels due to the lost instrumentation. Therefore, low pressure emergency pumps could not feed water into the primary system for core cooling. The primary coolant system heated up and soon exceeded its design pressure, the core became uncovered by coolant and the fuel rods started to melt down. Hydrogen was produced because the fuel rod claddings (Zirconium) exceeded temperatures of 1,200 °C and reacted with steam.

$$\mathrm{Zr} + 2\,\mathrm{H_2O} \rightarrow \mathrm{ZrO_2} + 2\,\mathrm{H_2}.$$

The pressure in the reactor pressure vessel was soon relieved into the primary containment because core melt penetrated the lower bottom wall by small holes or by a break of a low elevation pipe of the pressure vessel or by opening of a safety/ relief valve. The radioactive noble gases and the volatile fission products such as

Cs-134, Cs-137, Sr-90 etc. were released into the pressure vessel and into the primary containment. The water in the primary containment with pressure suppression chamber heated up and its design pressure was soon exceeded (Fig. 9.9). Leakage paths within the primary containment vessel led to hydrogen release into the upper part of the reactor building (Fig. 9.10). About 1 day after the earthquake and the impact of the tsunami wave on unit 1 a hydrogen detonation occurred in the upper part of the reactor building. It destroyed the upper structures of the reactor building. Four technicians were injured [20, 21].

The records did not show any deliberate attempt by the operation crew to depressurize the reactor pressure vessel during the accident course. This would have been necessary to add water by emergency pumps. Only about 2 h after the hydrogen detonation, when the primary coolant system had depressurized itself, the operators could begin feeding in fresh water using fire pumps [21]. However, the longer term water level in the reactor pressure vessel did not recover to more than midplane, regardless of the make-up water quantity being added. This indicates a low elevation leak in the pressure boundary of the reactor pressure vessel.

The accident developed in an almost similar pattern in units 2 and 3, though with a larger shift in time. The reactor core isolation cooling system worked longer (for 70 h in unit 2 and 20 h in unit 3). When this emergency cooling system failed in units 2 and 3 the operators tried to depressurize the reactor pressure vessel in order to inject water using the fire extinguisher lines. Problems occurred, however, due to the lack of electricity for the solenoid valves and lack of pressurized nitrogen to open the safety/relief valves. Therefore, water could not be injected for about 6.5 h in unit 2 and for about 7 h in unit 3. The core fuel became uncovered in both units. The fuel heated up, was significantly damaged, hydrogen was produced and volatile fission products and radioactive noble gases were released into the primary containment. A longer term water level in the reactor pressure vessel could not be restored to higher than about midplane in both units 2 and 3, indicating also a low elevation leak in the pressure boundary of the reactor pressure vessel. The primary containment pressure increased. Hydrogen was released probably through leakage paths as the containment vent lines could not be opened. This was due to not high enough pressure to break a rupture disk. As a consequence a hydrogen detonation occurred also in unit 3. In unit 2 no hydrogen detonation happened [21].

Also in unit 4 the total emergency electricity supply (diesel generators and batteries) was lost. This lead to an increase of the coolant water temperature in the fuel storage pool of unit 4. At 6 AM on March 15 a hydrogen explosion also occurred in this reactor building (unit 4), severely damaging its upper structures. At first it was thought to be due to hydrogen production from fuel heat up and coolant uncovery in the spent fuel pool. Later, photographs indicated that there was no overheat damage of that fuel in the spent fuel pool. The source of hydrogen was traced to be a backflow through the standby gas treating system shared as common piping with unit 3 [21].

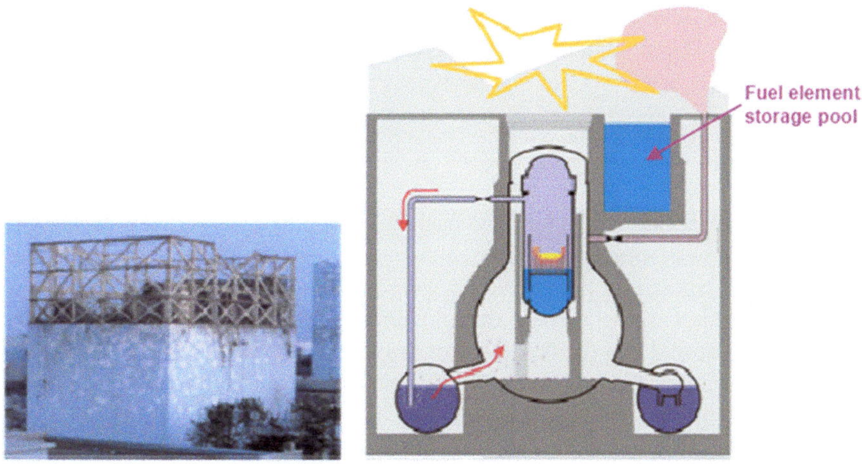

Fig. 9.10 Hydrogen explosions within the reactor auxiliary systems building destroyed the upper steel structure and the roof [12, 13]

9.3.1 Spent Fuel Pools of the Fukushima Daiichi Units 1–6

9.3.1.1 Unit 1

When the hydrogen detonation occurred and damaged the upper building structure, material might have been falling into the spent fuel pool. There is, however, no evidence that the fuel was damaged. A concrete pumping truck was used to provide makeup water inventory. An alternative cooling water system was put in service soon afterwards. The cooling water temperature has been maintained <35 °C [21].

9.3.1.2 Unit 2

Using existing piping water addition to the unit 2 spent fuel pool was possible. No fuel was damaged. A dedicated system using a heat exchanger was put in service afterwards. The cooling water temperature has been maintained <35 °C [21].

9.3.1.3 Unit 3

After the upper building structure had been destroyed by the hydrogen detonation water cannons were used for spraying water and helicopters dropped water into the spent fuel pool of unit 3. Then concrete pumps provided water addition to the spent fuel pool. The use of existing piping to restore water cooling started soon afterwards. Photographs showed that parts of the building structures had fallen into the pool. It is likely that no damage has occurred to the spent fuel. The cooling water temperature had been maintained <35 °C [21].

9.3.1.4 Unit 4

After the upper building structure had been destroyed by the hydrogen detonation, initially, water was sprayed by water cannons and concrete pumps. The structures supporting the spent fuel pool of unit 4 were improved by steel support pillars to provide protection against damage that might result from additional seismic events. Photographs showed that the fuel racks of the spent fuel pool of unit 4 were intact. A cooling system for the spent fuel pool was put in service. The coolant temperature of the spent fuel pool had been maintained <40 °C [21].

9.3.1.5 Units 5 and 6

No damage occurred to the fuel in the spent fuel storage pools 5 and 6. One Diesel generator of the units 5 and 6 could be restored soon enough such that emergency electric power was available for the cooling of the spent fuel [21].

9.3.2 Measurement of the Radioactivity Released

Large parts of the radioactivity released were initially carried out to the sea by the prevailing winds. Measurements of the radioactivity released over land (radioactive noble gases, radioactive I-131 (halflife 8 days) and radioactive aerosols, such as Cs-134 (halflife 2 years) and Cs-137 (halflife 30 years)), were carried out by specially equipped aircraft of the American National Nuclear Security Agency (NNSA). These measurements (Fig. 9.11) showed a particularly pronounced distribution of radioactivity towards the northwest of Fukushima. The measurements were evaluated in order to determine the radioactive exposure, which must be known for decisions about evacuation of the public. The Japanese government then evacuated the population (roughly 200,000 persons) in the vicinity of the nuclear power plants. The initial evacuation zone was soon expanded to a radius of 20 km. Afterwards, also some places situated beyond the zone of 20 km were evacuated because the annual dose to the population there had been estimated to run up too high [18, 19].

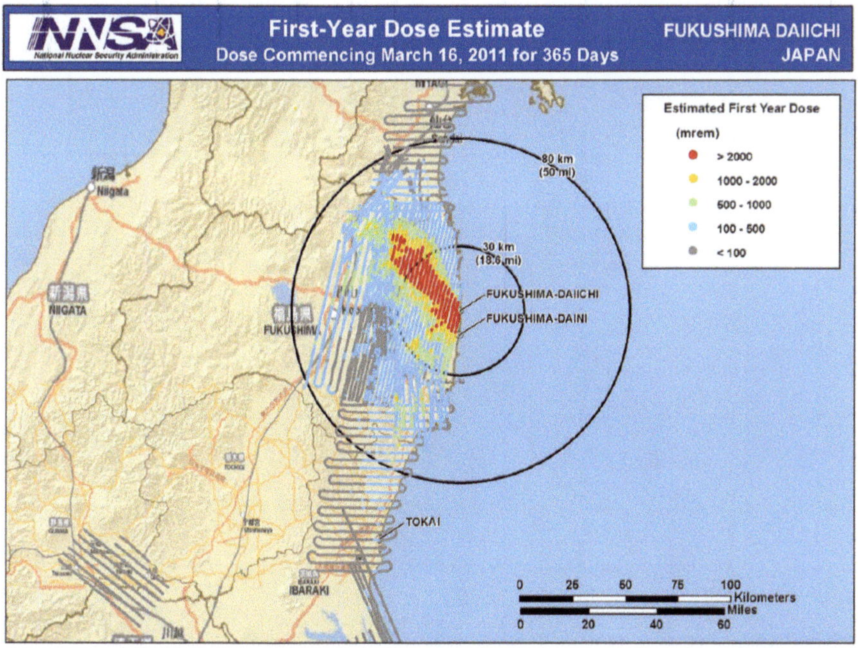

Fig. 9.11 Measured radioactivity and calculated dose levels from radioactive exposure within the first year after the accident of March 11, 2011 [16–19, 21]

9.3.3 Damage to Health Caused by Ionizing Radiation

Three months after the accident, the IAEA in Vienna found that the population had not suffered any measurable damage to health as a result of ionizing radiation [16].

The lifetime baseline risk (probability of having a specific cancer over the lifetime of 89 years of a person) was reported by the World Health Organization (WHO) [19] in a detailed analysis for children and adults.

Four members of the operating crew were killed by the earthquake and the following tsunami wave. Some 20 staff members were injured by the hydrogen explosions. There were in total 23,172 emergency and mitigation workers working at the Fukushima Daiichi reactor plants. According to a report of the World Health Organization (WHO) [19] most of them received <10 mSv total effective radiation dose. 75 workers received up to <200 mSv and 12 workers received up to 700 mSv total effective radiation dose (two of them had sustained β^--radiation exposures of the legs from contaminated water). The level laid down by the Japanese government was a maximum radiation exposure of rescue workers of 250 mSv [16–19, 21] (see also Sect. 4.4.3).

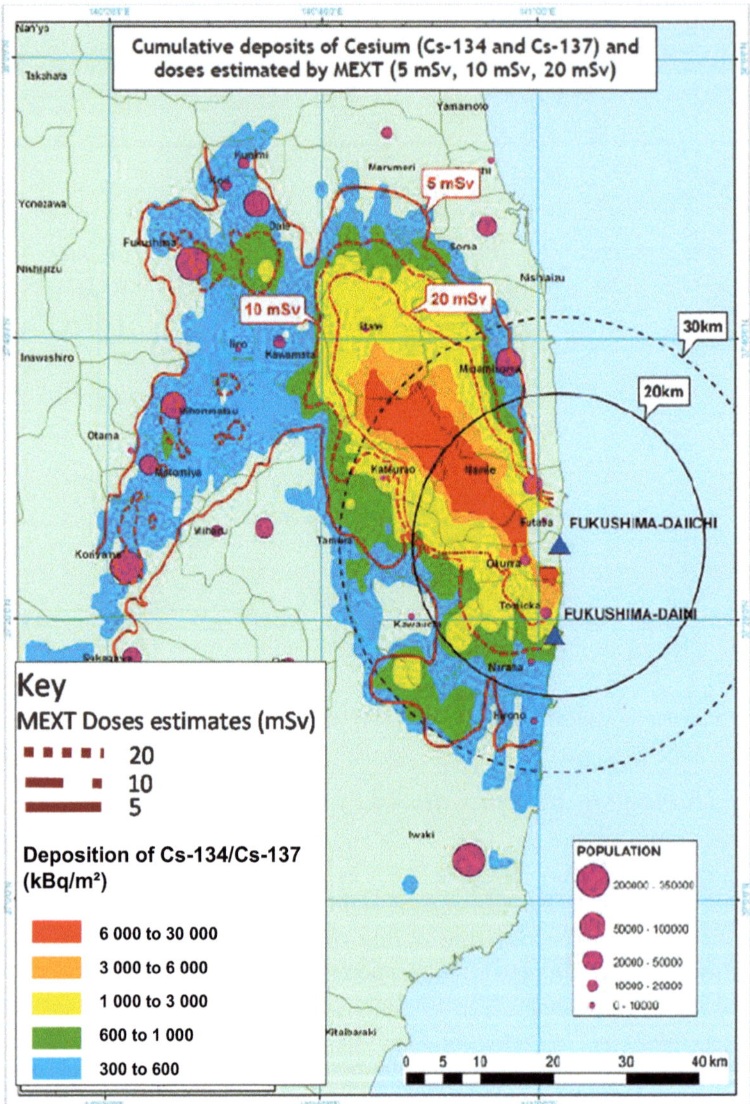

Fig. 9.12 Areas of Cs-134 and Cs-137 deposits in kBq/m^2 as well as estimates for radiation exposure of the Japanese public in certain locations during the first year after the Fukushima-Daiichi accident [18, 21]

9.3.4 Contamination by Cs-134 and Cs-137

While iodine-131, with a halflife of 8 days, had decayed within a month, depositions of Cs-134 stay on the ground for about 10 years, and depositions of Cs-137 remain for roughly 100 years. Japanese, French, and American measurements of ground depositions of cesium are shown in Fig. 9.12.

The results indicate the most highly exposed zone due to Cs-134 and Cs-137 with 6×10^6 to 30×10^6 Bq/m^2 to be in the northwestern direction from the Fukushima-Daiichi plant. Other zones show lower soil contamination levels due to cesium. Evaluations indicated regions in which persons, if they lived there and consumed only food available locally, would suffer a radiation exposure of 5 mSv/year and 10 mSv/year or 20 mSv/year, respectively, within the first year after the accident (This must be compared with the world wide effective annual radiation dose 2.4 mSv/year) [19] and the 8–100 times higher annual natural radiation exposure in Kerala, India or Brazil (Sect. 4.3). In the regions more highly exposed to radioactive cesium, 3,100 persons, if they were to return there, would suffer 50 mSv/year, and 2,200 persons would not be allowed to enter the "no-entry" zone. They would be exposed there to 100–500 mSv/year [17].

The Fukushima accident was classified in top category 7 of the international event scale of reactor accidents drafted by IAEA.

9.3.5 Lessons Learned

Despite the earthquake of Richter scale 9, the reactor units of Fukushima-Daiichi were shut down automatically as planned. The emergency power diesel generators supplied the internal grid as planned for almost 1 h until the tsunami hit.

The disaster occurred because the reactor facility had been designed only against a tsunami wave of 5.7 m height. The tsunami wave of 14 m height caused the emergency power diesel generators and direct current batteries to fail. Unfortunately, they had been installed at the lowest point (the basement of the turbine hall). The compartments of this turbine building could not be shut watertight.

Due to the failure of the diesel generators and the loss of the direct current batteries the reactor pressure vessel could not be depressurized (safety/relief valves could not be opened). The high pressure water injection system failed. The isolation condenser system worked only partially.

The efforts of the operators to start emergency core cooling by means of low pressure fire pumps failed. Water could not be injected because of the high pressure in the reactor. This caused the water level in the reactor pressure vessel to drop, the temperature of the fuel claddings to rise, hydrogen to be produced, the fuel elements to melt down, and fission products to be released into the pressure suppression chamber. The absence of hydrogen recombiners in the inner containment, and the non-availability of means for the depressurization of the hydrogen mixed with steam and fission products to be passed through aerosol filters in the stack of the plant, led after leaks out of the inner containment to hydrogen explosions and destruction of the relatively lightweight roof structure. The fuel element storage pool was uncovered. Radioactive iodine and, above all, radioactive cesium were released into the atmosphere.

9.3.5.1 Comparison with the Safety Design of Other Nuclear Power Reactors

The question whether the Fukushima-Daiichi accident scenario is also representative for other nuclear power plants in the world can be answered as follows:

Tsunami waves must only be expected for nuclear power reactors built near the ocean. LWRs in Europe and the USA are designed against floods occurring once in about 10,000 years, including waves, hurricanes, failure of dams, etc. (Chap. 7).

Most of the presently operating nuclear power reactors in Europe and the USA have similar safety design characteristics as pressurized water reactors and the boiling water reactors described in Chap. 3. These have a second emergency diesel power grid system protected against airplane crash and other external events, which ensures both cooling of the reactor core by way of the steam generators and cooling of the fuel element pools. The emergency power building, like the regular emergency power diesel buildings, **is protected against flooding**. The air intake openings of the diesel generators are located in the upper region of the building [20]. German pressurized and boiling water reactors—as an example—are equipped with hydrogen recombiners (backfitting) which recombine the hydrogen produced in accident situations within the inner containment. Boiling water reactors have inner containments which are inertized by nitrogen. The containments of pressurized water reactors in Germany—as an example—will resist to large scale hydrogen detonations (Chap. 10).

Boiling water reactors have the appropriate emergency buildings with the same functions as pressurized water reactors [20]. First of all, steam-driven turbopumps are available for emergency core cooling. Then pressure relief is initiated, and the inventory of the feed water tank—as an example in Germany—is passively fed into the reactor pressure vessel. The reactor core can be cooled by means of mobile pumps kept in the emergency building. There are several possibilities of feeding the reactor core with water by mobile pumps up to and including pumps of the firefighting system. The feed water reservoirs available include the demineralized water tanks (tanks for water of very high purity), the drinking water system, internal wells, and river water (severe accident management measures (Chap. 10)).

9.3.6 Recommendations Drawn from the Fukushima Accident

9.3.6.1 Recommendations of the American Nuclear Society Special Committee on Fukushima

A committee of the American Nuclear Society with safety experts from the USA and Japan made the following recommendations after thorough analysis of the Fukushima accidents [21].

- Flooding protection of diesel generators and direct current batteries is essential. Independent direct current connection should be provided for critical instrumentation, critical valve operation and control functions
- Ensure adequate dike heights against flooding of the emergency diesel generators and the direct current batteries. Provide diversity for both alternate and direct emergency power supply
- Provide robustness of the reactor core isolation cooling system (steam driven pumps and generators) in Boiling Water Reactors
- Improve the reliability to depressurize the reactor pressure vessel and maintain it depressurized during station black out (loss of all electrical power)
- Improve the reliability to vent the primary containment in Boiling Water Reactors
- Improve the instrumentation in the reactor pressure vessel to provide the operator with more knowledge about the course of a core accident.
- Provide the inner containment of Boiling Water Reactors with hydrogen recombiners
- The possibility of an earthquake damaging the wall of the liner of the spent fuel pool causing cooling water to be lost should be considered. A hardened strong pipe—as already realized in US reactor plants—should be installed which allows water to be fed to the spent fuel pool from the outside.

9.3.6.2 Additional Recommendations Drawn from the Fukushima Accident

The emergency power supplies (diesel generators, gas turbines, fuel cells direct current batteries) should be arranged in a building protected against tsunamis, flooding, earthquakes, hurricanes etc. An emergency operation room with the essential instrumentation power supply and the ability to operate the plant in case it cannot be operated from the main operator room should be available.

9.4 Comparison of Severe Reactor Accident on the International Nuclear Event Scale

The International Atomic Energy Agency (IAEA) introduced in 1990 the International Nuclear and Radiological Event Scale (INES) as a measure to compare severe nuclear accidents and their radiological impact on international scale. It represents seven increasing levels. Each level about a factor more severe than the previous level [22]. Figure 9.13 shows the INES scale in form of a pyramid.

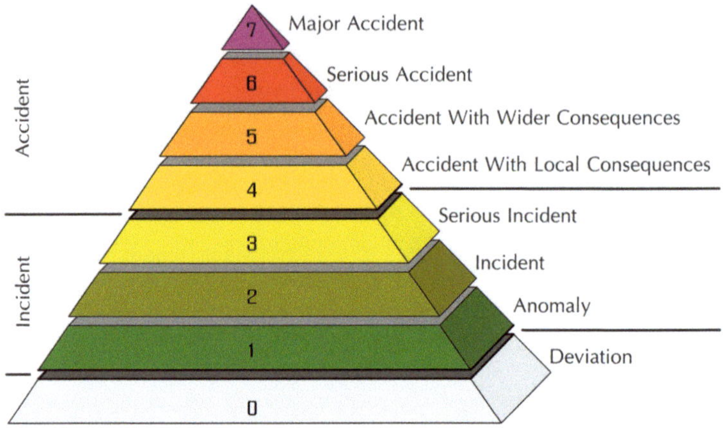

Fig. 9.13 International nuclear event scale [22]

Level 0: Deviation

- Classifies: Deviations of no safety significance, e.g. leakage from a primary coolant circuit

Level 1: Anomaly

- Classifies: impact on the defense in-depth, e.g. damaging of a fuel element during unloading process

Level 2: Incident

- Classifies: impact on radiological barriers, on people or environment, e.g. radiation levels in an operating area of more than 50 mSv.

Level 3: Serious incident

- Classifies impact on radiological barriers, on people or environment, e.g. radiation levels in an operating area of more than 1 Sv/h.

Level 4: Accident with local consequences

- Classifies: impact on radiological barriers, on people or environment, e.g. SL-1 accident (USA). Experimental reaction SL-1 reached prompt criticality killing three operators.

Level 5: Accident with wider consequences

- Classified: impact on radiological barriers, on people or environment, e.g. Windscale accident in the United Kingdom in 1957, Three Mile Island Accident near Harrisburg (USA) in 1979.

Table 9.1 INES classification of the most severe nuclear reactor accidents

Reactor	Windscale	Three Mile Island	Chernobyl	Fukushima Daiichi
INES-classification	Level 5	Level 5	Level 7	Level 7

Level 6: Serious accident

– Classified: impact on people and environment, significant release of radioactivity to require planned countermeasures, e.g. Kyshtym disaster at Mayak, Russia. A failed cooling system at a military nuclear waste reprocessing facility caused a steam explosion which led to release of 70–80 tons of highly radioactive material into the environment.

Level 7: Major accident

– Classified: impact on people and environment

Major release of radioactive material with wide spread health and environmental effects requiring planned and extended countermeasures.

Examples are:

– The Chernobyl disaster in 1986, Ukraine,
– The Fukuschima Daiichi nuclear disaster in 2011, Japan

Following this classifications of the Instrumental Nuclear and Radiological Event Scale (INES) the severe accidents of Windscale, Three Mile Island, Chenobyl and Fukushima are listed in Table 9.1.

References

1. Ireland JR et al (1981) Three Mile Island and multiple-failure accidents. Los Alamos Sci 3 (Summer/Fall):75–85
2. Knief RA (1992) Nuclear engineering, theory and technology of commercial nuclear power. Hemisphere Publishing Corporation, Washington, DC
3. Kemeny JG (1979) Report of The President's commission on the accident at Three Mile Island: the need for change: the legacy of TMI. The Commission, Washington, DC. ISBN 0935758003. http://www.threemileisland.org/downloads/188.pdf
4. Chernobyl (1995) Ten years on. Radiological and health impact. Nuclear Energy Agency (OECD), Paris Cedex
5. Chernobyl (2002) Update of Chernobyl: ten years on. Nuclear Energy Agency (OECD), Paris Cedex
6. (2011) Chernobyl: answers to long standing questions. International Atomic Energy Agency (IAEA), Vienna. http://161.5.1.75/newscenter/focus/chernobyl/faqs.shtml
7. (2011) Deaths due to the Chernobyl disaster. http://en.wikipedia.org/wiki/Deaths_due_to_the_Chernobyl_disaster
8. Chernobyl disaster. http://en.wikipedia.org/wiki/chernobyl_disaster
9. Shikalov N (1991) Kurchatov Institute Moscow. Personal communication
10. Kovan D (2011) Chernobyl 25 years on: time for a giant leap forward. Nucl News (Am Nucl Soc) 54(5):57

11. IAEA (1991) The international Chernobyl project, technical report, assessment of radiological consequences and evaluation protective measures report by an International Advisory Committee, Vienna, Austria
12. Mohrbach L (2013) Fukushima two years after the tsunami – the consequences worldwide. Atomwirtschaft 58(3):152
13. Mohrbach L (2011) Unterschiede im gestaffelten Sicherheitskonzept: Vergleich Fukushima Daiichi mit deutschen Anlagen, Sonderdruck aus Jahrgang 56 (2011), Heft 4/5|April/Mai. Internationale Zeitschrift für Kernenergie
14. DOE-NNSA Fukushima Survey March 27–28 (2011) PNG. http://en.wikipedia.org/wiki/File:DOE_NNSA_Fukushima_Survey_27-28.PNG
15. Henry P (2011) Das Megabeben in Japan, Spektrum der Wissenschaft, Aug 2011, 68–74, Spektrum der Wissenschaft Verlagsgesellschaft, Heidelberg
16. (2011) IAEA international fact finding expert mission of the nuclear accident following the great East Japan Earthquake and Tsunami. Preliminary summary, 24 May–1 June 2011
17. IRSN (2011) Assessment on the 66th day of projected external doses for the populations living in the North-West fallout zone of the Fukushima nuclear accident. Report DRPH/2011-10, IRSN, France
18. Nuclear News Special Report (2011) Fukushima Daiichi after the earth-quake and tsunami. Nucl News 54(4):17
19. World Health Organisation (2013) Health risk assessment from the nuclear accident after the 2011 Great East Japan earthquake and tsunami, based on a preliminary dose estimation, WHO Press, World Health Organisation, Geneva, Switzerland
20. Mohrbach L et al (2011) Earthquake and tsunami in Japan on March 11, 2011 and consequences for Fukushima and other nuclear power plants. http://www.vgb.org/vgbmultimedia/News/Fukushima_VGB_rev16.pdf
21. American Nuclear Society, Fukushima Daiichi: ANS Committee Report, March 2012, Revised June 2012. http://fukushima.ans.org/report/Fukushima_report.pdf
22. International Nuclear Event Scale. http://en.Wikipedia.org/International_Nuclear_Event_Scale

Chapter 10
Assessment of Risk Studies and Severe Nuclear Accidents

Abstract In Chaps. 7 and 9 the results of the risk studies for LWRs and the facts and consequences of the severe nuclear accidents of Chernobyl and Fukushima were assessed. The severe nuclear accidents of Chernobyl and Fukushima resulted in contamination of large areas and in the evacuation of the population of large areas.

It is concluded that LWRs in Europe are built and operated in densely populated areas with cities of several 100,000 inhabitants. It is impossible to evacuate cities of that size in due time and their contamination is beyond anybody's imagination. Therefore, the KHE safety concept was proposed around 1990 to build future LWRs such that the main accident path of the risk studies leading to the high number of deaths and to the large contamination of land should be eliminated by the safety design concept of the plant. The probabilistic safety studies leading to low frequencies of core melt in the range of 10^{-6} per year should be retained. However, the present risk concept should not be applicable any more to LWR plants built in future in Europe. It was concluded that the consequences of the most severe accidents in risk studies should be managed by the inner and outer containment of the reactor plant. A severe accident research program of the Research Center Karlsruhe, Germany, lasting for about two decades showed that it is possible to fulfill the requirements of the KHE concept. The KHE concept was essentially accepted by the German and French Safety Commissions and also incorporated in the German Atomic Law in 1994.

The main findings of the Karlsruhe research program for severe accidents were as follows:

- A large scale steam explosion in the reactor pressure vessel of a PWR would not jeopardize the mechanical integrity of the reactor pressure vessel. Consequently a steam explosion followed by failure of the integrity of the reactor pressure vessel and of the containment as assumed in WASH-1400 and the German Risk study can be considered impossible.
- A large scale hydrogen detonation after core melt in the spherical containment of a KWU PWR-1300 reactor plant cannot jeopardize the integrity of the steel

G. Kessler et al., *The Risks of Nuclear Energy Technology*, Science Policy Reports, DOI 10.1007/978-3-642-55116-1_10, © Springer-Verlag Berlin Heidelberg 2014

containment. Nevertheless hydrogen recombiners in the containment are useful for preventing or mitigating some accident sequences.

- Core melt down after a break of a pipe of the residual heat removal system in the annulus of the containment system or core melt down after an uncontrolled large scale steam generator tube break can be avoided by proper design.
- Containment failure after core melt down under high primary coolant pressure, as assumed in WASH-1400 and the German risk study, can be avoided by manual or automatic opening of pressure relief valves (ADS systems of Chap. 3) or—as a limiting case—reinforced anchorage of the pressure vessel cover.
- Core melting through the bottom part of the reactor pressure vessel can be counteracted by flooding the reactor pressure vessel on the outside with water (severe accident measure). Similar results can be obtained by installing a molten core spreading and cooling device (core catcher). Appropriate core catcher designs were developed after research.
- A rising steam pressure in the inner containment after a core melt can be avoided by water spray systems.
- A mechanically intact double containment, where the inner containment is either a steel containment or a concrete containment with an inner steel liner, having a leak rate of 0.3–1 % per day after a core melt accident and where the leaking radioactive gases and aerosols are passed from the annulus through aerosol filters to a stack can fulfill the requirements of the KHE safety concept. In this case the contamination by radioactive fission products is essentially limited to the site of the reactor plant. No evacuation of the population is necessary.

In summary all most severe accident consequences found in the WASH-1400 or the German risk study can be either controlled or eliminated or managed in future LWRs by appropriate design measures. Examples for such designs are EPR and KERENA (SWR-1000). Concluding remarks compare the KHE safety concept with the safety concept of presently operating reactors and more recent designs for LWRs.

10.1 Introduction

The findings reported in Chaps. 6 and 8 show nuclear power generation (for the operating about 370 nuclear reactors in 2012) in the range of damage (deaths) of similar magnitude as other technical or power generation systems. Yet, the severe reactor accidents of Chernobyl and Fukushima (Chap. 9) have shown that

- large areas were contaminated with Cs-134 and Cs-137 such that food production either had to be restricted for long periods of time (the half-life of Cs-134 is ~2 years and of Cs-137 is ~30 years) or, where contamination was lower, must be monitored for radioactivity over long time;
- the population of large areas had to be evacuated or even to be resettled.

While the economic losses arising, e.g., from major tanker accidents or oil drilling platforms were estimated to run up to several billion dollars, the economic consequences of the Chernobyl and Fukushima accidents are higher by about two orders of magnitude. Besides the damage to health resulting from the radioactivity to which workers and the population were exposed it is the contamination of land by Cs-137 as well as the ban on food over prolonged periods of time which are a problem very specific to the use of nuclear power [1–4].

In discussions about the risk of technical systems or especially by nuclear energy the following argument is often stressed:

The large damage, caused by severe nuclear accidents, is associated with extremely low probabilities of occurrence per year. Therefore the risk as the product of damage times probability of occurrence is small.

Around 1989, Kessler-Hennies-Eibl (KHE) [5–8] raised the question whether this risk argumentation and the associated findings of the risk studies had to be accepted as unavoidable for future LWR safety concepts or whether they could be improved upon.

European light water reactors are often located by large rivers passing through densely populated regions with cities of more than 100,000 inhabitants and large industrial plants. Cities of that size are impossible to evacuate fast enough in Europe. Radioactive contamination of such cities and surrounding densely populated areas is beyond anybody's imagination.

From 1990 on, this thinking led to deeper research into the sequence and consequences of severe core meltdown accidents in LWRs and to the proposal of an extended safety concept in Europe.

10.2 Principles of the KHE Safety Concept for Future LWRs

The safety concept for future LWRs as proposed by KHE at the Research Center and Technical University of Karlsruhe, Germany, is based on these considerations [5–8]:

– The major consequences of accidents as determined in WASH-1400 [9] and the German Reactor Risk Study Phase A [10] assume that

  • a steam explosion also called fuel coolant interaction (FCI) after core meltdown or
  • core melt-through under high primary pressure or
  • a major hydrogen detonation

  damage the outer reactor containment such that a large leak will release considerable amounts of airborne radioactivity (100 % radioactive noble gases, 50–90 % of the radioiodine, Cs, and Tc), roughly within 1 h after core melt down;

– a leak in the annulus between the outer containment and the concrete shell, or an uncontrolled steam generator accident with steam blowdown valves getting stuck in the open position, release only somewhat smaller amounts of airborne radioactivity;

The different contributions of these most severe accident sequences to the overall damage consequences, i.e. early deaths (fatalities), are shown in Fig. 10.1 as the result of the German Safety Study, Phase A [10–12]. The results are normalized to a 1.2 GW(e) PWR plant. These results are similar in terms of early deaths to the results of the US Reactor Safety Study WASH 1400 [9] as shown in Fig. 6.6 for 100 LWRs in the USA.

Figure 10.2 shows the areas which would have to be evacuated as a function of the different release categories during a severe core melt accident with subsequent containment failure. The abscissa of Fig. 10.2 shows the fraction of radioactive aerosols released and also the probability of occurrence for the radioactivity release categories (steam explosion, high pressure core melt through, hydrogen detonations etc.) as already presented in Chap. 6. On the ordinate the areas for evacuation are given, which become necessary as a function of the different weather conditions which are possible in the surroundings of the nuclear plant. The curves shown account for 50 % (dotted line) and 95 % (full line) of all weather conditions occurring [11–14].

The reasoning of the KHE safety concept is as follows:

If it can be demonstrated that the above accident phenomena (and their consequences detailed in Figs. 10.1 and 10.2) occurring after a severe core meltdown accident result in neither early nor late failure of the outer containment i.e. the outer containment retains its integrity, then the

– radiological consequences to the population,
– need for evacuation and resettlement of the population,
– hazard of contamination of large areas

are reduced to a minimum, i.e. limited to about the area of reactor plant itself. The curves shown by Fig. 10.1 are shifted to the left (close to the ordinate) [8].

If in addition the reactor core can be prevented from melting through the concrete baseplate into the ground below the reactor containment, there would be no danger of contamination of the groundwater over long periods of time.

The safety requirements of the outer reactor containment as outlined in the Karlsruhe KHE safety concept then are as follows:

– The consequences of severe core meltdown accidents must be managed by the structures of the inner and outer reactor containments. The inner reactor containment should retain the leaktightness of smaller than 0.3–1 % leakage rate even after a severe core meltdown accident. The molten reactor core must not melt through the bottom of the outer containment.

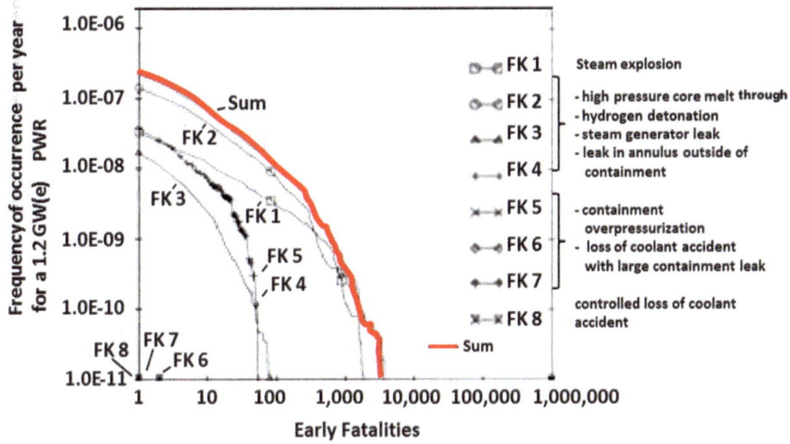

Fig. 10.1 Results of the German Reactor Safety Study, Phase A, shown with different contributions of those accident sequences resulting in the highest damage consequences normalized to a 1.2 GW(e) plant [10–12]

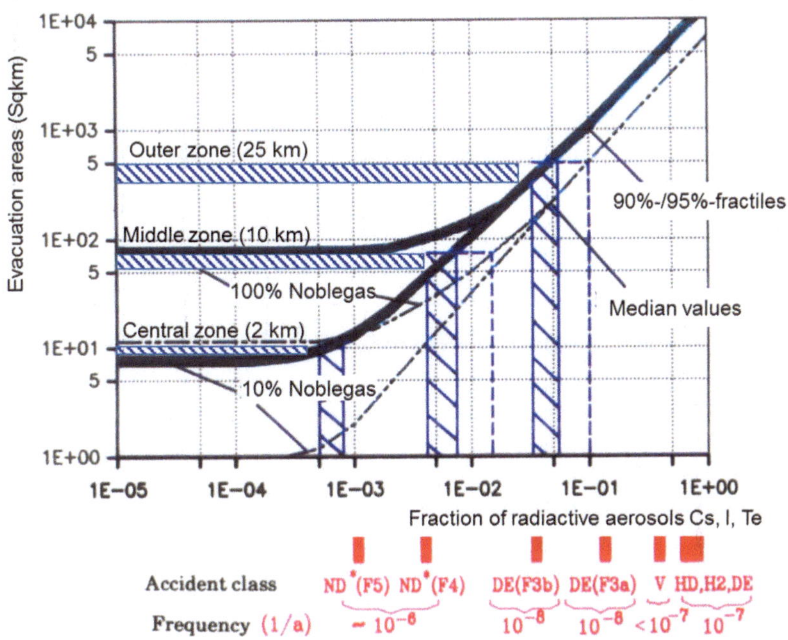

Fig. 10.2 Areas of evacuation required after a severe core melt accident [12–14]. *Dotted line* accounts for 50 % (mean values) and *full line* for 90–95 % of all weather conditions occurring

- The frequency of occurrence of core meltdown is determined by the diversity and number of redundancies of the cooling circuits, safety systems, emergency core cooling systems, emergency power supply systems. The overall frequency of occurrence for a core meltdown should be around 10^{-6} per reactor year. It can hardly be reduced below this level for practical reasons (common mode failures etc.).
- Probabilistic safety analyses only can serve for the optimization of safety systems and for demonstrating that the overall frequency of occurrence is in the range of 10^{-6} per reactor year. However, probabilistic safety analyses no longer shall support the argument that risk—as a product of frequency of occurrence multiplied by amount of damage—is so low that it can be accepted.

The results of the research program conducted at the Karlsruhe Research Center on the basis of this KHE safety concept are presented in the following sections of this chapter.

This KHE safety concept was essentially accepted by the German and the French Reactor Safety Commissions for future LWRs after thorough discussions [15, 16].

10.3 New Findings in Safety Research

This section will contain the new findings of safety research into the accident phenomena, described in the section above, which, in WASH-1400 [9] and the German Reactor Risk Study Phase A [10], still resulted in major accident consequences.

10.3.1 Steam Explosion (Molten Fuel/Water Interaction)

A steam explosion is the explosion-like evaporation of a cool liquid, such as water, in contact with very hot liquid fuel [17, 18]. This contact must allow a very fast heat input from the molten fuel to the coolant (water) in a very short time. Peak pressures may be in excess of 10 MPa. Steam explosions can also occur due to interaction of other liquids, e.g. contacts between

- hot oil and water,
- molten aluminum and water,
- water and liquefied gas at cryogenic temperatures,
- hot lava (volcanic eruption) and water.

In an LWR, an accident sequence can include mainly two contact modes for a steam explosion:

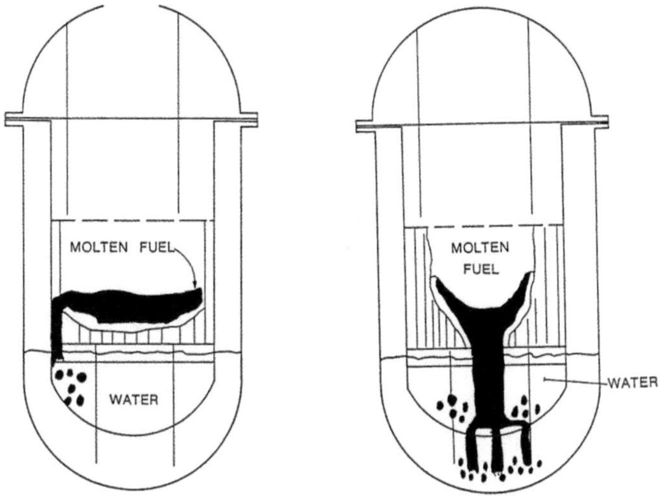

Fig. 10.3 Molten fuel pouring near the edge of a core (*left*) and molten fuel pouring through the lower grid plate [17, 18]

- During a very fast superprompt critical power transient leading far beyond the nominal reactor power, the fuel rods may burst and the molten fuel will be injected as very finely dispersed fuel (with very large heat transfer area) into the cooling channel under high pressure and mixed with the cooling water. This happened, e.g., to a certain extent in the so-called SL-1 accident (water-cooled experimental reactor in USA) [19–21] and in the Chernobyl accident, Ukraine, (Chap. 9);
- in the second mode of contact, molten fuel of the reactor core after a core meltdown accident can come into contact with remaining water (Fig. 10.3) either within the reactor pressure vessel, after melting through the grid plate, or outside the reactor pressure vessel, after melting through the bottom hemispherical head of the reactor pressure vessel [17, 18].

10.3.1.1 Mechanically Released Energy in a Steam Explosion

The maximum mechanical energy which can be converted from the thermal energy of the molten fuel in a steam explosion is obtained in the case of heat transfer at constant volume and rising pressure and ensuing isentropic expansion of the steam [22]. Heat transfer from the fuel melt to the water must occur roughly within 1 ms. The ratio of volumes of the fuel melt and water in that case should be around 1. Theoretically, this would allow an efficiency of roughly 40 % to be attained for the conversion of thermal into mechanical energy [23–25]. However, the efficiency measured in experiments with a simulated core melt (corium) on average nearly always is below about 1 %. The maximum efficiency of conversion in some

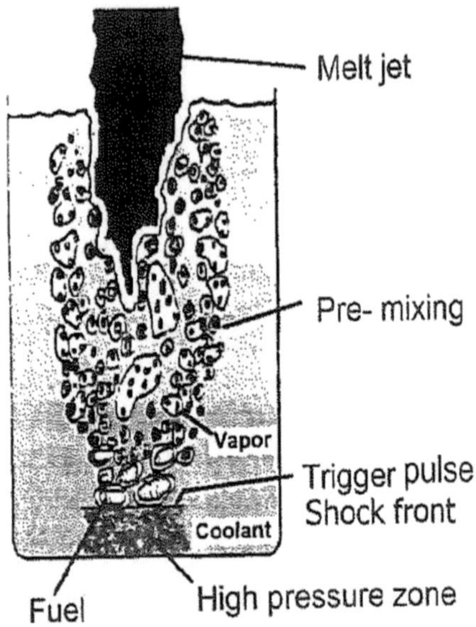

Fig. 10.4 Melt jet injected into water with different zones for premixing and fragmentation [17, 18]

experiments was 2–3 % [22, 26, 27]. Only experiments with iron—aluminum—thermite and water resulted in efficiencies roughly a factor of 2 higher.

Estimates of the damage resulting in the SL-1 accident led to an efficiency of 10–15 % [19]. However, the SL-1 accident is to a certain extent in the category of the first mode of contact between the molten fuel and water, which is initiated by a (superprompt critical) power transient with the fuel rods rupturing.

10.3.1.2 Description of a Steam Explosion Sequence

One of the most important preconditions of a steam explosion is fragmentation of the melt into many particles of about 0.1 mm size to create the area necessary for fast heat transfer.

Estimates show that this kind of fragmentation from a large molten core mass into many particles of roughly 0.1 mm in size is not possible in one single step, as the energy input would be too large. The melt first can be split up only into larger droplets of cm in size (premixing). In this process, Rayleigh-Taylor or Kelvin-Helmholtz instabilities play a major role when the molten jet flows into the water (Fig. 10.4). These larger melt droplets roughly cm in size are surrounded by a vapor film.

A pressure pulse is required as the trigger initiating a steam explosion. The pressure pulse causes the vapor film to break down and very fine melt particles (Fig. 10.5) to be produced in a process of fine fragmentation. As a consequence, there is very fast evaporation and a steam explosion, respectively [17, 18, 23, 24, 28].

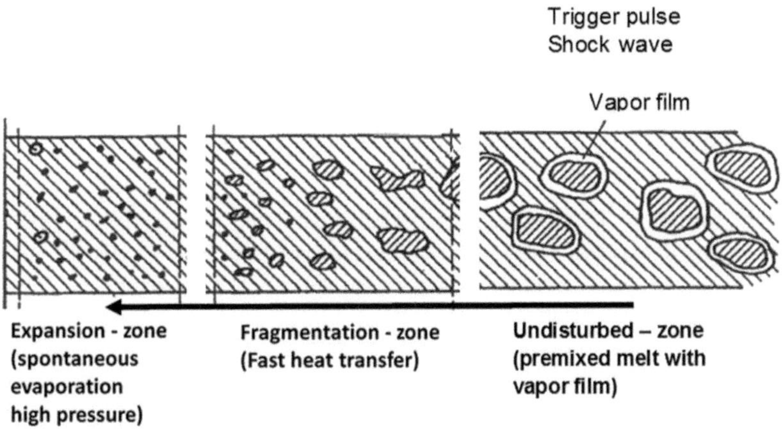

Fig. 10.5 Theoretical model for steam explosion with premixing, fragmentation (trigger pulse) and spontaneous evaporation (expansion zone) [17, 22]

However, this ideal concept of the theoretical chain of events in a steam explosion will only occur in a random process in experiments including dissipation effects. Moreover, the core melt will never contact water as a bulk substance because melt-through processes will always proceed in an incoherent manner in terms of time and location [27].

10.3.1.3 Steam Explosion in the Reactor Pressure Vessel

Accounting for the above described discrepancies between experimental results and the ideal theoretical models, it is now **postulated** that the molten reactor core melting through the gridplate and falling as a molten jet into the water-filled region below the gridplate would give rise to a steam explosion producing a maximum amount of stress acting on the reactor head and its bolts. WASH-1400 [8] had assumed that a steam explosion would cause the head of the reactor vessel to be blown off and penetrate the outer reactor containment as a bullet (α-mode failure). This maximum accident was the subject of many research programs between 1980 and 2010.

These are the most important findings of the associated Karlsruhe Safety Research Program [22, 29]:

- Studies of an assumed core meltdown accident by means of the SCADP/RELAP [30, 31] or MELCOR [32] computer codes led to a molten core mass of 110 ton at a temperature of approx. 3,000 K and a pressure of 0.25 MPa above the baseplate (the 110 ton corresponding to 85 % of the core mass of a KWU-1300 PWR). Additional studies using the MC3D [33, 34] and MATTINA [35] computer codes, which had been verified against experiments, showed that the core melt, after having molten through the gridplate (Fig. 10.3), would flow out in a molten jet of about 0.2 m^2 cross section. There would be premixing with a water volume fraction of 0.5–0.6. The **maximum** content of thermal energy in the larger melt droplet as premixing zone amounts to **roughly 3 GJ**. (On the average of all possible cases, the thermal energy content would be only 0.5–2.0 GJ with the corresponding water volume fractions of 0.2–0.5.)

In the further course of the analysis it was postulated that fine fragmentation to 0.2–0.3 mm, size as measured in experiments, would occur and a steam explosion would be initiated. After careful inspection and assessment of all experimental findings and theoretical analyses against theoretical models available internationally, **a conservative value of 15 % was selected** as the efficiency of conversion of thermal into mechanical energy. This results in a maximum mechanical energy release by the steam explosion of

$$3\ \text{GJ} \times 0.15 = \mathbf{0.45\ GJ}$$

(as the average of all possible cases, the result would only be 0.075–0.3 GJ).

10.3.1.4 Dynamic Mechanical Analysis of the Pressure Vessel

These results about the release of mechanical energy in the course of a steam explosion in the lower plenum (bottom hemispherical head) of the reactor pressure vessel were used to conduct dynamic mechanical stability analyses accompanied by 1:10 scale experiments (BERDA experiments) [36–39]. Theoretical models of similarity theory and accompanying strength analyses allowed the results of the 1:10 scale experiments to be transferred to the dimensions of the reactor pressure vessel.

As the steam explosion is initiated already while the melt jet is discharged, some 80 ton of core melt would still be present on the baseplate. After rupture of the mechanical anchorage (break) of the gridplate, this volume must be accelerated upward together in the reactor pressure vessel. The upper part of the reactor pressure vessel of a PWR contains the internal structures with the guide tubes for the control and shutdown rods and for the in-core instrumentation (Fig. 10.6). These must be compressed by the core melt, accelerated upward so as to be able to transfer the dynamic forces of the core melt to the head structures and head bolts. These internal structures and the head structures were simulated in great detail in the 1:10-scale experiments (BERDA) [37, 40].

This is the overall finding of all Karlsruhe BERDA experiments and theoretical analyses:

- Acceleration of the gridplate, with the remaining core melt resting on it, up to the top internal structures of the head requires at least approx. 2 GJ.
- Another 0.8 GJ would be necessary to compress the internal structures and elongate the head bolts by a few mm. The compression of the upper internal

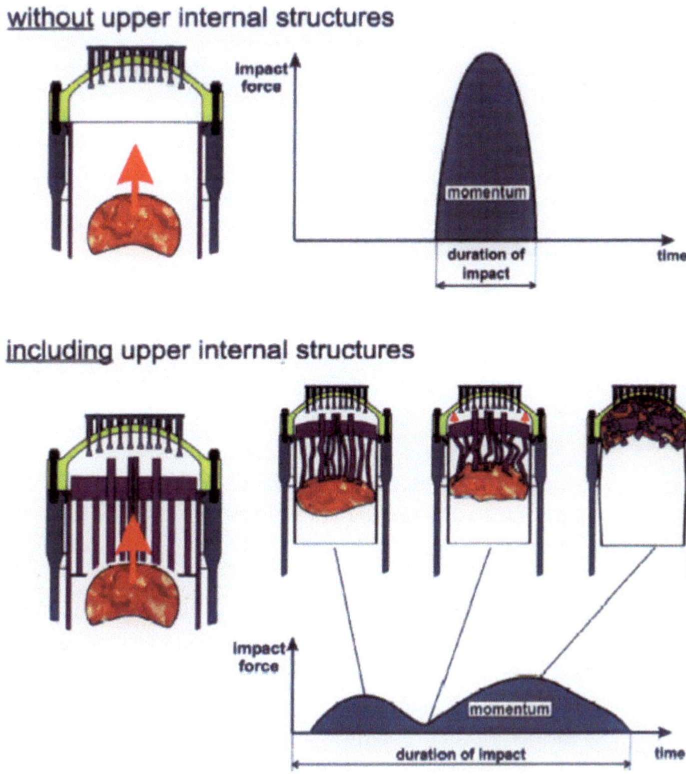

Fig. 10.6 BERDA experiments for melt slug impact on head of the reactor pressure vessel (without and with internal structures) [37, 40]

structures results in a decrease of the impact forces and an extension of the time period for action of these impact forces (Fig. 10.6). The reactor head would remain intact in this process.

– On the whole, at least **2.8 GJ** of mechanical energy would have to be released in the steam explosion to accelerate the core melt plus the gridplate upward, compress the internal head structures of the reactor pressure vessel, and expand the head bolts by a few mm.

This 2.8 GJ must be compared to the 0.45 GJ occurring in the steam explosion at conservatively assumed efficiency of 15 % for the conversion of thermal into mechanical energy. Even an efficiency of 40 %, i.e. 1.2 GJ, for the conversion of thermal into mechanical energy in a steam explosion could not jeopardize the mechanical integrity of the reactor pressure vessel [40–44].

Consequently, the steam explosion with failure of the reactor pressure vessel and outer containment (α-mode failure) as assumed in WASH-1400 [9]

and in the German Risk Study Phase A [10] **can be considered impossible on grounds of physics.**

A demonstration originally proposed in USA [45, 46] of the non-existence of the α-mode failure of the reactor containment was brought to a scientifically successful conclusion by these detailed Karlsruhe experiments and theoretical analyses.

10.3.2 Hydrogen Detonation

WASH-1400 [9] and the German Risk Study Phase A [10] postulated conservatively, with little detailed scientific and technical analysis, that a large-volume hydrogen detonation in the outer reactor containment would cause the containment to rupture and radioactivity to be released to the environment. This was doubted after some first theoretical estimates by KHE [5–7]. Appropriate containment design concepts were proposed which would be able to withstand a very conservatively assumed large-volume hydrogen detonation [7, 47]. This demonstrated that containment design concepts can be conceived which can resist to such large-volume hydrogen detonations.

These considerations were followed by many years of theoretical code development, such as GASFLOW [48], DET-3D [49], and COM3D [50] and experimental investigations, such as the RUT experiments in Russia [51–55]. The conclusion can be drawn from these theoretical and experimental efforts that a large-volume hydrogen detonation following a core meltdown accident can be managed by the reactor containment of existing modern PWRs, like the Konvoi-PWRs of Kraftwerk Union [56].

As a first severe accident management measure, PWRs were equipped with so-called passive autocatalytic recombiners [57] able to reduce slow release-rates of hydrogen release of approx. 0.5 kg H_2/s during core meltdown accidents. However, also core meltdown accidents must be accounted for with higher rates of hydrogen release of up to 7 kg H_2/s. The related H_2-steam-air mixtures produced are capable of detonating [55, 56].

10.3.2.1 Load Carrying Capacity of a KWU-1,300 PWR Containment in a Hydrogen Detonation

Analysis of the release of hydrogen in a core meltdown accident initiated by a small leak in the primary system will be described here as an example of the kind of analysis performed [56]. A small leak in the primary system with a delayed pressure drop in the secondary steam system results in water/steam and hydrogen release into the containment with a maximum release rate of 7 kg H_2/s over a certain period of time during the accident sequence [56]. The hydrogen is produced during the accident sequence by overheating of the Zircaloy claddings above 1,300 °C and their chemical reaction with steam according to

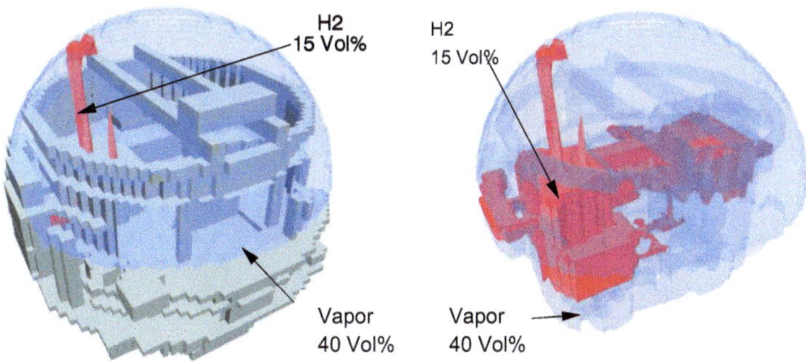

Fig. 10.7 Hydrogen-Steam-air mixtures in the outer containment (two different side views) of a German PWR after an assumed core melt accident at the time of 7,950 s after begin of the accident sequence [56]

$$Zr + 2H_2O \rightarrow ZrO_2 + 2H_2.$$

This hydrogen release is accompanied by steam release. The aggregate volume of hydrogen released eventually into the outer containment then corresponds roughly 855 kg H_2 [56]. In the further course of the accident sequence, the hydrogen concentration in the air and steam mixture exceeds 15 vol% in parts of the reactor containment. The spatial hydrogen distribution and the hydrogen concentration within the mixture of air, steam, and hydrogen within the outer containment is shown for the time of 7,950 s after begin of the accident sequence in Fig. 10.7. This mixture of air and steam with 550 kg of H_2, which is able to detonate, can be ignited, e.g., by an overloaded hydrogen recombiner. This was demonstrated in the RUT experiments [55, 56]. The calculations with the three-dimensional-time-dependent detonation code DET3D [49], which also takes into account shock wave reflections within the reactor containment, resulted in short time pressure peaks of 2–5 MPa and impulse-type loads of 10–30 kPa·s. When the transient phase of the detonation is over, there remains a quasi-steady-state pressure of approx. 0.58 MPa and temperatures around 1,064 K over hours. This does not jeopardize containment integrity of the Konvoi-PWRs of Kraftwerk Union in Germany.

10.3.2.2 Structural Dynamics Response of the Spherical Steel Containment

The results of the detonation with respect to the impulse and pressures acting on the containment wall were used for analyses with the ABAQUS [58] and PLEXUS [59] codes. Figure 10.8 shows the deformations arising in the spherical outer steel containment of the reactor in the steel shell (magnified by a factor of 5) [56]. The largest plastic strains of approx. 4.6 % occur in the vicinity of the materials transfer lock and close to the upper pole (2.4 %). None of these plastic strains will cause the steel shell of the spherical reactor containment to fail [60–62].

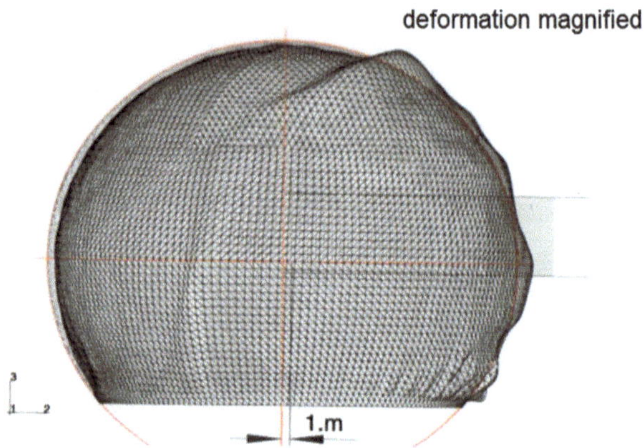

Fig. 10.8 Deformations (magnified) in the outer containment walls of a Kraftwerk Union PWR-1200 during a hydrogen detonation [56]

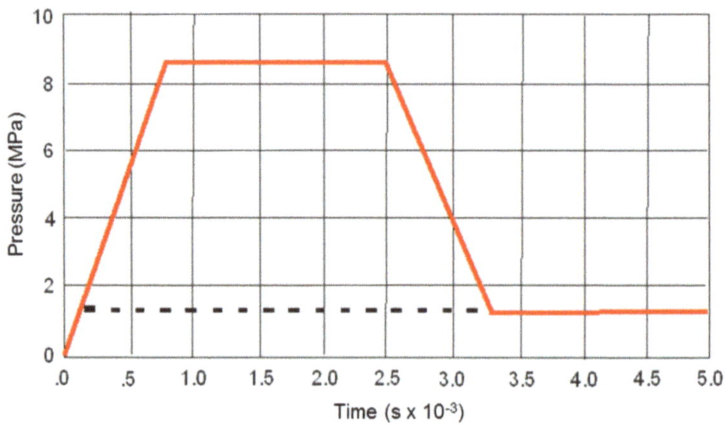

Fig. 10.9 Enveloping conservative pressure time curve determined from space- and time-dependent hydrogen detonation calculations for EPR [63]

10.3.2.3 Load Carrying Requirements of the EPR Containment in Case of Large Scale Hydrogen Detonations

Similar analyses as for the KWU-1,300 PWR containment were also performed for the EPR containment. Figure 10.9 displays the respective pressure-time curve [63] covering all results of the hydrogen detonation analyses for EPR. On the basis of this pressure-time curve similar structure dynamics response calculations can be done for the prestressed concrete containment of EPR.

10.3.2.4 Hydrogen Release in BWRs

The inner containments of modern BWRs (Chap. 3) are inertized by nitrogen in order to avoid hydrogen ignition. Hydrogen recombiners decrease the released hydrogen concentration and spray systems or steam condensers can decrease steam pressure buildup [102]. The inner containment of BWRs must also be equipped at key locations with instrumentation for the measurement of the hydrogen concentration and gamma radiation.

10.3.3 Break of a Pipe of the Residual Heat Removal System in the Annulus of the Containment by Steam

The reference case for loss-of-coolant accidents with leaks in primary pipes in the annulus outside the containment as referred to in the German Risk Studies Phase A [10] and Phase B [64] was the assumed break of a pipe or the failure of valves of the residual heat removal system. This can ultimately cause core meltdown at low primary pressure. The radioactivity from the core melt in this case would bypass the leaktightness function of the outer containment and escape directly into the annulus and, through filters, on into the environment.

This weak spot in the design of existing early pressurized water reactors must be avoided in future PWRs by appropriate design measures, according to the KHE Safety Concept. To avoid such possible bypasses, the function of the multiple barrier system (containment) of retaining the radioactivity must be maintained for all pipes connected to the primary cooling system (double containment function) [8, 64, 65]. This is technically feasible.

10.3.4 Core Meltdown After an Uncontrolled Large Scale Steam Generator Tube Break

In the highly unlikely case of a large scale steam generator tube break, primary coolant can flow to the secondary side. The loss of primary coolant causes the primary pressure to drop and the high-pressure safety feed system to be automatically initiated. Should the necessary shutdown of the high pressure safety feed systems fail, this would ultimately cause overfeeding of the steam generators, and the main steam relief valves would open. When these main steam relief valves do not close again, primary coolant will flow straight into the environment. In the further course of the accident, there could be core meltdown, and the radioactivity released would escape directly into the environment.

Also this very rare core meltdown accident, which could become possible as a result of the present design of steam generators and main steam pipes, must be solved technically as a requirement of the KHE Safety Concept by appropriate routing of the steam pipes, use of the proper valves, which can close during such accidental situations, and by accident management measures such that it will no longer be of importance in modern PWRs, e.g. KWU PWR-1,300 and EPR (Chap. 3) as it was the case in the early risk studies e.g. the German Risk Studies Phase A and B [10, 64].

10.3.5 Core Meltdown Under High Primary Coolant Pressure

WASH-1400 [9] and the German Risk Study Phase A [10] had assumed core meltdown under high primary coolant pressure to lead to failure of the outer containment, followed by a major early radioactivity release, as a result of the reactor pressure vessel acting as a bullet. Core meltdown under high primary pressure could occur in an uncontrolled emergency power case (station blackout) or uncontrolled failure of the main feedwater supply [9, 10, 64]. In both cases, the ultimate consequence is heating of the primary coolant plus primary pressure increase, thus causing the pressurizer relief valves to open. The reactor pressure vessel will be voided. The water level in the reactor core will drop, the fuel rods will heat up, and there will be a zirconium—steam reaction at the fuel rod claddings. This heats the reactor core still further, causing it to start melting. After roughly 1 h, the core will have molten more than 80 %. After 3 h, the core will melt through the grid-plate in the reactor pressure vessel.

Molten fuel flows into the water contained in the bottom hemispherical head of the reactor pressure vessel. This water evaporates quickly. However, the coolant cannot be removed fast enough through the pressurizer relief valves, which causes the coolant pressure to rise. After approximately 3.5 h, the core will melt through the bottom of the reactor pressure vessel at a high primary pressure.

Very high buoyancy forces will arise which can accelerate the reactor pressure vessel upward.

In the German Risk Study Phase B [64] there had already been estimates of mechanical resistance offered by the anchorage of the reactor pressure vessel and the primary piping. The outcome had been that a primary internal pressure of >3 MPa during melt-through would cause the anchorage of the reactor pressure vessel to fail. However, the integrity of the outer reactor containment would be jeopardized by the reactor pressure vessel accelerated upward only above a primary internal pressure of 8–10 MPa.

As a conclusion drawn from all these findings, it was proposed for existing PWRs to reduce the primary pressure in the reactor pressure vessel by timely pressure relief by opening of a pressurizer relief valve (accident management measure), thus allowing the core to melt through at a low pressure as in a loss-of-coolant accident (LOCA). Failure of this accident management measure results in a

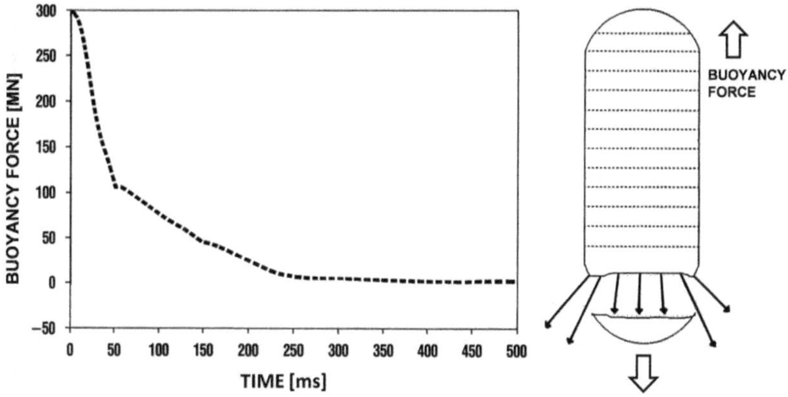

Fig. 10.10 Buoyancy force on pressure vessel lower head because of core melt through under operational pressure of 15.5 MPa [66]

frequency of occurrence roughly one order of magnitude lower per reactor year than for core meltdown under high pressure [64].

Within the framework of the KHE Safety Concept, the RELAP-MOD-3 code [30, 31] was used to determine the buoyancy forces arising from failure of the bottom hemispherical head at the full primary pressure of 15.5 MPa in the RPV [66].

This buoyancy force is shown in Fig. 10.10 as a function of time. It starts with 300 MN and decreases down within about 250 ms. A technical concept was proposed for reinforced anchorage of the reactor pressure vessel [5–7]. This stronger mechanical anchorage of the reactor pressure vessel—as an ultimate technical safety solution—would prevent the vessel from moving upwards even in the case of the failure of depressurization of the primary coolant system (opening of the pressure relief valves). The integrity of the containment would be preserved even in this hypothetical accident (core meltdown under high primary pressure). In this way it provides high flexibility in safety design to avoid the core melt down under high primary pressure.

In the EPR or AP1000 safety concepts (Chap. 3), the possibility was chosen to install two or three blowdown valves, with very high discharge capacities of 900 ton/h (EPR) which is to ensure depressurization to at least 2 MPa (EPR) with a high reliability within a short time period. These blowdown valves can be actuated, e.g. from the control room (EPR) when the coolant outlet temperature exceeds 650 °C. Depressurization to <2 MPa avoids significant melt dispersal (direct containment heating Sect. 10.3.8) in case of core melt break through the bottom of the reactor pressure vessel [67, 68]. The three tandem pressure relief and safety valves as well as the two blowdown valves (depressurization devices) for EPR are shown in Fig. 10.11. For AP1000 and US-APWR the depressurization systems were described in Sects. 3.2.5.2 and 3.2.6 and displayed in Figs. 3.13 and 3.16.

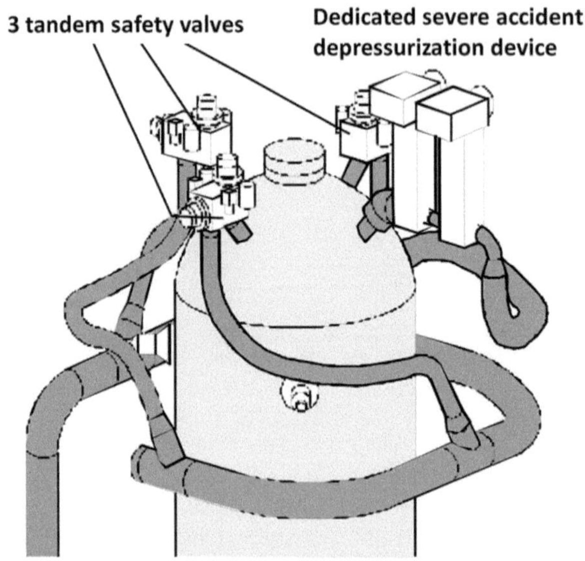

Fig. 10.11 Safety valves and dedicated depressurization devices of EPR [67]

This changes the core meltdown accident sequence under high pressure into a core meltdown under low pressure [67, 68].

10.3.6 Core Melt Down Under Low Coolant Pressure

Even if no steam explosion occurs at low pressure in the bottom hemispherical head of the reactor pressure vessel (Sect. 10.3.1.3) during the influx of molten fuel, the residual water remaining there will evaporate very quickly and the core will melt down. There are then two possibilities for the progress of the accident evolution and two safety design options:

- Cooling the reactor pressure vessel by flooding with water from the outside as an accident management measure [69–73],
- No cooling of the reactor vessel by flooding with water from the outside because of the hazard of a steam explosion when the molten core would melt through. Instead installation of a molten core fuel retention and cooling device (core catcher), accommodating and retaining the core melt after core meltdown [68].

The molten core will always remain subcritical. This was explained by Fig. 2.10 in Sect. 2.10.2. As the water in the core is evaporated and the lattice structure of the fuel elements is destroyed the molten core becomes subcritical to a k_{eff} of less than 0.9. Even the meltdown and displacement of the absorber material of the control and shutdown rods (silver, indium, cadmium or boron carbide) cannot jeopardize this subcriticality of the molten core.

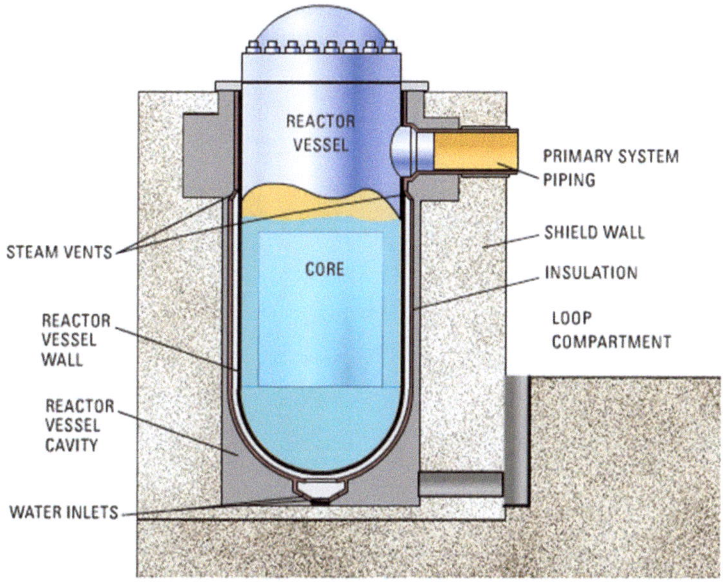

Fig. 10.12 Cooling of AP-1000 vessel from the outside in case of core melting by flooding with water [73]

10.3.6.1 Possibility of Cooling the Molten Reactor Core by Flooding the Reactor Pressure Vessel with Water from the Outside

In this variant applied, e.g. for AP600 and AP1000 the reactor pressure vessel is to be flooded with water from the outside (Fig. 10.12) as the reactor core is melting [69–73]. There are a number of theoretical and experimental investigations in the wake of the Three Mile Island accident in the USA which make this accident management measure appear successful. The heat fluxes from the melt to the wall of the bottom hemispherical head, heat conduction through the wall of the pressure vessel, the temperatures in the wall of the pressure vessel, and the stability of steel as a function of the wall temperature are taken into account.

These research investigations in the USA demonstrated that the molten core will not melt through the wall when the reactor pressure vessel is flooded with water from the outside. This is valid also for the case that the thermal insulation of the pressure vessel remains intact on the outside of the reactor pressure vessel. However, for PWR's with a power output of 1,400 MW(e) or more this variant of flooding the reactor pressure vessel with water from the outside needs more research efforts [74].

Molten core cooling by flooding the reactor pressure vessel from the outside is also proposed for the SWR-1000 (KERENA) design (Fig. 10.13) [75]. The SWR-1000 bottom head of the pressure vessel has many penetrations (welded tubes for control rods etc.). Therefore, BWRs have a higher surface to volume ratio in this bottom part than PWRs. In addition, BWRs have a lower power density

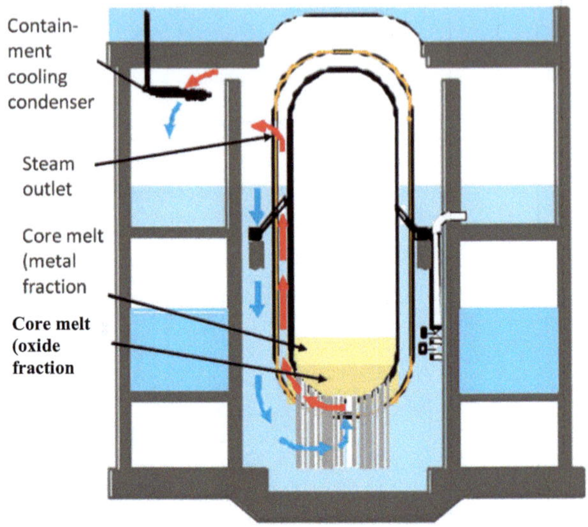

Fig. 10.13 Cooling of molten core by flooding the reactor pressure vessel with water on the outside [75]

in the core melt than PWRs (see e.g. Tables 3.1 and 3.3) [76]. The reactor pressure vessel of SWR-1000 can be flooded passively by water from the flooding pools. In addition to cooling of the molten core in the lower head by outside flooding of the pressure vessel, there exists the backup possibility of retaining molten fuel (if it would melt through the bottom of the reactor pressure vessel) by a special steel support plate underneath. This support plate fixes the control rod drives in the bottom hemispherical head of the reactor pressure vessel. It can act as a back-up core catcher in case parts of the core would melt through the bottom head [75, 77].

The decay heat (afterheat) of the core melt can be transferred by evaporating water. The resulting steam can be cooled and condensed by the containment cooling condenser [75, 77] (Fig. 10.14).

10.3.6.2 Penetration of the Core Melt Through the Bottom Head of the Reactor Pressure Vessel

The second safety design option e.g. realized in EPR, is to keep the reactor cavity dry (no flooding with water) and to install a molten core fuel retention and cooling device (core catcher) outside of the reactor pressure vessel [67, 68]. This design will be described for EPR in Sect. 10.3.7.

Core Melt Through in Case of Presently Operating LWRs

For presently operating LWRs, having no molten core retention device core catcher the situation will be different. This will described e.g. for the PWR1,300 design.

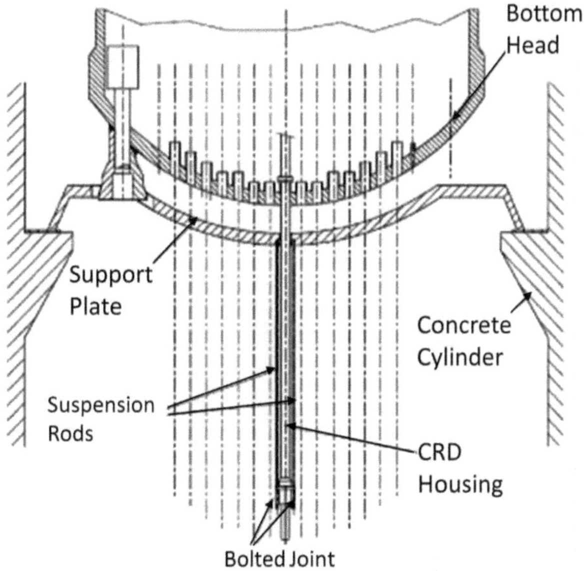

Fig. 10.14 Steel support plate acting as a back-up core catcher [75]

In most presently, operating PWRs the reactor pressure vessel is not flooded with water from the outside. Then this would be the further course of accident events:

The molten mass of the reactor core will melt through the bottom of the reactor pressure vessel and drop into the dry reactor cavity. This is where the molten fuel will react with the concrete and its water of crystallization, giving rise to water vapor, hydrogen, carbon monoxide, and carbon dioxide. While these gases flow into the compartments above the reactor cavity, fuel melt moves downward into the concrete foundation (Figs. 10.15 and 10.16). The fuel melt also spreads in radial direction and is able (as in the containment design of the Konvoi-PWR of Kraftwerk Union) to melt through the lateral biological shield. In this process, the core melt can contact the water of the building sump (Fig. 10.16). The sump water will then evaporate. This will cause the pressure in the outer reactor containment to rise gradually. After roughly 5–6 days, the fuel melt can eventually penetrate the concrete foundation of approximately 6 m [64, 78–80]. These theoretical predictions are supported by experiments and theoretical analyses (WECHSL code [79, 80, 103–105]). After several days, the steam pressure can reach the design pressure of the outer reactor containment of 0.6 MPa [9, 10, 64].

In the German Risk Study, Phase A [10] it is concluded that the outer containment would develop a leak when exceeding this design pressure of 0.6 MPa. The radioactive aerosols and radioactive gases would escape into the environment.

More recent investigations [81, 82] showed, however, that the outer containment of KWU-1,300 PWRs would develop a leak at the materials transfer lock not below some 1.2 MPa. In addition it was shown that dangerous overpressures in the reactor containment cannot be reached if a pressure reduction by a so-called ex-venting

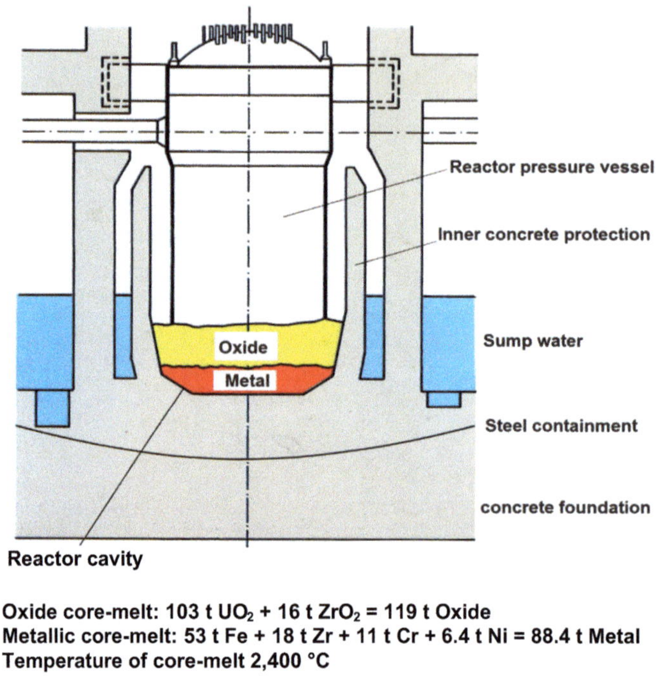

Fig. 10.15 Core melt in the reactor cavity after melting through the lower head of the reactor pressure vessel [78]

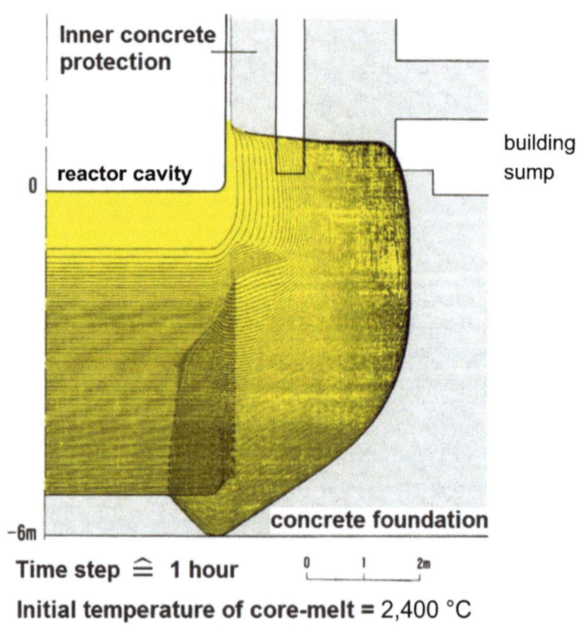

Fig. 10.16 Penetration of core melt into the concrete of the base-mat of the reactor building (calculational results of the WECHSL-code) [64, 79]

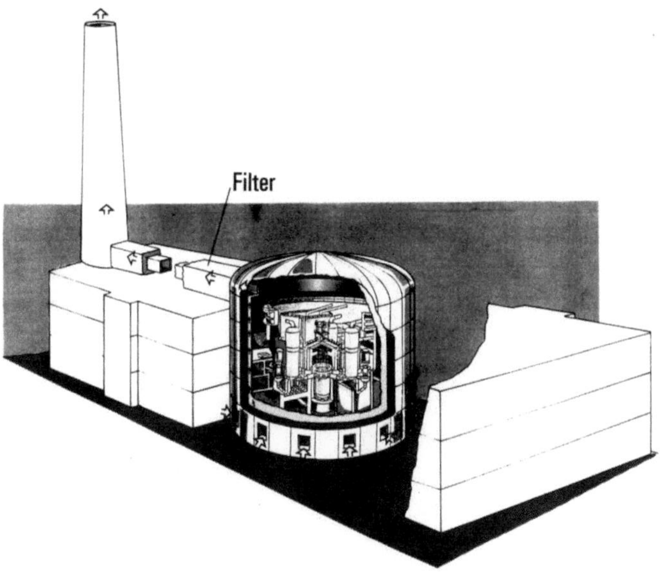

Fig. 10.17 Exventing filter of a German PWR for pressure relief in the outer containment [78]

filter is achieved. Therefore, all German LWRs were equipped after the German Reactor Safety Study Phase B [64] with so-called aerosol ex-venting filters having a filter efficiency of 99.9 % [78, 83, 84]. The pressure will be relieved by opening a valve to this exventing filter and radioactive gases and radioactive aerosols are emitted through this filter (Fig. 10.17).

During pressure buildup over several days, however, most of the radioactive aerosols will have deposited already in the outer reactor containment by sedimentation and thermophoresis etc.. This decreases the amount of radioactive aerosols released to the environment by orders of magnitude [78].

The Role of Exventing Filters and of Double Containment with Low Leak Rate

The efficiency of exventing filters can be demonstrated in Fig. 10.18. In this case the exventing filter would be opened 84 h after the initiation of the core melt accident sequence [85]. The radioactive aerosols would be retained with a certain filter efficiency within the filter. Radioactivity passing through would be released through a 120 m high stack into the environment. Figure 10.18 shows the 70 year committed effective dose in mSv for individuals living in certain distances from the nuclear plant. The results include 95 % of the possible weather conditions in the surroundings of the reactor plant. The effective doses are calculated from all exposure pathways (inhalation, ingestion, skyshine, groundshine) (Chap. 4).

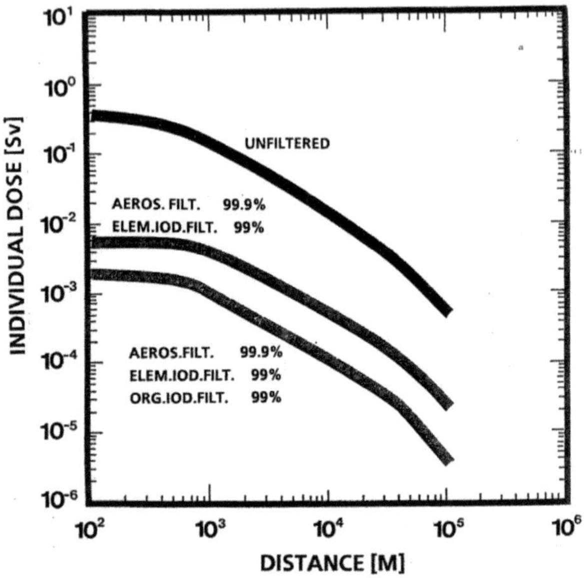

Fig. 10.18 Potential committed effective doses for different filter systems in case of radioactivity release through exventing filters [85]

A pure aerosol filter causes only relatively small dose reductions. Therefore the combination of aerosol filters with elemental and organic iodine filters (both with 99 % filter efficiency) is much more efficient.

A much higher quality of the containment function against radioactivity release after a severe core melt accident can be achieved if a double containment with an inner steel containment or inner concrete containment with steel liner is designed. If the inner containment has a leak rate of 0.3–1 % during or after the severe core melt accident, then the leak rate from the inner containment into the annulus between inner containment and outer concrete containment can be only 0.3–1 % of the radioactivity released in the annulus, i.e. 0.3–1 % of the noble gases and 0.3–1 % of the radioactive aerosols (iodine, cesium, strontium etc.). This leaking radioactivity is collected in the annulus and can be released only through the exventing filters and filters for elemental and organic iodine into a stack and from there into the environment. This is shown in Fig. 10.19 by the double containment of the EPR. In this case the effective radioactive doses in the environment around the plant are decreased by at least a factor of 100 compared to the case described above (exventing filters for presently operating plants (Figs. 10.17 and 10.18) [85]. Such a double containment fulfills the requirement of KHE safety concept. No evacuation is necessary outside the reactor plant.

In this way the objective of the KHE safety concept defined in Sect. 10.2 is achieved.

In the EPR and other LWRs (Chap. 3) safety concepts a dome spray system can spray water from the dome of the containment and condense vapor in the containment. In this way a decrease of the containment pressure is possible from 0.65 MPa to about 0.2 MPa within 1 day. The water for the spray system is taken from the

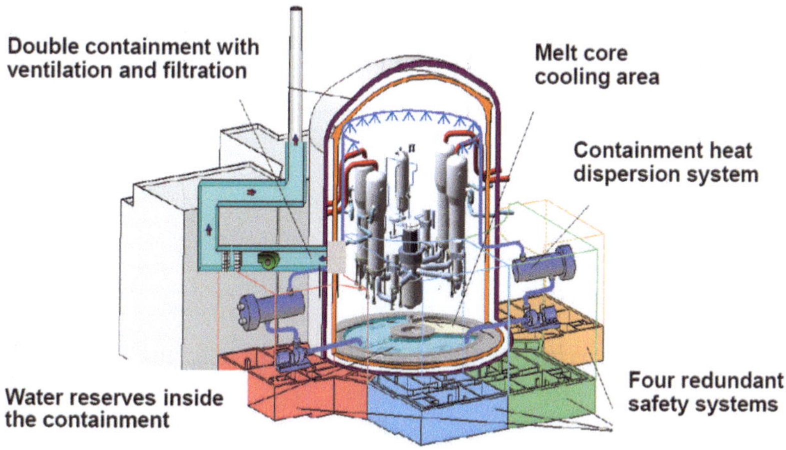

Fig. 10.19 EPR double containment system with reactor pressure vessel, cooling systems and core catchers [86]

in-containment refueling water storage tank (IRWST) and be pumped through a heat exchanger to the spray system (Chap. 3) [86].

The Core Melt of Present LWRs Penetrating into the Subsoil Underneath the Reactor Building

Further penetration of the core melt into the ground below the outer reactor containment was neither studied in the WASH-1400 [9] nor in the German Risk Studies [10, 64]. In the Three Mile Island accident the core melt did not penetrate through the bottom head of the reactor vessel, probably due to the fact that enough water was available timely enough for cooling. In the Chernobyl accident, dropping sand and lead from helicopters on the destroyed reactor core created a molten mass which ultimately did not melt through the bottom foundation of the reactor building despite fears that this might happen. In the Fukushima accident the core melt caused some small holes in the lower part of the reactor pressure vessel but the core melt did probably not penetrate further, because cooling could be provided early enough.

All experimental and theoretical investigations culminate in the conclusion that, in a PWR for instance, the core melt—after having molten through the bottom head of the reactor vessel and through the concrete base plate—would move further into the subsoil below the foundation of the reactor building [79, 87, 88]. In a matter of roughly 200 days, it would expand to a radius of approx. 12 m (Fig. 10.20). It would comprise a volume of 1,000 m^3 and consist of UO_2, ZrO_2, CrO_2, FeO_2, SiO_2, Al_2O_3 and CaO. The SiO_2 fraction would amount to roughly 75 %. Ground-water would cool this enlarged molten mass and slow down and stop eventually its further penetration into the subsoil [87].

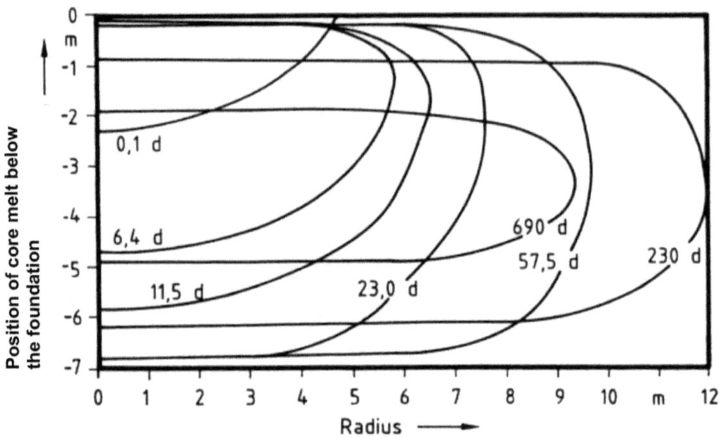

Fig. 10.20 Core melt penetrating into the subsoil below the foundation of the reactor building [87]

Slowly, the groundwater could dissolve fission products out of this originally molten and subsequently solidified mass. In a study [88] leaching rates roughly a factor of 100 higher than those applied for vitrified HLW in a deep geological repository waste were assumed. The leaching rate from the porous mass of the melt is determined by processes such as molecular diffusion, adsorption, desorption, ion exchange and colloid buildup. Further transport of the key radionuclides, Sr-90, Tc-99, and Cs-137, requires consideration of the hydrodynamic transport equations for advection and dispersion in the groundwater [87, 88]. The radionuclides could be carried through the groundwater to a well or into a river and then would move downriver.

Radiation exposure of the public mainly from Sr-90 and Cs-137 would then be possible through the intake of drinking water from the groundwater in the environment and downriver of the location of the core melt. Moreover, flooding by the river could cause the flooded regions to be contaminated as a consequence of sedimentation of radionuclides and drying up of the flooded regions.

10.3.6.3 Possible Countermeasures Against Core Melt into Subsoil

Countermeasures against the spreading of radionuclides would be possible [87] by

- installing sealing walls extending deep down to the contaminated groundwater,
- sinking wells to pump off radioactively contaminated groundwater.

If no such countermeasures would be taken, the groundwater of a relatively large area, and over long periods of time, would not be fit for use as drinking water or for irrigation of agricultural areas.

10.3.7 *Molten Core Retention and Cooling Device (Core Catcher)*

The countermeasures listed above for the period after melting of the reactor core through the concrete baseplate of the reactor containment, and the radioactive contamination of the groundwater and rivers in the vicinity, can be rendered superfluous by

- flooding the reactor pressure vessel on the outside vessel with water in case the core is going to melt down (accident management measures Sect. 10.3.6.1)
- a molten core cooling and retention device (core catcher) underneath the reactor pressure vessel. This cooling device for molten core masses is part of the EPR concept (Fig. 10.21), but not of any other PWR safety design concepts known up to now.

In the EPR design, the core melt is first kept in the reactor cavity for a short period of time so that core masses dropping slightly later than the initial bulk of molten fuel can also be collected. After penetrating a melt plug (steel plate covered by a layer of concrete), the core melt flows through an inclined canal onto a dispersion area of approx. 170 m^2. The core melt is allowed to spread there evenly to a thickness of roughly 30 cm. Then flooding of the melt with water is initiated passively by water flow from the IRWST. In this way, the melt is cooled from the top and solidifies in part. From the bottom, the melt is cooled by active bottom cooling and stabilized in this way (Fig. 10.21).

10.3.7.1 Other Core Catcher Designs

As a result of research programs, a number of other concepts were developed to cool core melts [89–94]. At this point, only the COMET concept developed at Karlsruhe will be explained briefly [90, 93–95]. In the COMET concept, the melt is to be flooded with water from below after erosion of a sacrificial layer. The rapidly evaporating water disrupts the melt, cooling its interior as a water-steam mix. Figure 10.22 shows the COMET concept. The melt is collected below the reactor pressure vessel and then first erodes a dry sacrificial layer of concrete 15 cm high. Afterwards the melt can spread completely and melt any cooling channels in the concrete layer from the top. This allows water to enter under the pressure of an overflow tank located high up. This is followed by an effective phase of melt cooling and fragmentation. The melt solidifies within a short period of time and can be flooded fully and cooled. The steam produced is cooled in a heat exchanger and condensed.

Both the EPR core catcher and the COMET concept were tested at Karlsruhe and developed in many years of pilot experiments (KAPOOL, KATS, COMET) [93–96].

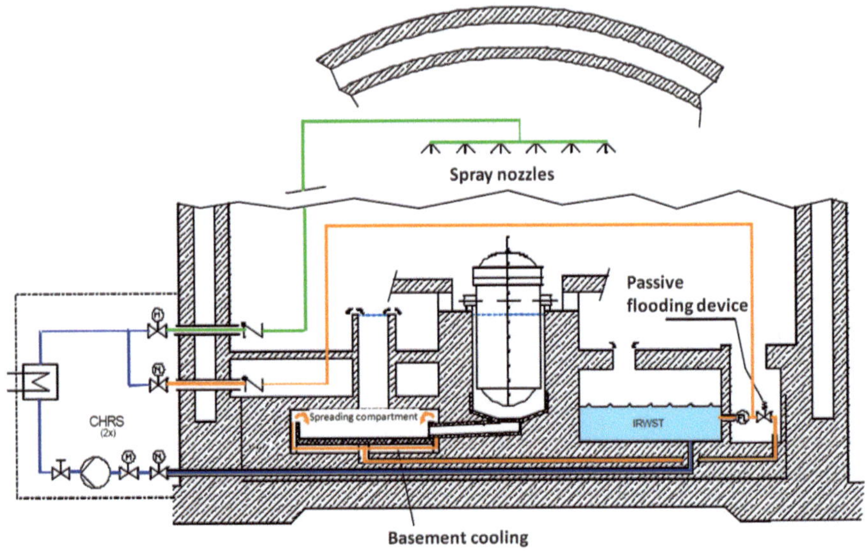

Fig. 10.21 Molten core fuel retention and cooling device (core catcher) together with spray nozzle system for containment atmosphere cooling [68]

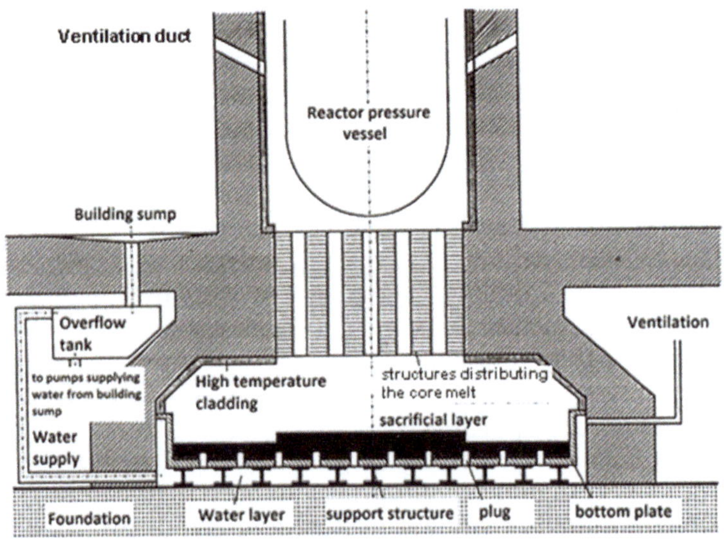

Fig. 10.22 The COMET core catcher and molten fuel heat removal concept [90]

10.3.8 Direct Heating Problem

As the reactor core is melting through the bottom hemispherical head of the reactor pressure vessel under high pressure, the core melt, in the form of a water vapor—melt spray, is driven into other compartments and into the outer reactor containment, respectively. Depending on the design of the inner and outer containments of a pressurized water reactor, this makes it possible for fine particles of the core melt to be distributed over large volumes of the reactor containment. As the droplets and particles of the melt at the same time carry the residual heat, this accident sequence in present PWRs is also referred to as direct containment heating. The Zr-particles can interact with steam and generate hydrogen. The hot fuel particles heat the containment atmosphere and increase the pressure.

These phenomena were studied in experimental programs which, in turn, allow the appropriate design proposals to be made for the reactor cavity and compartments below the reactor pressure vessel [97, 98].

In this way, problems of direct containment heating can be excluded for future reactors, such as the EPR, by a properly enforced flow of water vapor and melt spray into specific compartments.

10.3.9 Summary of Safety Research Findings About the KHE Safety Concept

The findings of recent safety research as outlined in this chapter differ decisively from the results of the early WASH-1400 [9] and German Reactor Risk Study [10]. The accident phenomena,

- steam explosion,
- hydrogen detonation,
- high-pressure core meltdown,
- containment bypass in the annulus and uncontrolled steam generator tube failure associated with radioactivity release through the main steam relief valves

which had the most severe accident consequences (Figs. 10.1 and 10.2), are either partially controlled in present PWR designs (KWU PWR-1,300) or can be eliminated or managed in future LWRs by appropriate design measures. Examples of such future LWRs are the EPR and the SWR-1,000 (KERENA) (Chap. 3). Both reactor lines were designed on the basis of the research findings reported in this chapter and along safety recommendations laid down by the French and German reactor safety commissions [16, 99, 100]. They still need to undergo the licensing procedures required in those countries in which they will be built and operated.

The results of the German Risk Study Phase A [10] for earlier German pressurized water reactors, which findings are outlined individually in Fig. 10.1, are not valid anymore for future LWR designs following the KHE safety concept, e.g. EPR and SWR-1000 (KERENA).

With the above new research results the curves of Fig. 10.1 **shrink to very small damage consequences close to the ordinate** [8]**. Core melt down accidents will no longer lead to large scale contamination of areas outside the reactor plant. Figure** 10.2 **shrinks to areas which belong to the vicinity of the nuclear plant containment only.**

While these statements apply to the safety concept of future LWRs, e.g. EPR or SWR-1,000 (KERENA), they also will apply to a very large extent to the new LWR designs in the USA (AP1000) and Japan (ABWR-II) described in Chap. 3.

The above statements, however, apply only partially to the majority of LWRs operating around 2013 in the world. Many of these LWRs were built from 1970 on and are constructed on safety design concepts and principles as they were analyzed in WASH-1400 [9] and the German Safety Studies Phase A [10] and Phase B [64].

10.3.9.1 Applicability of the Above New Research Results to Presently Operating LWRs and Future

The research results of Sect. 10.3.1 (steam explosion) can—after thorough examination—also be applicable to other PWR pressure vessels, if their design is similar to the design of KWU-1,300 pressure vessels (experiments described in Sects. 10.3.1.3 and 10.3.1.4 were only performed for the KWU-1,300 pressure vessels). Results for hydrogen detonations (Sect. 10.3.2) were only reported for the KWU-1,300 PWR containment design up to now (spherical steel containment 56 m inner diameter and 5.8 cm thickness). Similar conclusions can certainly also be drawn after appropriate analysis for the EPR containment during still ongoing licensing processes.

However, most of the new PWR designs—described in Chap. 3—and of presently operating PWRs appear to rely on the installation of hydrogen recombiners only. This is not sufficient as reported in Sect. 10.3.2 for the KWU-PWR. There exist accident sequences, where despite the installation of hydrogen recombiners, large scale hydrogen detonations can occur, because the hydrogen production rate is too high. Therefore, this case must be examined for each PWR design separately.

Most of the presently operating BWRs have inner containments which are inertized by nitrogen to avoid hydrogen ignition. They are also equipped with hydrogen recombiners to avoid pressure build up by hydrogen. In case of pressure buildup above the limiting pressure of the inner containment exventing filters could relieve the containment pressure.

If the reactor cavity of PWRs, e.g. AP600 or AP1000 can be flooded in case of threatening core melting then a core catcher can be avoided. If the flooding of the reactor pressure vessel with water on the outside is not considered sufficient and the decision is taken to equip a new PWR with a core catcher, then research is necessary for such a new core catcher design [95, 96].

10.4 Severe Accident Management Measures

The publication of the US Reactor Safety Study [9] and the German Reactor Safety Studies A and B [10, 64] as well as the occurrence of the severe reactor accidents of Chernobyl and Fukushima prompted the introduction of a number of severe accident management measures which can be applied by the operation team during the course of a severe accident. This improves the safety of presently operating LWRs. Many of these accident management measures—described below—will also be applied to the new LWR designs described in Chap. 3.

10.5 Plant Internal Severe Accident Management Measures

The safety concept of plant internal severe accident management measures was introduced around 1990. It has been elaborated by safety experts with the assistance of the operational staff of LWR plants [64, 83, 84]. They will be applied in the case of accident events which lead to beyond design basis accidents, i.e. when essential safety systems failed to act and core melt is to occur.

The planned procedures of internal severe accident management measures are described, e.g. in Germany, in a severe accident management handbook which complements the operational handbook.

The procedures laid down in the operation and in the severe accident management handbook are trained by operators, e.g. in Germany, in special reactor simulation centers.

Severe accident management procedures enable the use of especially prepared components and systems of the reactor plant. The main objectives to be accomplished by the severe accident management procedures are to ensure the required protection goals as described in Sect. 5.2 (shut down and subcriticality of the core, long term cooling of the core, no release of radioactivity).

10.6 Examples for Severe Accident Management Measures for LWRs

10.6.1 Examples for Severe Accident Management Measures for PWRs

10.6.1.1 Securing the Emergency Electrical Power in PWRs

In addition to the regular emergency diesel generators the following systems can be applied: gas turbines, mobile motor engines with generators or pumps, especially equipped fuel cells and batteries. Also underground electrical cables to other power plants can be applied. It is important that the required connections are prepared at special locations **in advance** for all voltage levels and AC or DC current needed in the reactor plant. The required emergency electrical power for a 1,300 MW(e) plant is in the range of 700 kW(e) [84].

10.6.1.2 Securing the Feedwater Supply for the Steam Generators in PWRs

In the KWU-PWR, as an example, the feedwater tank contains about 300–400 m^3 of water and after the depressurization of the steam generators about 350 m^3 of water is available also in the feedwater lines. The preheaters contain about 150 m^3 of water. In the demineralized water tanks there are about 1,440 m^3 available.

In modern PWRs, e.g. AP1000, US-APWR, EPR the IWRST contains a very large amount of water (Chap. 3).

In addition, water can be supplied—as described above—by independently powered pumps from deep wells on the site of the reactor plant. Water can also be taken from a river, a lake or the ocean.

The transfer of the afterheat in a 1,300 MW(e) plant by boiling the coolant water and steam release needs a supply of about 20 kg/s of water [84].

10.6.1.3 Securing Core Cooling in PWRs

The most important part of severe accident management measures in PWRs is the **depressurization of the steam generators and of the primary cooling system**. Present PWRs, e.g. in Germany, were equipped with large cross section safety/relief valves. The safety/relief valves can be opened by the operator staff from the operation room—if needed. For future reactor designs e.g. AP1000 and EPR the automatic depressurization system (ADS) were described in Chap. 3. The electrical energy supply or the supply of pressurized nitrogen or air for the safety relief valves must be secured redundantly or even diversely in advance. It is important that the safety/relief valves can be closed again (see Three Mile Island accident).

10.6.2 Examples for Severe Accident Management Measures for BWRs

Core melting under high system pressure must be avoided by rapid depressurization through a number of parallel safety/relief valves into the pressure suppression chambers (Chap. 3). It is important that the supply of electrical energy and pressurized nitrogen or air—as described for PWRs Sect. 10.6.1.3 above-is always available in redundant and diverse ways for the action of the safety/relief valves [101].

10.6.2.1 Securing the Emergency Electrical Power Supply for BWRs

In BWRs, in addition to the options described above for PWRs, there is the possibility to use the steam generated by the afterheat in the core in order to drive small steam turbines combined with water pumps or generators. The water can be taken from the condensation chamber or other water supplies. This option represents a diverse and redundant possibility to the diesel generators, or combustion turbines etc. Cooling the control rod guide tubes by applying additional diverse emergency power and diverse coolant water supply appears to be promising [76].

10.6.2.2 Securing the Feedwater Supply in BWRs

In case of complete failure of the feedwater supply in BWRs, water can be taken from deep wells located on the site of the plant or it can be supplied by mobile pumps from water sources like a lake, river or the ocean. The power supply of these pumps must be secured as described above.

Similar possibilities can be used for the cooling of the pressure suppression chamber.

10.6.2.3 Overpressure in the Inner Containment of BWRs

In case of rising pressure in the nitrogen inerted inner containment of BWRs the hydrogen can be consumed by catalytic hydrogen recombiners and steam can be condensed by water spray cooling or special containment coolers (Chap. 3).

10.7 Emergency Control Rooms

German LWR plants, e.g. are equipped with a special emergency control room, sheltered against external events and poisonous or radioactive gases. The emergency control room is equipped with the critical instrumentation and control mechanisms to control the plant during the course of a severe accident. It is important that independent emergency power supply is available for critical instrumentation and control mechanisms in the emergency control room.

10.8 Flooding of the Reactor Cavity Outside of the Reactor Pressure Vessel

For all presently operating LWRs it must be investigated whether flooding of the reactor pressure vessel is feasible as an accident management measure. This is extremely important for the case that accident progression cannot be stopped and core cooling is endangered.

If the above described severe accident management measures are successfully applied, the safety concept of presently already operating LWRs can approach to a certain extent the safety concept of new reactor designs described in Chap. 3. It is important that these severe accident management measures are well planned in advance and trained by the operational staff.

10.9 Mobile Rescue Teams

For the case of a severe accident in an LWR plant mobile rescue teams must be available which can reach the plant in short time. They must be able to provide:

- mobile radiation measurement and decontamination equipment,
- remotely controlled dredging machines or cranes,
- mobile robotic machines with radiation instruments and tools to enter the internal rooms of the containment,
- mobile emergency power generators.

10.10 Concluding Remarks

Almost all safety concepts and licensing requirements for the operation of present LWRs in the world are still based on the **probabilistic approach** described in Chaps. 5 and 6.

The new research results described in Sect. 10.3 show that the deterministic approach, e.g. applied for EPR and SWR-1,000 (KERENA), is feasible in parallel to the application of the probabilistic approach (for optimizing the internal plant systems). Future LWRs can be built such that their containment can withstand large scale steam explosions and hydrogen detonations. High pressure core melt through the bottom part of the reactor pressure vessel or core melt through the concrete basement of the containment can be avoided by design. Leakages into the annulus between inner and outer containment as well as leakages after steam generator tube failures followed by stuck open steam relief valves can be avoided by proper design.

In BWRs, hydrogen detonations can be avoided by inertization of the inner containment with nitrogen, by equipment of the inner containment with hydrogen recombiners and cooling of the inner atmosphere by spray cooling with steam condensation.

The outer concrete part of the outer double containment can be designed against the impact of large commercial airplanes. As described in Chap. 7 LWR plants can also be designed against relatively high intensity earthquakes and against high level tsunamis or floods and hurricanes.

However, the LWR plants must be designed following such requirements **from the very beginning**. (For already operating reactor plants the buildup of dikes with adequate height for protection against tsunamis or floods is feasible.)

New reactor designs aim at an operation time of 60 years for LWR plants. Many of the presently operating reactors, e.g. in the USA and France, – originally designed for about 35–40 years—have obtained a license for an operation time of 60 years after additional amendments. While this prolongation of operation time is helpful for the economics of these plants, it also prevents new LWR plants with improved safety concepts (Chaps. 3 and 10) to be built in time.

The KHE safety concept described in Sect. 10.2 (no evacuation, no relocation etc. in case of a severe core melt accident) which is applied, e.g. to EPR and SWR-1,000 (KERENA), was required mainly in Europe until now by licensing organizations and safety advisory commissions [15, 16]. These requirements had been incorporated already in an amendment of the German Atomic Law for future nuclear power reactors [100].

Acknowledgment The authors are grateful to Dr. E. Kiefhaber for critically reviewing the contribution and for his scientific advice.

References

1. Stan S. Costs and consequences of the Fukushima Daiichi disaster. http://www.psr.org/environment-and-health/environmental-health-policy-institute/responses/costs-and-consequences-of-fukushima.html
2. Fukushima cleanup could cost up to $250 billion. http://newsonjapan.com/html/newsdesk/article/89987.php
3. Rose M (2013) Major nuclear accident would cost France $580 billion: study. http://www.reuters.com/article/2013/02/07/us-france-nuclear-disaster-cost-idUSBRE91603X20130207
4. Samet JM et al (2013) Selected health consequences of the chernobyl disaster; a future systematic literature review, focus group findings, and future directions. http://www.greencross.ch
5. Hennies HH, Kessler G, Eibl J (1989) Improved containment concept for future pressurized water reactors. In: 5th International conference on emerging nuclear energy systems (ICENES), Karlsruhe, Germany, 3–6 July 1989
6. Hennies HH, Kessler G, Eibl J (1992) Containments and core catchers in future reactors. Atomwirtschaft 37:238–247
7. Eibl J et al (1992) How to eliminate containment failure in tomorrows PWRs (pressurized water reactors). Nucl Eng Int 37(453):51–55

8. Kessler G (2002) Requirements for nuclear energy in the 21st century. Nuclear energy as a sustainable energy source. Prog Nucl Energy 60(3–4):309–325
9. (1975) Reactor safety study: an assessment of accidents risks in US Commercial Nuclear Power Plants. In: Rasmussen NC (ed) US Nuclear Regulatory Commission, WASH-1400 (NUREG-75/014), Washington
10. Deutsche Risikostudie Kernkraftwerke Phase A (1980) Gesellschaft für Reaktorsicherheit (GRS). Verlag TÜV Rheinland
11. Ehrhardt J et al (1982) Der Einsatz des Unfallfolgenmodells der "Deutschen Risikostudie Kernkraftwerke" bei Risikoabschätzungen zu verschiedenen Reaktortypen, KfK-Nachrichten, Jahrg. 14-4:269–277
12. Bayer A, Heuser FW (1981) Basic aspects and results of the German Risk Study. Nucl Saf 22:695–709
13. Erhardt J et al (1989) Radioactivity release and health consequences by filtered containment venting. In: SMIRT seminar: containment of nuclear reactors, Los Angeles, USA, 10–15 August 1989
14. Qu J (1992) Konsequenzen und Wirksamkeit von Umsiedlungsmaß-nahmen nach kerntechnischen Unfällen, KfK 4990
15. Gemeinsame Empfehlung von RSK und GPR für Sicherheitsanforderungen an zukünftige Kernkraftwerke mit Druckwasserreaktoren, Bundesanzeiger Nr. 218, 20. November 1993
16. IPSN-GRS Proposals for the development of technical guidelines for future PWRs (1998) vol 5. Structuring GPR-RSK Recommendations as Guidelines, Common report IPSN/GRS No. 42. Institut de Protection et de Sureté Nucléaire, Saclay, France, Gesellschaft für Reaktorsicherheit, Garching, Germany
17. Corradini ML et al (1988) Vapor explosion in light water reactors: a review of theory and modelling. Prog Nucl Energy 22:1–117
18. Magallon D (2005) FCI phenomena uncertainties impacting predictability of dynamic loading of reactor structures (from OECD SERENA programme). In: PSA-2 Workshop, Aix-en-Provence, 7–9 November 2005
19. Berman M, Beck DF (1989) Steam explosion triggering and propagation: hypothesis and evidence. In: Proceedings of the 3rd international seminar on containment and nuclear reactors, UCLA, Los Angeles, 10–11 August 1989 (SNL report SAND89-1878C)
20. The SL-1 Reactor Accident. http://www.radiationworks.com/photos/sl1reactor1.htm; http://en.wikipedia.org/wiki/SL-1
21. Consideration of BORAX-type reactivity accidents applied to research reactors. IRSH Report 2010/128. Institut de Radioprotection et de Sûreté Nucléaire, 08/2011
22. Jacobs H et al (1994) Untersuchungen zur Dampfexplosion, PSF Statusbericht 23. März 1994 KfK 5326, Kernforschungszentrum Karlsruhe, 214–232
23. Cronenberg AW, Benz R (1980) Vapour explosion phenomena with respect to nuclear reactor safety assessment. Adv Nucl Sci Technol 12:247–335
24. Fletcher DF, Anderson RP (1990) A review of pressure-induced propagation models of the vapour explosion process. Prog Nucl Energy 23:137–179
25. Board SJ et al (1975) Detonation of coolant explosions. Nature 254(3):319–321
26. Diab A et al (2000) Long-term validation of the molten fuel-moderator interactions model. Nucl Technol 169:114–125
27. Kolev MI (1999) In-vessel melt water interaction caused by core support plate failure under molten pool. In: 7th International conference on nuclear engineering, Tokyo, Japan (ICONE 7360)
28. Berthoud G (2000) Vapor explosions. Annu Rev Fluid Mech 32:573–611
29. Struwe D et al (1999) Consequence evaluation of in-vessel fuel coolant interaction in the European Pressurized Water Reacto, FZKA 6316, Forschungszentrum Karlsruhe
30. Allison M et al (1993) SCDAP/RELAP5 mod3.1 code manual, vol I–IV. NUREG/CR-6150, EGG-2720

31. Coryell E et al (1997) SCDAP/RELAP5 mod3.2 code manual, vol I–V. NUREG/CR-6150, INEL-96/0422
32. Summers RM et al (1995) MELCOR computer code manuals, vol 1–2 (Vers: 1.8.3). NUREG/CR-6119, SAND93-2185
33. Valette M (1997) MC3D V3.0 directions for use. Commissariat a l'energie atomique Grenoble, STR/LTEM, STR-LTEM-96-52
34. Berthoud G, Valette M (1994) Development of a multidimensional model for the premixing phase of a fuel coolant interaction. Nucl Eng Des 149:409–418
35. Jacobs H et al (1995) Multifield simulations of premixing experiments. In: Proceedings of "A multidisciplinary Intern. Seminar on intense multiphase interactions", Santa Barbara, CA, USA, 9–13 June 1995, pp 56–69
36. Krieg R (1995) Missiles caused by severe pressurized-water reactor accidents. Nucl Saf 36:299–309
37. Krieg R et al (1995) Slug impact loading on the vessel head during a postulated in-vessel steam explosion in pressurized water reactors – assessments and discussion of the investigation strategy. Nucl Technol 111:369–385
38. Hirt A (1998) Rechenmodell zum Aufprall von Kernschmelze auf die oberen Einbauten und den Deckel eines Reaktordruckbehälters, FZKA 6054, Forschungszentrum Karlsruhe
39. Malmberg T (1995) Aspects of similitude theory in solid mechanics, Part I: Deformation behavior, FZKA 5657, Forschungszentrum Karlsruhe
40. Krieg R et al (2003) Load carrying capacity of a reactor vessel under molten core slug impact. Final report including recent experimental findings. Nucl Eng Des 293:237–253
41. Stach T (1997) Zur Skalierung von Modellversuchen zum Aufprall flüssiger Massen auf deformierbare Strukturen, FZKA 5903, Forschungszentrum Karlsruhe
42. Krieg R et al (1980) Transient, three-dimensional potential flow problems and dynamic response of the surrounding structures, Part I. Comput Phys 8(2):139
43. Krieg R et al (2000) Load carrying capacity of a reactor vessel head under a corium slug impact from a postulated in-vessel steam explosion. Nucl Eng Des 202:179–196
44. Krieg R (1997) Mechanical efficiency of the energy release during a steam explosion. Nucl Technol 117:151–157
45. Theofanous B et al (1997) An assessment of steam explosion-induced containment failure, Part I: Probabilistic effects. Nucl Sci Eng 97:259–281
46. Amarasooriya WH et al (1987) An assessment of steam-explosion-induced containment failure. Part III: Expansion and energy partition. Nucl Sci Eng 97:296–315
47. Eibl J (1994) Zur bautechnischen Machbarkeit eines alternativen Con-tainments für Druckwasserreaktoren – Stufe 3, KfK 5366, Kernfor-schungszentrum Karlsruhe
48. Travis JR et al (1998) GASFLOW-II: a three-dimensional-finite-volume fluid-dynamics code for calculating the transport, mixing, and combustion of flammable gases and aerosols in geometrically complex domains, theory and computational model, vol 1, FZKA-5994 and LA-13357-MS
49. Redlinger R (1999) DET3D: a code for calculating detonations in reactor containments. Proc. Jahrestagung Kerntechnik 99, Kerntechnische Gesellschaft e.V. Deutsches Atomforum e.V. Annual Meeting on Nuclear Technology 99, Karlsruhe, 18.–20. Mai 1999
50. Kotchourko AS et al (1999) Reactive flow simulations in complex 3D geometries using the COM3D code. Proc. Jahrestagung Kerntechnik 99, Kerntechnische Gesellschaft e.V. Deutsches Atomforum e.V. Annual Meeting on Nuclear Technology 99, Karlsruhe, 18.–20. Mai 1999
51. Veser A et al (1999) Experiments on turbulent combustion and COM3D verification. Proc. Jahrestagung Kerntechnik 99, Kerntechnische Gesellschaft e.V. Deutsches Atomforum e.V. Annual Meeting on Nuclear Technology 99, Karlsruhe, 18.–20. Mai 1999
52. Dorofeev SB et al (2001) Evaluation of limits for effective flame acceleration in hydrogen mixtures. J Loss Prev Process Indus 14:583–589

53. Dorofeev SB et al (1999) Effect of scale and mixture properties on behavior of turbulent flames in obstructed areas, FZKA 6268, Forschungszentrum Karlsruhe
54. Kuznetsov M et al (1999) Effect of obstacle geometry on behaviour of turbulent flames, FZKA 6328, Forschungszentrum Karlsruhe
55. Breitung W et al (2005) Innovative Methoden zur Analyse und Kontrolle des Wasserstoffverhaltens bei Kernschmelzunfällen, FZKA 7085, Forschungszentrum Karlsruhe
56. Krieg R et al (2003) Assessment of the load-carrying capacities of a spherical pressurized water reactor steel containment under a postulated hydrogen detonation. Nucl Technol 141 (2):109–121
57. Rohde J et al (1997) Selection of representative accidents and evaluation of H2-control measures in PWR containments. In: 14th Session of RSK light water reactor safety committee, January 1997
58. ABAQUS (1989) A general purpose linear and nonlinear finite element code, user manual standard 5.8. Hibbit, Karlson & Sorenson Inc., Providence, RI
59. Bung H et al (1993) A new method for the treatment of impact and mechanics in reactor technology (SMIRT12), Stuttgart, Germany
60. Krieg R (2005) Failure strains and proposed limit strains for a reactor pressure vessel under severe accident conditions. Nucl Eng Des 235:199–212
61. Jeschke J et al (2011) Critical strains and melting phenomena for different steel sheet specimens under uniaxial loading. Nucl Eng Des 241:2045–2052
62. Ostermann D et al (2011) Critical strains and necking phenomena for different steel specimens under biaxial loading. Nucl Eng Des 241:2045–2052
63. Breitung W et al (2005) Innovative Methoden zur Analyse und Kontrolle des Wasserstoffverhaltens bei Kernschmelzunfällen, Abschlußbericht zu Teilprojekt 1 des HGF-Strategiefondsprojekts 98/07
64. Deutsche Risikostudie Kernkraftwerke Phase B (1990) Verlag TÜV Rheinland, Köln
65. Smidt D (1979) Reaktorsicherheitstechnik, Sicherheitssysteme und Stör-fallanalyse für Leichtwasserreaktoren und Schnelle Brüter. Springer, Berlin
66. Jacobs G (1995) Dynamic loads from reactor pressure vessel core melt through under high primary pressure. Nucl Technol 111:351–356
67. Plank H et al (2009) Severe accident management measures for future NPPs. http://sacre.web.psi.ch/ISAMM2009/oecd-sami2001/Papers/p20-Plank/SAM-Paper-b.pdf
68. Czech J et al (1999) European pressurized water reactor: safety objectives and principles. Nucl Eng Des 187:25–32
69. Tong LS (1968) Core cooling in a hypothetical loss of cooling accident. Estimate of heat transfer in core meltdown. Nucl Eng Des 8:309–312
70. Henry RE et al (1993) External cooling of a reactor vessel under severe accident conditions. Nucl Eng Des 139:31–43
71. Rempe JL et al (1993) Light water reactor lower head failure. Idaho National Engineering Laboratory, NUREG/CR-5642, EGG-2618
72. Thinnes GL et al (1989) Comparison of thermal and mechanical responses of the Three Mile Island Unit 2 vessel. Nucl Technol 87:1036–1049
73. Cummings WE et al (2013) Westinghouse AP1000 advanced passive plant. In: Proceedings of ICAPP 2003, Paper 3235, Cordoba, Spain
74. Park JW (2012) Investigation of core melt coolability inside the large evolutionary advanced power reactor APR1400, atw 57, Heft 1
75. Stosic ZV et al (2008) Boiling water reactor with innovative safety concept: the generation III + SWR-1000. Nucl Eng Des 238:1863–1901
76. Ma W et al (2009) On the effectiveness of CRGT cooling as a severe accident management measure for BWRs. In: OECD workshop on the implementation of severe accident management measures (SAMI-2009), Schloß Böttstein, Switzerland

77. Kolev NI (2004) External cooling – the SWR-1000 severe accident management strategy. In: 12th International conference on nuclear engineering – ICONE-12, Arlington, VA, USA, 25–29 April
78. Kuczera B (1991) Aktueller Stand der Reaktorsicherheitsforschung dargestellt anhand von Ergebnissen aus der deutschen Risikostudie Kernkraftwerke – Phase B, Radioaktivität – Risiko – Sicherheit, Herausgeber Kernforschungszentrum Karlsruhe (2. veränderte und aktualisierte Auflage 1991)
79. Reimann M (1987) Verification of the WECHSL-code on melt/concrete interaction and application to the core melt accident. Nucl Eng Des 103:127–137
80. Reimann M et al (1981) The WECHSL-Code: a computer program for the interaction of a core melt with concrete, KfK 2980, Kernforschungszentrum Karlsruhe
81. Krieg R et al (1987) Failure pressure and failure mode of the latest type of German PWR containments. Nucl Eng Des 104:381–390
82. Göller B et al (1988) Failure pressure and failure mode of the bolted connection for the large component port in German PWR containments. Nucl Eng Des 106:35–45
83. Birkhofer A (1989) Anlageninterner Notfallschutz, Achtes deutsches Atomrechtssymposium 1.–3. März 1989 München, Carl Heymann Verlag KG, Köln
84. Schenk H (1990) Maßnahmen zum anlageninternen Notfallschutz. Atomwirtschaft 11:514–520
85. Kessler G et al (1995) Confinement of the radiological source term during beyond-design-basis events in future pressurized water reactors. Nucl Technol 111:305–318
86. EPR – European Pressurized Water Reactor, the 1600 MWe Reactor. http://www.areva-np.com/common/liblocal/docs/Brochure/EPR_US_%20May%202005.pdf
87. Tromm W et al (1991) Radionuclide dispersion after core-concrete melt leaching by ground-water. Kerntechnik 56(6):7–12
88. Al-Omari I (1990) Abschätzung der Strahlenexposition infolge störfallbe-dingter Radionuklideinleitungen von kerntechnischen Anlagen in Fließge-wässer unter Berücksichtigung der Zeitabhängigkeit relevanter Parameter, KfK 479
89. Alsmeyer H et al (1987) BETA-experiments in verification of the WECHSL-code: experimental results on the melt-concrete interaction. Nucl Eng Des 103:115–125
90. Alsmeyer H (1989) Containment loadings from melt-concrete interaction. Nucl Eng Des 117:45–50
91. Turricchia A (1992) How to avoid molten core/concrete interaction (and steam explosions). In: Proceedings of the 2nd OECD(NEA) CSNI Spec. meeting on molten core debris-concrete interaction, KfK 5108, NEA/CSNI/R(92)10, Ed. H. Alsmeyer, p 503
92. Seiler JM et al (1992) Conceptual studies of core catchers for advanced LWRs. In: Proceedings of the international conference design and safety of advanced nuclear power plants, Tokyo, October 1992, vol III, p 23.3-1
93. Alsmeyer H et al (1998) Beherrschung und Kühlung von Kernschmelzen außerhalb des Druckbehälters. Nachrichten Forschungszentrum Karlsruhe. Jahrg 29(4):327–335
94. Tromm W et al (1993) Fragmentation of melts by water inlet from below. In: 6th International topical meeting on nuclear reactor thermal hydraulics (NURETH-6), Grenoble, 5–8 October 1993
95. Fieg G et al (1996) Simulation experiments on the spreading behavior of molten core melts. In: Proceedings of the 1996 national heat transfer conference, Houston, TX, 3–6 August 1996. American Nuclear Society, La Grange Park, IL, vol 9:121–130
96. Lewis BJ (2008) Overview of experimental programs on core melt progression and fission product release behaviour. J Nucl Mater 380:126–143
97. Meyer L et al (2003) Low pressure corium dispersion experiments with simulant fluids in a scaled annular cavity. Nucl Technol 141:257–274
98. Meyer L et al (2009) Direct containment heating integral effects tests in geometries of European nuclear power plants. Nucl Eng Des 239:2070–2084

99. Köberlein K (2001) Bewertung des Unfallrisikos fortschrittlicher DWR in Deutschland. GRS-175, Gesellschaft für Reaktorsicherheit
100. Bundesgesetzblatt, Gesetz zur Sicherung des Einsatzes von Steinkohle in der Verstromung und zur Änderung des Atomgesetzes und des Stromeinspeisungsgesetzes (Artikel 4, 7. Gesetz zur Änderung des Atomgesetzes, § 7 Absatz 2a), Nr. 46, 19.7.1994
101. American Nuclear Society, Fukushima Daiichi: ANS Comittee Report, revised June 2012
102. Kersting E et al (1993) Safety analysis for boiling water reactors. A summary, GRS-98. Gesellschaft für Anlagen- und Reaktorsicherheit, Garching
103. Nazaré S et al (1975) Über theoretische und experimentelle Möglichkeiten zur Bestimmung der Stoffwerte von Corium, Abschlußbericht Teil II, KFK 2217
104. Schneider H et al (1975) Zur Bestimmung der Zusammensetzung ver-schiedener Corium-Schmelzen, KFK 2227
105. Sehgal B (2012) Nuclear safety in light water reactors, severe accident phenomenology. Elsevier, Academic Press, Waltham, USA

Part II
Safety of German Light-Water Reactors in the Event of a Postulated Aircraft Impact

Franz-Hermann Schlüter

Chapter 11
Introduction

Abstract The safety of German nuclear power plants in the event of a postulated aircraft impact is addressed. The related requirements increased during the last decades and led to adequate constructive design measures.

The safety of nuclear power plants in the event of a postulated aircraft impact has been discussed by the general public and experts in Germany for many years (see [1]). This led to the design and implementation of safety measures for such incidents beginning in the 1970s of the previous century. The basis for initial considerations was the accidental crash of a fast flying military aircraft. Due to the special situation in West Germany with the stationing of NATO partners, a high number of low-flying military aircraft movement could be observed, fraught with numerous crashes. Since 1977 new nuclear power plants and comparable facilities in Germany have been required to comply with design standards to take into account the possibility of an airplane crash (see [2] to [4]). Load assumptions were defined by the German Reactor Safety Commission guidelines (*RSK-Guidelines*) in 1981 based on the impact of a Phantom with a mass of 20 ton and a speed of 215 m/s. A design according to these requirements aimed simultaneously to the protection against the hazard of third party actions.

The attacks on the World Trade Center in New York on September 11, 2001 were cause for the evaluation of the safety standards of German nuclear power plants in the case of a comparable terrorist attack. The question arises whether the power plants are adequately protected against a deliberate forced crash with large commercial aircraft and what dangers this may pose to the general public. The decision to turn off the eight oldest reactors after the events in Fukushima in 2011 was also significantly influenced by the factor of a possible airplane crash. For this reason, a special contribution in this book deals with this issue. An emphasis is placed on possible crash scenarios and the determination of the load approaches.

G. Kessler et al., *The Risks of Nuclear Energy Technology*, Science Policy Reports,
DOI 10.1007/978-3-642-55116-1_11,

While the enveloping structures of the first nuclear power plants served only as weather protection and the load-bearing framework for the technical components, in course of time they also had to take the function of a barrier, especially from external events. The requirements were developed further to what is referred as inherent protection, which means that the outer shell of a containment must be capable of preventing the penetration of a military aircraft according to the RSK guidelines such that no relevant concrete spalling occurs on the interior side of the structure. In addition to this, all resultant consequences such as induced vibrations and fuel fires must be mastered with respect of the safety objective: the save enclosure of the radioactive material. The constructive design of the outer reactor containment is of great importance to achieve this.

References

1. BfS, Bundesamt für Strahlenschutz (2014) Handbuch Reaktorsicherheit und Strahlenschutz (RS-Handbuch), version 01/2014. http://www.bfs.de/en/bfs/recht/rsh.html
2. BMI, Der Bundesminister des Inneren (1977) Sicherheitskriterien für Kernkraftwerke. 21 October 1977. BAnz. Nr. 206 from 03.11.1977
3. BMU (1981) RSK-Leitlinien für Druckwasserreaktoren, 3rd edn. 14 October 1981
4. Institut für Bautechnik (IfBt) (1974) Richtlinie für die Bemessung von Stahlbetonbauteilen von Kernkraftwerken für außergewöhnliche äußere Belastungen (Erdbeben, äußere Explosion, Flugzeugabsturz). Berlin, 1974 (einschließlich ergänzende Bestimmungen Version November 1975)

Chapter 12
Overview of Requirements and Current Design

Abstract The possible actions on nuclear power plants caused by different aircraft crash scenarios are described. To reduce the remaining risks of a possible crash on the safety-relevant buildings special requirements have to be taken into account during the design. Three groups of German nuclear power plants can be distinguished with respect to their resistance against the external event of an airplane crash according to the corresponding start of construction. The development of the structural design is described, from the first generation up to the current design.

12.1 Possible Actions

A differentiation must be made between an accidental and a deliberate forced impact when considering a postulated aircraft impact. The impact scenarios associated therewith have to be viewed in a differentiated way. This is considered in more detail in Chap. 13.

Before the requirements for the design are specified, the possible actions against which the facility must be protected are described.

Special emphasis lies on the mechanical actions. The impact of an aircraft or wreckage on the outer shell of the building creates an impact force of a very short duration. In the first load phase the hit structure is only locally loaded. Depending on the size, mass and speed of the projectile in comparison to the mass and strength of the hit structure, the structural element will be subjected to shear, be penetrated or even perforated. To prevent the penetration of the projectile into the structure, the local resistance to punching shear is decisive. In this phase concrete fragments can spall off (scabbing) the rearward side of the wall, which can act as projectiles, endangering the mechanical components behind it.

In a second phase, the structure is globally involved in the load transfer and excited to vibration. The structural elements of the building deform and experience stress. Depending on the design, this can result in permanent deformations and cracking.

G. Kessler et al., *The Risks of Nuclear Energy Technology*, Science Policy Reports, DOI 10.1007/978-3-642-55116-1_12, © Springer-Verlag Berlin Heidelberg 2014

As a result of the structural vibrations, the mechanical components and systems are excited to vibration according to the coupling with the structure of the building, similar as in case of an earthquake. These induced vibrations can influence the integrity and functionality of the facility.

A consequence of an aircraft impact can also be the load due to flying wreckage or falling debris. In the case that a structure is not designed to withstand an airplane crash, the structure can collapse and endanger safety-related facility components. An example of this is the exhaust air chimney, which could fall on the reactor structure. Internal structures can also collapse and create debris loads.

Further resultant consequences are fuel fires (kerosene) or explosions. Fire, great heat, smoke and/or pressure waves can compromise important safety-related facility components.

12.2 Design Requirements

To reduce the remaining risks of a possible airplane crash on the reactor building of a power plant as well as on buildings which have to be protected with respect to the safe residual heat removal (referred to as safety-relevant buildings), the following requirements have to be taken into account during the design:

- Protection against penetration of the outer shell
- Assurance of the global structural stability
- Protection against the effects of vibrations
- Protection against the effects of wreckage and debris
- Protection against kerosene fires and their consequences

Explicit requirements were already specified in the RSK-Guidelines [1] (referred to as RSK-LL in the following) in 1981. Among other things, a load-time function is described for the impact of an airplane upon which the design must be based. The global structural stability of the safety-related buildings must be assured, taking into account the possible protective effect of the other surrounding building structures. Perforation has to be avoided to prevent consequential damages. In addition to this, no concrete fragments can be allowed to spall from the rear side of a target structure, so long as safety-relevant mechanical components could be compromised and their failure cannot be assured through the principle of spatial separation. This case is called inherent protection. The effects of wreckage and debris, as well as those of kerosene fires and the induced vibrations caused by an aircraft impact must be considered during the design process according to the RSK-LL. Neither the simultaneous occurrence of a single error and an airplane crash nor the simultaneous maintenance case must be postulated in the design.

Within the framework of consultations for the revision of the nuclear regulations, a defense-in-depth concept was suggested by the RSK in 2005 (cf. [2]). The load case of an airplane crash is not a classical design load case for incident management even after the attacks on the World Trade Center on September 11, 2001, but is rather — comparable to the requirements according to the RSK-LL—allocated to security level 4. These are very rare events which the design of the facility must account for to limit the effects of accidents. Considerable releases to the environment are to be avoided. The protection goal of the enclosure of radioactive materials is to be ensured by maintenance of the integrity of the reactor coolant pressure systems and the inner steel containment. All systems necessary to control the event sequences must be appropriately designed.

All structures of the buildings must meet the minimum requirement for load bearing capacity. Unlike with security level 3 (e.g. earthquakes), the building is not required to be fully functional after an aircraft impact. Cracking and irreversible deformation are permitted, so long as safety-relevant issues are not affected. For reasons of structural stability or functional capability of facility components, additional requirements which go beyond the minimum requirements for loading bearing capacity (e.g. limitation of deformation and width of cracks) might be necessary at defined points.

According to the BMU safety criteria (cf. [3]), the effects of debris, kerosene fires, kerosene explosions and additional consequences are to be respected when considering a postulated airplane crash. In particular, this includes:

- Kerosene fire on the facility compound
- Explosion of the kerosene inventory (partially or entirely) outside of buildings
- Fire or explosion of kerosene (liquid or vaporized) which has entered buildings due to permanent existing openings or those caused by the crash
- Entry of combustion products and supply air with reduced oxygen content as a consequence of combustion processes into the ventilation system with respect to human actions, electrical equipment and supply air for diesel generators.

The requirements correspond to the BMU safety requirements [4] and are compiled in the "Handbook for reactor safety and radiation protection" (RS-Handbook) ([1] of Chap. 11) which is continuously updated.

12.3 Development of the Design in Germany

Dependent on the chronological order of their construction German nuclear power plants were, with the exception of the very first ones, designed with increasing higher standards with respect to an accidental "random" airplane crash of a certain type. Specific aircraft models were chosen to represent the particular effects and were the basis of the design standards for structures and components.

Three groups of German nuclear power plants can be distinguished with respect to their resistance against the external event of an airplane crash according to the corresponding start of construction:

- a first group, which does not feature a systematically planned resistance to airplane crashes

Table 12.1 Design of German nuclear power plants in power operation against military aircraft crash

	Power plant	Type	Generation/ building line	First criticality	Military aircraft crash design
1	Grafenrheinfeld (KKG)	PWR	3	09.12.1981	Phantom
2	Gundremmingen B (KRB B)	BWR	72	09.03.1984	Phantom
3	Grohnde (KWG)	PWR	3	01.09.1984	Phantom
4	Gundremmingen C (KRB C)	BWR	72	26.10.1984	Phantom
5	Philippsburg 2 (KKP 2)	PWR	3	13.12.1984	Phantom
6	Brokdorf (KBR)	PWR	3	08.10.1986	Phantom
7	Isar 2 (KKI 2)	PWR	4 (convoy)	15.01.1988	Phantom
8	Emsland (KKE)	PWR	4 (convoy)	14.04.1988	Phantom
9	Neckarwestheim 2 (GKN 2)	PWR	4 (convoy)	29.12.1988	Phantom

Table 12.2 Design of nuclear power plants in post-operational phase against military aircraft crash

	Power plant	Type	Generation/ building line	First criticality	Military aircraft crash design
1	Biblis A (KWB A)	PWR	2	16.07.1974	Not explicit
2	Biblis B (KWB B)	PWR	2	25.03.1976	Starfighter
3	Neckarwestheim 1 (GKN 1)	PWR	2	26.05.1976	Starfighter
4	Brunsbüttel (KKB)	BWR	69	23.06.1976	Not explicit
5	Isar 1 (KKI 1)	BWR	69	20.11.1977	Not explicit
6	Unterweser (KKU)	PWR	2	16.09.1978	Starfighter
7	Philippsburg 1 (KKP 1)	BWR	69	09.03.1979	Not explicit
8	Krümmel (KKK)	BWR	69	14.09.1983	Phantom

- a second group, which is protected against the accidental crash of a military aircraft of model "Starfighter"
- and finally a third group, which is designed to resist the accidental crash of a fast flying military aircraft of type "Phantom" with the worldwide highest protection in this regard.

After the shutdown of the eight older nuclear power plants in 2011 in Germany, only facilities of the third group remain in operation (cf. Table 12.1). The eight disconnected facilities, which are in what is called a post-operational phase, belong to the first and the second group (cf. Table 12.2). A large amount of radioactive inventory remains in the fuel assembly storage pools and as the case may be in the reactor vessels, even in the disconnected facilities, which must be reliably cooled for several years. The risk of an aircraft impact is therefore still relevant for these power plants as well, until all inventory has been loaded into the Castor containers after the decay time.

The crash of a large commercial aircraft, whether accidental or deliberate, is not part of the regular design plan of any German nuclear power plant. Nevertheless, due to the design of the enclosure, especially for the third group of power plants however, a high degree of protection is expected.

In the Tables 12.1 and 12.2 the design of German nuclear power plants with respect to an airplane crash is summarized.

References

1. BMU (1981) RSK-Leitlinien für Druckwasserreaktoren, 3rd edn. 14 October 1981
2. RSK-Empfehlung (2005) Gestaffeltes Sicherheitskonzept (386. Sitzung am 8.09.2005). http://www.rskonline.de/English/statements—recommendations/index.html
3. BMU (2009) Sicherheitskriterien für Kernkraftwerke. Revision D. April 2009. http://www.bmu.de/files/pdfs/allgemein/application/pdf/sicherheitskriterien_kernkraftwerke_revisiond.pdf
4. BMU (2012) Sicherheitsanforderungen an Kernkraftwerke vom 22. November 2012 (BAnz AT 24.01.2013 B3). http://www.bfs.de/de/bfs/recht/rsh/volltext/3_BMU/3_0_1_1112.pdf

Chapter 13
Impact Scenarios

Abstract The possible impact scenarios are considered from accidental aircraft impacts to deliberate forced impacts. The different relevant aircraft models and crash scenarios are discussed, including possible approach speed and angle.

13.1 General

To determine the specific effects resulting from an aircraft impact, the possible scenarios must be considered: which aircraft under which conditions are relevant for an impact. Beginning in the early 1970s, safety measures were introduced in the design of nuclear power plants in Germany to take into account the crash of an aircraft. Before September 11, 2001 it was assumed, however, that the threat of an airplane crash could only come about accidentally. A deliberate act was not considered in the design. Only after the attacks on the World Trade Center a scenario of a deliberate crash of large passenger airplane was considered and its consequences were examined.

13.2 Accidental Aircraft Impact

An accidental aircraft impact can be put down to an accident resp. a technical or human error. The probability of such an impact can be seen as very low and is determined by the quantity of aircraft movement close to the nuclear power plant. With respect to the effects and the frequency, fast flying military aircraft must be predominately considered in this case.

The territory of the Federal Republic of Germany is characterized by a dense net of civil flight routes. On top of this there was a very high density of air traffic due to

G. Kessler et al., *The Risks of Nuclear Energy Technology*, Science Policy Reports, DOI 10.1007/978-3-642-55116-1_13,

German and allied military aircraft until reunification in 1989 and the improvement of relationships between east and west. In the 1970s and 1980s many crashes occurred during military training flights, especially during low-level flights. In this time frame 15–25 accidental crashes occurred per year, excluding those that occurred during takeoff and landing manoeuvres close to airports (radius approx. 5 km). The possible scenarios and crash frequencies were discussed in detail in context of the German risk study of nuclear power plants [1,2]. Fast flying military aircraft often with free flight paths and flying at low levels where considered a potential for danger.

The number of crashes, their spatial distribution, impact speed, impact angle and the different types of aircraft with their specific mass distributions were analyzed and probabilistically evaluated in the threat investigation. The possible armament of the military aircraft was not considered. Figure 13.1 shows cumulative frequency curves of different parameters.

The distribution curve of the mass shows that only few crashed military aircraft had a takeoff weight higher than 20 t. A further focal point of the distribution curve are aircraft around 10 t. This is the result of the large number of Starfighters, weighing about 10 t, and Phantom F4, weighing about 20 t, used in the 1970s and 1980s.

The cumulative frequency curve of the impact speed v shows that crashed military aircraft with a speed higher than 225 m/s barely contribute to the sum. The impact angle ϑ is nearly equally distributed between 0° and 90°.

The endangering due to accidentally crashing large commercial aircraft is significantly lower than with military aircraft. This is explained by the fact that large commercial aircraft follow predetermined flight paths and do not operate in free air traffic, in addition to not performing low-level flying with the exception of takeoff and landing. The crash probability is much lower than with fast flying military aircraft. The pilot of a military aircraft will in the case of a major fault leave the plane by an ejector seat, leaving the aircraft to itself. Due to their high operating altitude of large commercial aircraft usually maneuverability options are available in case of a fault. The pilot and co-pilot cannot leave the aircraft and will usually attempt a controlled emergency landing. The accidental crash of a large commercial aircraft is seen as very unlikely and is not considered for the design of nuclear power plants.

Civilian light aircraft are not bound to a flight path and fly at low altitudes much like military aircraft. In comparison to large commercial aircraft they have a considerably higher crash frequency. However, the mass and speeds are lower than those of the military aircraft mentioned above, and the effects are therefore covered within the scope of the military aircraft. Falling wreckage of a possible explosion of large commercial aircraft in flight are also covered by this case.

Starting with a uniform distribution of the crash frequency over the area of the Federal Republic of Germany of 10^{-10} crashes/m^2/year yields the value of 10^{-6} crashes/year over the area of a nuclear power plant. This value is a conservative estimate and means that a power plant can be hit by an accidentally crashing military aircraft with a probability 10^{-6}/year, independent of the possible damages.

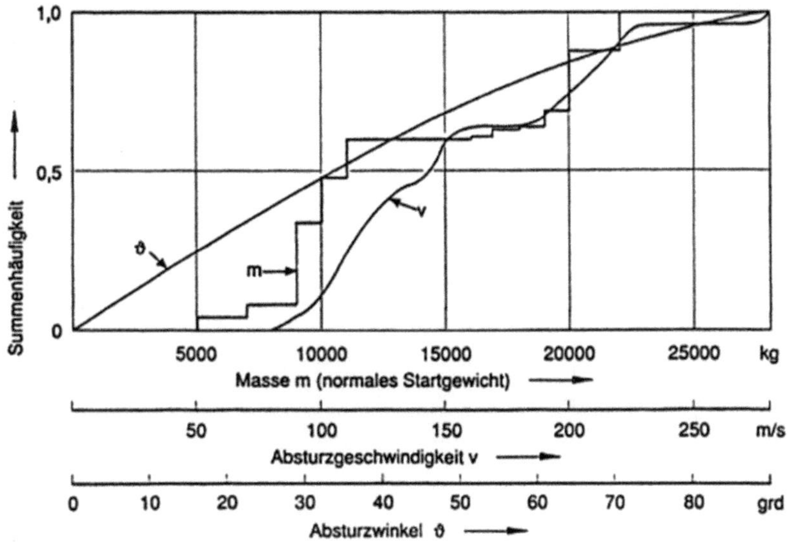

Fig. 13.1 Cumulative frequency distribution (Summenhäufigkeit) of the mass m (Masse), the impact speed v (Absturzgeschwindigkeit) and impact angle ϑ (Absturzwinkel) of crashed military aircraft (according to [2])

Hits that result in relevant damages to safety-related buildings have an even smaller probability of occurrence.

Numerous studies were performed with regards to possible scenarios of an accidental airplane crash in the beginning of the 1970s. The result for the design of a nuclear power plant was the crash first of a Starfighter and later of a Phantom, representing many possible events (Table 13.1).

For the older nuclear power plants in Germany (cf. Table 12.2) the crash of a military aircraft of the model Starfighter (Fig. 13.2) with a mass of 13 t and an impact velocity of mach 0.30 (=102 m/s or 367 km/h) was used as a design basis.

For newer nuclear power plants and nuclear facilities in Germany (cf. Tables 12.1 and 12.2) the crash of a military aircraft of the type Phantom (cf. Fig. 13.3) with a mass of 20 t and an impact velocity of mach 0.65 (=215 m/s or 774 km/h) was assumed as covering event. A load function was established for this case in 1973 (cf. [12]), which was finally included in the RSK-LL in 1981 [6]. The possible impact angles are assumed between 0 and 90°. Normally the worst load case perpendicular to the impact surface is assumed for design.

An analysis shows that the load parameters (mass, speed) of over 90 % of the observed accidental crashes lie beneath the load assumptions (Phantom) of the newer German nuclear power plant design requirements.

Table 13.1 Major impact data for the military aircraft Starfighter and Phantom

	Dimension	Unit	Starfighter	Phantom
1	Mass, m	t	13	20
2	Speed, v	m/s	102	215
3	Max. impact force, F_{max}	MN	17	110
4	Impact period	ms	~120	70
5	Impact area	m^2	2.1	7.0
6	Momentum	kg (m/s)	$1{,}326 \times 10^6$	$4{,}300 \times 10^6$
7	Kinetic energy	kg (m^2/s^2)	$67{,}626 \times 10^6$	$462{,}250 \times 10^6$

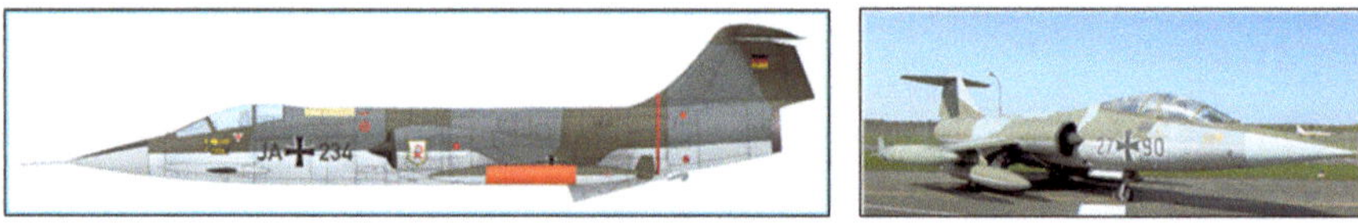

Fig. 13.2 Fast flying military aircraft of type Starfighter F-104G; until withdrawal from the German armed forces in 1991 a number of 292 aircraft of this type crashed (from [3,4])

Fig. 13.3 Fast flying military aircraft of the model Phantom F-4 (Boeing [5])

13.3 Deliberate Forced Aircraft Impact

After the terrorist attack in the USA on September 11, 2001 the question arose to what extent nuclear power plants would withstand an impact of a large commercial airplane or rather which damage scenarios are to be expected. As a result many national and international studies were conducted, in which possible scenarios for a deliberate forced crash were considered and the load functions for the crash of a large commercial airplane under consideration of the spatial load distribution were developed. Among others, the author was involved in several of these studies. Since most of these studies are confidential and therefore not open to the general public, however, only general references and information can be discussed here.

13.3.1 Relevant Airplane Models

Civil and military aviation consists of a multitude of different aircraft models which pose very different threats with respect to the case of a deliberate forced crash into a nuclear facility. With regard to this problem the first step is to collect the necessary data for the different aircraft types. An emphasis was put on passenger aircraft which are currently in service in large numbers. In addition, military aircraft are not considered for the case of a deliberate crash.

Even though many aircraft models are used in civilian aviation it is possible to limit the types with respect to their threat potential to a few which are considered relevant. Due to the similarities between the aircraft it is possible to concentrate the analysis on the representative large manufacturers Airbus and Boeing. Several examples of typical aircraft models are presented below.

13.3.1.1 Boeing 747

The Boeing 747, also known as the "Jumbo Jet," was for years the largest passenger aircraft used in commercial aviation (cf. Fig. 13.4). Since its first flight in 1969 over 1,400 airplanes of this model have been produced, many of them are still in service worldwide to this day. With a carrying capacity of 300–500 passengers it has to be considered as a possible instrument of terrorist activity. The B747 exists in different configurations. The maximum take-off weight of the version B747-400 is approx. 395 t. The length is approx. 71 m and the wingspan 64.5 m. The Boeing 747 has approx. 170 t of kerosene on board at take-off, of which one third each is stored in the two wing tanks and in the center tank. This corresponds to a volume of about 220,000 l. Due to the prevalence and the fact that the dimensions and weight represent the upper boundaries of commercial aircraft, the Boeing 747 is considered a relevant aircraft model with threat potential.

13.3.1.2 Boeing 777

The Boeing 777 is a long-range aircraft with a length between 64 and 74 m and a wingspan between 61 and 65 m. The maximum take-off weight amounts to 250–340 t. The weight share of kerosine amounts to a maximum of just under 150 t. Commissioned in 1995, the Boeing 777 was developed to accommodate the wish for an aircraft with a capacity between the Boeing 767 and 747. It possesses a seating capacity of between 300 and 550 passengers.

The Boeing 777 is constructed as an aircraft with only two jet engines. To be licensed for transatlantic flights, it had to comply with higher reliability requirements. In the case of the failure of one engine, the continuation of the flight has to be assured for 3 h. The engines used are therefore some of the strongest that have ever been used in aircraft. In comparison, the outer diameter of the engines is nearly as

Fig. 13.4 Boeing 747-400 with a take-off weight of 395 t (Photographs [5])

large as the diameter of the body of a Boeing 737. Although the differences to the Boeing 747 with respect to mass and dimensions are not considered relevant to the threat, the engines are interesting for a separate examination due to their size and dimensions.

13.3.1.3 Airbus A380

The Airbus A380 is currently the largest civilian commercial aircraft that has entered serial production. It is a four-engined wide-body aircraft with two continuous decks with a capacity for approx. 525 passengers and is in use since 2007. Approximately 320 aircraft have been commissioned to date. The length of the aircraft is 72.30 m and the wingspan 79.80 m. It has an elliptical cross section of the fuselage of 7.15 m in width and 8.40 m in height. The maximum take-off weight is close to 560 t including 320,000 l (=256 t) of kerosene.

This aircraft model was not considered in detail in the examinations directly after September 11, 2001 because the aircraft was not in use and no specific details were available neither from the manufacturer nor from the airline companies at the time. The first examples of this aircraft have only been in use since 2007 (Fig. 13.5).

13.3.1.4 Airbus A340

The aircraft A340, built by Airbus, has a capacity of 260–340 passengers. The length of the aircraft is between 60 and 64 m, the wingspan approx. 60 m and the maximum take-off weight 270 t including 120 t of kerosene. A larger model with a fuselage that is elongated by 25 ribs has a capacity of 420 passengers. The maximum take-off weight of this larger variant increases to approx. 380 t, of which approx. 150 t are kerosene (Fig. 13.6).

The A340-600 corresponds roughly to the Boeing 747 with respect to total weight and its dimensions. The greater wing angle and the slightly lower weight result in a higher mass concentration in the middle of the aircraft where the wings

Fig. 13.5 Airbus A380 with maximum take-off weight of approx. 560 t, in use since 2007 (Photograph [7])

Fig. 13.6 Airbus A340-600 with a maximum take-off weight of 370 t (Photograph [8])

and therefore the kerosene tanks are located. Higher loads on the target structure are therefore to be expected.

13.3.1.5 Airbus A320

The Airbus A320 is a typical representative of two-engined narrow-body aircraft with more than 1,000 units sold. It transports 150–180 passengers. The length is 37.57 m, the wingspan 34.10 m. The maximum take-off weight is 78 t with approx. 19 t kerosene. This aircraft model is comparable to the Boeing 737 or the McDonnell Douglas MD-80. This model of aircraft can be seen as representative for this most widespread category of aircraft (Fig. 13.7).

Several specifications of the relevant passenger aircraft are summarized in Table 13.2.

Fig. 13.7 Airbus A320 with maximum take-off weight of 75 t (Photograph [8])

Table 13.2 Specifications of several relevant passenger aircraft

	Parameter	Unit	A380	B747	A340-600	A320
1	Max. take-off weight	t	560	395	380	78
2	Kerosene weight	t	256	170	150	19
3	Length	m	72.30	70.60	75.30	37.57
4	Wingspan	m	79.80	64.50	63.45	34.10
5	Height of fuselage	m	8.40	7.85	5.64	3.96
6	Engines	Quantity	4	4	4	2

13.3.2 Approach Angle and Approach Speed

To determine the decisive load of a structure such as the outer containment of a reactor building it is crucial to know at which angle and with which speed an attacker could reach a target.

For the definition of possible and realistic flight approaches intense discussions and studies were conducted after the attacks in the USA on September 11, 2001 (cf. e.g. [9]). A diverse number of aviation institutions, airline companies and experienced pilots were involved. Furthermore, possible approach scenarios were tested in flight simulators. A summary of the first examination results are described in [10], but the basic report is classified confidential as "VS-Vertraulich" and is therefore not freely available.

To determine the aeronautical parameters varying conditions such as altitude, wind and visibility conditions were considered under different approach angles for target approaches to similar objects. It is generally known that the approach angle for a planned landing is usually approx. 3° with respect to the horizontal plane. Under these ideal conditions it is not a great problem for a reasonably experienced pilot to hit a target the size of a reactor building very precisely. The accuracy decreases slowly as the pitch increases, at higher values it decreases drastically. The probability of a hit is markedly reduced for angles greater than approx. 10°. An important aspect with respect to the probability of a hit is that in comparison to the almost horizontal approach with optimal maneuverability to the over 400 m tall World Trade Center in New York, a reactor building must be hit with pinpoint accuracy in the horizontal and vertical direction, taking into account the loss of

Fig. 13.8 The Pentagon in Washington in a perspective view and as a satellite image after the attack (from [11]). The building complex is over 280 m wide and is therefore a "linear" construction

maneuverability due to descent. An approach that is too low might hit the ground in front of the reactor building or surrounding buildings, which results in a much lower threat potential. An approach that is too high would either graze the dome area and possibly be deflected, which leads to a significant reduction of the resulting load and therefore a reduction of the threat, or miss the reactor building completely. Similar considerations are valid for a sideways flyby.

The Pentagon was also hit during the attacks in the USA on September 11, 2001, which is a low building in comparison to the World Trade Center. This linear complex of buildings (cf. Fig. 13.8) with an edge length of approximately 280 m is a much easier target to hit, however, in comparison to a punctual reactor building with an outer diameter of approx. 66 m. For clarification, Fig. 14.1 shows the size comparison between the dimensions of a reactor building and the aircraft models Boeing 747, Airbus A320 and Phantom in a horizontal section as an example.

In addition to the possible approach angles the question of the possible impact speeds has to be addressed. The acceptable speeds set by aircraft manufacturers are about 900 km/h (= 250 m/s) for relevant aircraft models. This is a value for the optimal flight altitude however. Close to the ground these values are significantly reduced to about 650 km/h (=180 m/s). In the case of a deliberate forced crash the possibility has to be taken into account that the attacker will attempt to reach speeds above those that are permitted for "normal" flight. Modern aircraft have electronic measures, however, to prevent such an extreme flight situation. It is also important to note that at speeds that are significantly higher than the normal speed range the reaction time for course correction is decreased and the aircraft also becomes more difficult to steer. The attacker has a conflict of objectives: on one hand to achieve a speed as high as possible to maximize damage, on the other hand to maximize the accuracy by reducing the speed.

As shown in the summary of the German Federal Environment Ministry [10], an upper speed limit of $v = 175$ m/s (=630 km/h) was established for the determination of the mechanical load-time function (cf. Chap. 14) which is used in the further analysis under consideration of the available information and the considerations with respect to the accuracy above.

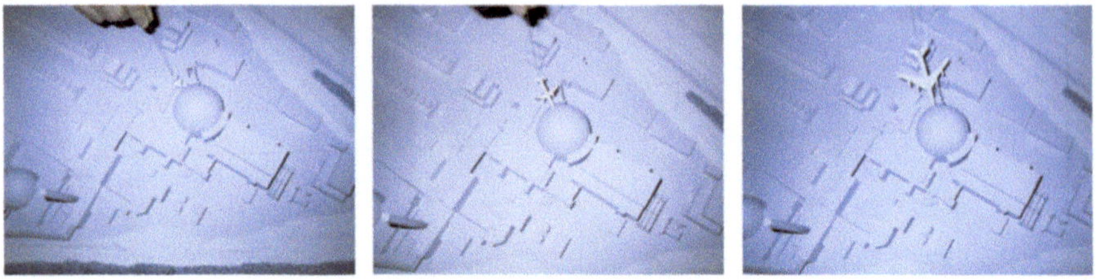

Fig. 13.9 Approach to a nuclear power plant with the three aircraft models Phantom, Airbus A320 and Boeing 747 in a size comparison

Fig. 13.10 Possible approach zone to the cylindrical part of a convoy containment; situation with a true to scale model of a Boeing 747

The special structural situations of nuclear power plant facilities also have to be taken into account for the target accessibility considerations. In order to clarify the relationships and to review the possible scenarios from the "bird's eye view," models of power plant facilities were manufactured among other thing. For example it was examined whether it is possible to approach the unprotected regions of the reactor building in realistic scenarios even with the encircling buildings. In the areas of a containment with buildings in front the resulting impact force is considerably reduced.

Figure 13.9 shows the reactor building of a convoy plant in a size comparison with three aircraft models in a possible approach situation for clarification. It shows that even with apparently advantageous terrain and buildings, an approach in the scenarios above is still possible (Fig. 13.10).

In summary with respect to the approach scenarios it is initially to be assumed that it is possible for a large commercial aircraft to hit the reactor building in the unprotected area in the case of a deliberate forced crash. A detailed examination could be carried out for all sites.

References

1. (1980) GRS: Deutsche Risikostudie Kernkraftwerke, Fachband 4 – Einwirkungen von außen. Verlag TÜV Rheinland, Köln
2. (1990) GRS: Deutsche Risikostudie Kernkraftwerke, Phase B. Verlag TÜV Rheinland, Köln
3. Photograph: Markus Kutscher, Frankfurt, Germany. http://en.wikipedia.org/wiki/de:Creative_Commons
4. Photograph: Lexikon der Flugzeuge von Wolfgang Bredow. http://www.bredow-web.de/Luftwaffenmuseum/Kampfjets
5. Boeing Image Gallery. http://www.boeing.com
6. (1981) BMU: RSK-Leitlinien für Druckwasserreaktoren. 3. Ausgabe
7. Photograph: Roger Green, Bedford, UK http://en.wikipedia.org/wiki/de:Creative_Commons
8. Photograph: Airbus Image Gallery. http://www.airbus.com/aircraftfamilies/passengeraircraft
9. Dräger P (2002) (KTG): Zur Widerstandsfähigkeit von Sicherheitsbehältern für Kernkraftwerke gegen Terrorattacken mit großen Verkehrsflugzeugen. München
10. (2002) BMU: Schutz der deutschen Kernkraftwerke vor dem Hintergrund der terroristischen Anschläge in den USA vom 11. September 2001. Zusammenfassung der GRS-Studie durch das BMU. Bonn
11. Photograph: David B. Gleason from Chicago, IL, USA. http://en.wikipedia.org/wiki/de:Creative_Commons
12. Drittler K, Gruner P, Sütterlin L (1973) Zur Auslegung kerntechnischer Anlagen gegen Einwirkungen von außen – Teilaspekt Flugzeugabsturz. Bericht IRS-W-7.

Chapter 14
Determination of a Load Approaches for Aircraft Impacts

Abstract As basis for the analysis and verification of building and mechanical components adequate load approaches for aircraft impacts have to be developed. Different approaches and mathematical models to derive load-time-functions for the contact force between airplane and hit structure are presented. Finally the shown impact load-time functions of the different types of aircraft are compiled for comparison.

14.1 General Information

The crash of an aircraft on a target is physically an impact problem with largely non-linear processes. The kinetic energy of a projectile, determined by its mass and speed, is changed into other forms of energy during impact. For the design of a structure the contact force between the striking and hit bodies are important. In principle this is dependent on the ductility of the participating structural elements.

In the case of an aircraft impact a so called soft impact takes place. This is characterized by the fact that the projectile is very ductile. The kinetic energy is largely absorbed by the striking body itself in the form of plastic deformation, such that only a fraction of the energy is introduced into the impacted body. Due to the large deformations of the projectile on impact the size of the resulting contact force is practically independent of the behaviour of the impacted structure. This allows a decoupled examination. In a first step the impact force-time function can be determined using only the characteristics of the projectile. The impacted structure can then be loaded with this load function in a dynamic analysis in a second independent step. The spatial distribution of the load over the whole impact area with respect to the time sequence is of special interest.

It is necessary to establish the load functions for the aircraft models and approach scenarios discussed in the previous chapter in order to continue the analysis of the problem. The first functions were created in the early 1970s: first

G. Kessler et al., *The Risks of Nuclear Energy Technology*, Science Policy Reports, DOI 10.1007/978-3-642-55116-1_14,

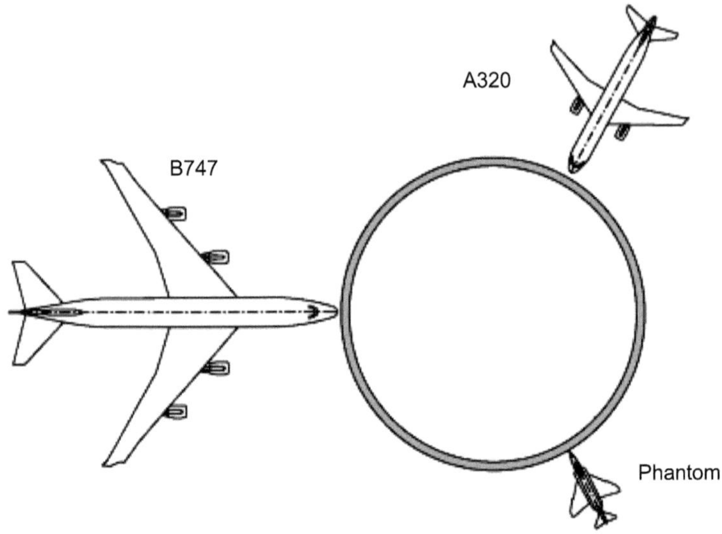

Fig. 14.1 Comparison of the geometric dimensions in a horizontal section: Phantom–Airbus A320–Boeing 747 on impact of a convoy containment with diameter of 66 m

the load function based on the "accidental" crash of a Starfighter, then of a Phantom. As previously mentioned, the latter became a part of the RSK-Guidelines of 1981. Furthermore, the document establishes that the load can be assumed to be equally distributed over an area of 7 m^2.

After the attacks on September 11, 2001 the deliberate crash of a large commercial aircraft was also considered. Different conditions apply in the case of such a crash compared to the load function of a Phantom. Figures 14.1 and 14.2 clarify the geometric relationships. While the impact of a phantom can be assumed to be almost punctiform, this assumption only serves as a rough estimate for the impact of a B747 when considering the relationship of the dimensions sketched here. While the wingspan of a Phantom only measures 11.77 m, that of a Boeing 747 is almost 65 m which roughly corresponds to the diameter of a convoy reactor building.

14.2 Mathematical Models to Determine an Impact Load-Time Function

An experimental determination of the load function for different aircraft models through real tests is not feasible with justifiable means. The determination was therefore carried out on the basis of theoretical models with support of limited experimentation.

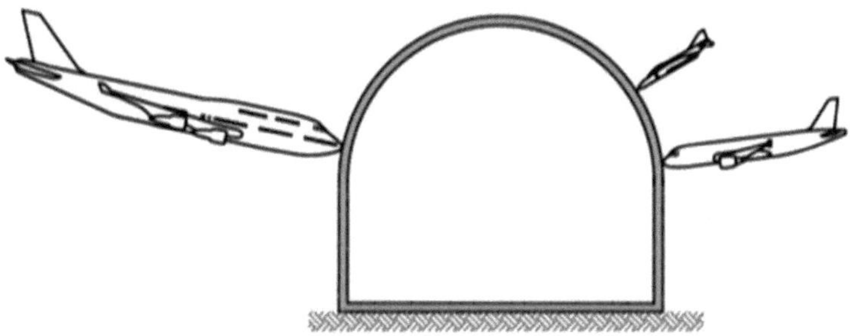

Fig. 14.2 Comparison of geometric dimensions in a vertical section: Phantom–Airbus A320–Boeing 747

To calculate the time-dependent contact force during the impact between two bodies, two different approaches are available, both of which have their roots in Newton's second law of motion (force = mass × acceleration) in addition to respecting the conservation of energy and momentum. Riera developed a comparatively simple method for the process of an aircraft impact in the late 1960s [1], which yields adequately applicable results. It is still used today amongst complex numerical simulations for crash situations and has been verified in numerous experiments. The so-called Riera-Model is a one dimensional mathematical model in the form of a thin walled tube with specific mass and bursting load distributions. The load on the rigid target body is therefore dependent on the mass M and the deformation speed v of the impacting body. As a matter of principle, the impact force is calculated as the change of momentum $M \cdot v$ in the region of the impact zone:

$$F(t) = \frac{\partial}{\partial t}(Mv) = v\frac{\partial M}{\partial t} + M\frac{\partial v}{\partial t}$$

This simple relationship used for the impact of an aircraft shows the influence of different effects. The first term on the right is determined by the change in mass of the projectile within a fixed control volume at the impact point. The second term corresponds to the force in the impact zone that is transmitted to the rest of the projectile, which is assumed to be rigid, and must be equal to the crushing load of the structure in front of the impact zone for reasons of equilibrium. For high impact speeds (>150 m/s), the first term and therefore the mass distribution of the aircraft is the dominant factor in determining the impact load-time function, whereas for lower speeds the bursting load has a significant influence. It is important to note that the mass distribution of an airplane is usually very well known, but that the bursting load distribution of an aircraft fuselage can only be stated within a wide range. This shows that the Riera-Model calculates the load function for high speeds relatively reliably, however only with larger uncertainties for smaller speeds.

Different conditions apply to the impact of a large commercial plane than to that of a military aircraft. Taking into consideration the geometric relationships the one-dimensional Riera-Model can only be seen as a rough approximation although this approximation should give a result on the safe side. In the examinations after September 11, 2001 the mathematical models were refined, in particular to take into account the time dependent spatial distribution on the target. In addition to modified Riera-Models complex computer simulations such as those used in the automobile industry to analyze crashes were used. While the complex analyses are verified by crash tests in the automobile industry such analyses are almost impossible to realize in the case of an airplane crash and have therefore not taken place.

Within the frame of the examinations to the consequences of an impact of large commercial aircraft, an enhanced mathematical model for the realistic compilation of the existing conditions was also applied by the author. In this model, the aircraft is broken down into discrete masses which are spatial distributed according to the geometry. In particular the masses are assumed to be distributed along the fuselage and main wings. The individual masses are connected by springs that are able to transmit axial forces, bending moments and shear forces using a special non-linear approach to the spring characteristics.

Figure 14.3 shows the basic layout of the mathematical model. The airplane is treated as a plane model in the x–y plane. The third axis can be neglected for the given problem. A row of nodes with assigned masses M_i are assumed in the flight direction in the axis of the fuselage (x-axis). These are connected via axial force springs R_i which describe the force-deformation relationship during the deformation. At the same time the masses are connected by special elements that describe the flexural behaviour of the fuselage in the x–y-axis. The two main wings are coupled to the fuselage as rigid elements at a certain angle to the longitudinal axis. Again different rows of individual masses with axial force springs are attached to these in the direction of flight. The jet engines are also represented by a group of springs and masses which are connected to the wings through special coupling springs.

The axial force springs used in the simulation have in principal a characteristic curve as shown in Fig. 14.4. Starting with an initial stiffness k_O a maximum force R_B can be carried in relation to the displacement u. This corresponds to the bursting load of the fuselage or the wing element. Beginning at a value u_V a hardening with a parabolic gradient takes place until a specified force R_C is reached at u_C. The unloading and reloading takes place with a stiffness of k_E. These parameters must be determined for each section i of the aircraft model. Once deformation of the normal force spring reaches the value u_C, the masses are considered to be rigidly coupled. At this point in time a balancing of momentum of the two concerned masses take place by calculating a new resulting speed.

The method of calculation consists of an integration in time of a coupled spring-mass system. In principle it is based on the finite element method. All masses are assigned a starting speed as an initial condition. This corresponds to the approach speed being considered. Upon impact with a rigid target this speed is gradually decreased though the effect of the springs. To determine the impact load-time function the forces in the respectively first facing spring is evaluated (cf. Fig. 14.3) and the spring force is plotted over time. Specifically these are the springs in the

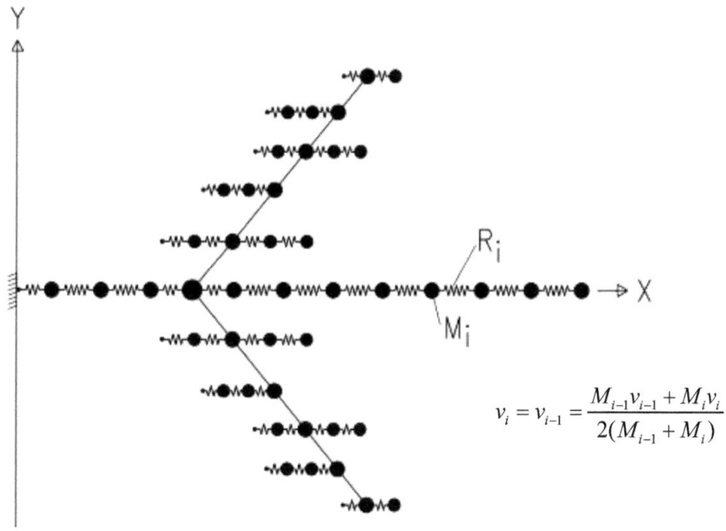

Fig. 14.3 Discrete spring-mass model for the calculation of the impact load-time function

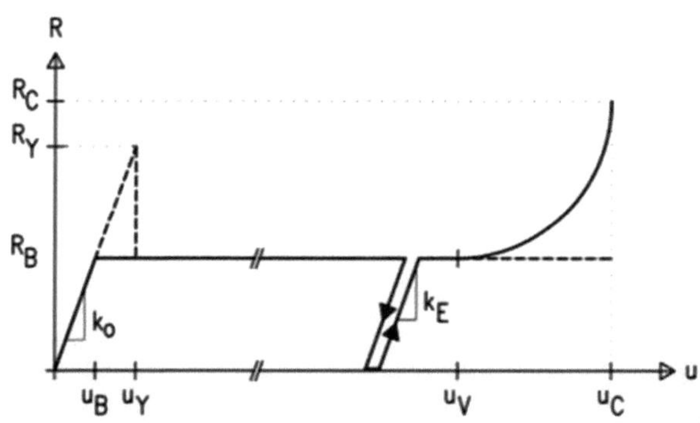

Fig. 14.4 Non-linear characteristic of normal force spring to describe the deformation

extension of the fuselage axis and the wing and jet engine masses. With this method the loads can be determined in the time dependent order of reaching the different impact locations. The sum of all that impact springs provides the integral impact load-time function in the examined case.

With the help of the mathematical model the different approach situations can be analyzed. The chronological order of the failure of individual components such as fuselage, wings and jet engines can be determined. The obstacle can be modelled as a planar, curved or offset structure. The accuracy of the results is decisively dependent on how realistic the parameters of the individual aircraft models are. In addition to the geometry and mass distribution the rigidity and bursting loads must be available for the different sections of the fuselage.

Table 14.1 Load approach for Starfighter F 104

Starfighter F 104		
1	Mass	13 t
2	Impact speed	102 m/s (= mach 0.3)
3	Max. impact force	17 MN
4	Impact period	120 ms
5	Time response	Half sine
6	Impact area	2.1 m^2

14.3 Load Approach for Fast Flying Military Aircraft

14.3.1 Load Approach for Starfighter

Due to the number of observed crashes of military aircraft of the model Starfighter load functions were first derived at the end of the 1960s and the beginning of the 1970s. After the examinations had been carried out under the assumption of a fully plastic impact the load approach detailed in Table 14.1 was established. For the design of the structures a static load of 17 MN was used as a substitute for the maximum impact force. The assumed impact area was 2.1 m^2. Using this procedure a minimal protection was to be ensure for similar effects.

14.3.2 Load Approach for Phantom

In the further development an intensive investigation of the danger of an accidental airplane crash was conducted in the beginning of the 1970s. A representative load approach was to be defined for the design that covered all relevant aircraft types. Starting with the Riera-Model of a deformable projectile Drittler, Gruner, and Sütterlin continued the development of a mathematical model in 1973 and suggested an impact load-time function for the accidental crash of a Phantom onto a rigid concrete wall. This load function is shown in Fig. 14.5 and corresponds to the function later adopted in the RSK-Guidelines in 1981. It was used for the design of all new nuclear power plants in Germany.

In the years after, i.e. the end of the 1970s and the beginning of the 1980s, numerous additional studies were conducted regarding the impact problem of deformable and rigid projectiles on rigid and deformable targets. The effects of wreckage and debris were given special consideration. Aside from the theoretical development extensive experiments were simultaneously carried out in different scales to verify the mathematical models. Especially worthy of mention are the so-called Meppen experiments (cf. [6]). Ballistic tests where deformable steel pipes were shot at thick reinforced concrete slabs were carried out on the exercise area of the German armed forces in Meppen within the frame of a research project supported by the German Federal Ministry of Research and Technology and with

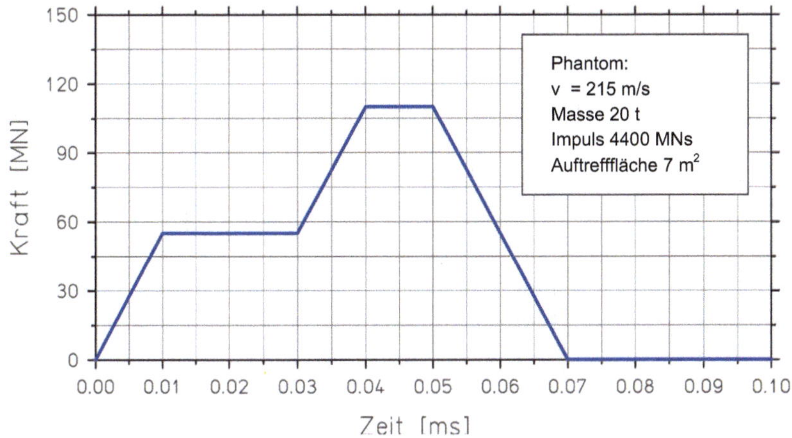

Fig. 14.5 Impact load-time function according to the RSK-LL for a fast flying military aircraft based on a Phantom with an impact velocity v = 215 m/s

the involvement of different institutions. Impact speeds of up to 215 m/s were reached. Extensive knowledge was gained about the impact load-time curve and the behaviour of impacted structures. Among other researchers the author also examined the findings of the Meppen tests in the frame of his dissertation [13].

Figure 14.6 shows the experimental setup in Meppen. Using a special accelerating device the approx. 6 m long steel pipes were shot at reinforced concrete slabs. After the impact the deformed projectiles typically looked like the one shown in Fig. 14.7. Figure 14.8 schematically shows the failure pattern of a reinforced concrete slab after an impact.

In a spectacular large scale test carried out in the USA in 1988 by the SANDIA National Laboratories on behalf of Japanese institutions (cf. [13]) a full scale verification of the theoretically determined load function for a Phantom was realized for the first time. In the experiment a Phantom with a mass of nearly 20 t and a speed of 215 m/s in accordance with the assumptions made in the RSK-LL function was launched against a rigid target (3 m thick concrete slab), which was suspended almost friction-free on an air cushion. During the experiment many measurement signals were taken on the aircraft fuselage as well as on the target body, which were then evaluated by Japanese and American teams. In addition, the data were reevaluated by the German side in a joint research project lead by the GRS and with the involvement of the author (cf. [15–20]). Through this many valuable insights were gained about the processes of an airplane crash. The resulting load function of the experimental data concurred relatively well with the function according to the RSK-LL. Figure 14.9 shows a sequence of pictures during the impact.

In addition to the test under realistic conditions performed with the Phantom within the framework of the Sandia tests experimental studies were also performed concerning the impact of jet engines and load functions determined accordingly.

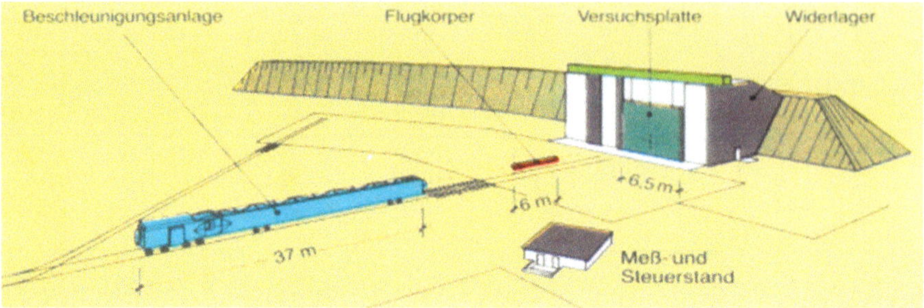

Fig. 14.6 Experimental setup in Meppen (cf. [6])

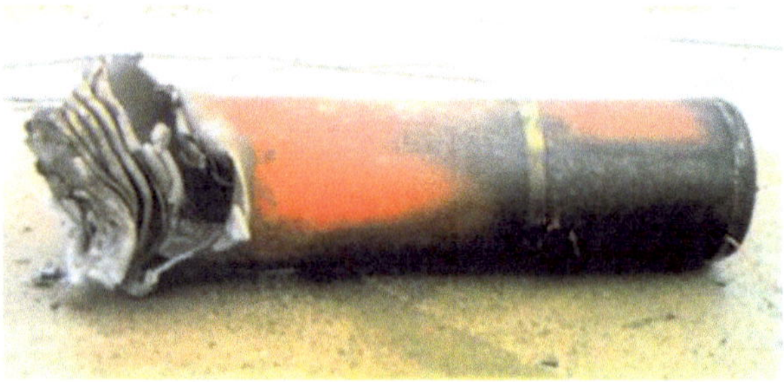

Fig. 14.7 Projectile after impact (soft impact, experiments in Meppen)

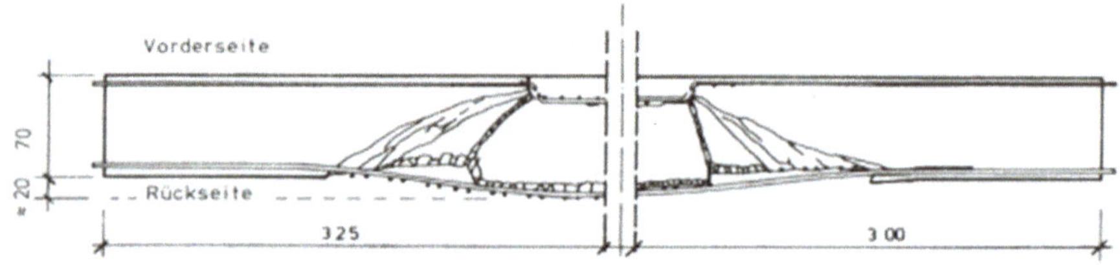

Fig. 14.8 Schematic representation of the failure pattern of a reinforced concrete slab after an impact (soft impact, experiments in Meppen)

The spectacular large scale experiment with the Phantom verified that the theoretically determined load approach according to the RSK-LL describes the loads upon impact relatively well. The design parameters used in Germany were therefore verified.

Using the tests carried out at the SANDIA National Laboratories other mathematical models to determine the load function could also be verified. Amongst others the author was able to verify the mathematical models using the discrete masses and springs described above. The modeling of fuselage of the approx. 17.6 m long military aircraft was carried out with 176 individual masses and springs. The bursting load and mass distributions were assumed analog to the

Fig. 14.9 Pictures of the large scale experiment with a Phantom at the Sandia National Laboratories in the USA in 1988; the target body after the impact on the *right*

Sandia experimental data according to Fig. 14.10. While the mass distribution can be determined relatively well, there is a large degree of uncertainty concerning the bursting load. It is determined by the strength and rigidity of the materials with simultaneous consideration to the complex processes due to the crushing such as stability, imperfections, the geometric non-linearity and the non-linearity of the materials, temperature, etc. It turns out that the contribution of the bursting load to the impact force is relatively low. The final result can therefore still be considered reliable even with the uncertainties in the bursting load.

Figure 14.11 shows the resulting deformation of the nodes at specific points in time as a result of the calculations. The deformation process is visible as a compression of the nodal masses. At $t = 70$ ms the fuselage is almost completely compressed in the impact zone. The temporal development of the contact force is shown in Fig. 14.12 and is compared with the impact load-time function according to the RSK-LL. Due to a slightly lower gradient the maximum value of the impact force of approx. 135 MN is slightly larger than the RSK-LL value. The analysis of the Sandia experiments also resulted in slightly higher values (cf. [13]). It can be seen that there is a fairly good agreement to the experimental results which shows the applicability of the used mathematical model.

14.4 Load Approaches for Large Commercial Aircraft

Load approaches for large commercial aircraft are not included in the general nuclear facility technical guidelines. Only after the attacks on September 11, 2011 studies were carried out worldwide and load functions were determined for various commercial aircraft classes. For the most part the information is not publicly available and can therefore not be discussed in detail here. Contributions to the determination of load approaches were presented in recent years, for example at the SMiRT conferences, cf. e.g. [21, 22]. The topic was also treated from different perspectives within the framework of general scientific research amongst

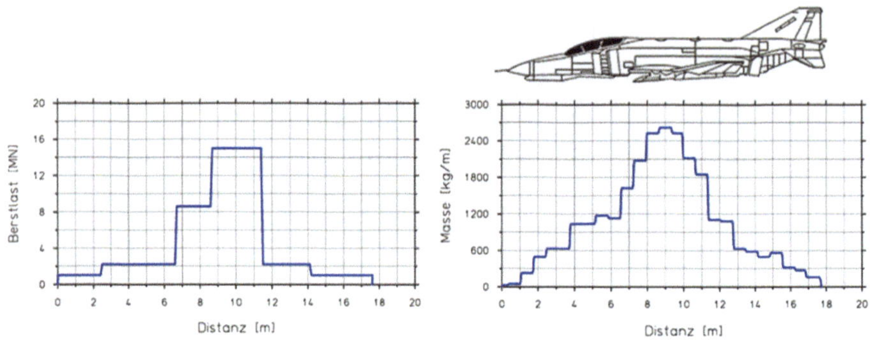

Fig. 14.10 Idealized bursting load and mass distributions of a Phantom (data corresponding to the Sandia test)

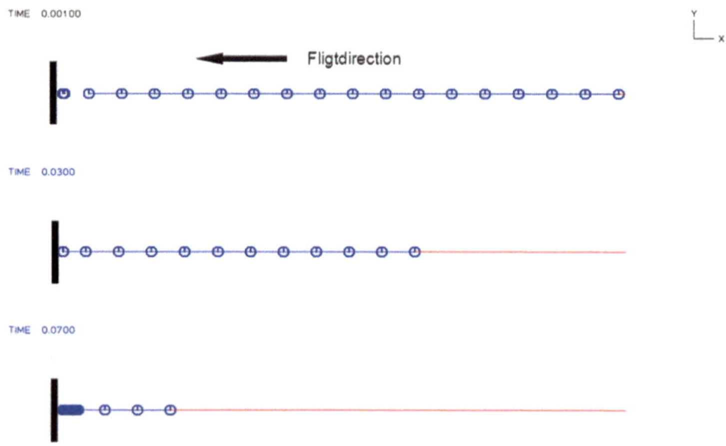

Fig. 14.11 Sequence of nodal deformation (buckling) along the fuselage axis at t = 1 ms, t = 30 ms and t = 70 ms

others to examine the influence of impact projectiles filled with liquids (cf. e.g. [23]). Using two aircraft models with chosen parameters as examples the general behaviour is shown below, once using the long-range aircraft Boeing 747 and once using the typical medium-range aircraft Airbus 320.

For a description of the process the deformation of the aircraft on impact or rather its step-by-step destruction is of interest. It is only in this matter that the surface loading of an impacted structure with temporally fluctuating load distributions as well as the temporal difference of the impact of different aircraft components can be determined. A computational analysis of the impact will be shown with a two-dimensional aircraft model which takes into consideration the spatial distribution of rigidity and mass as well as the destruction due to non-linear material behaviour.

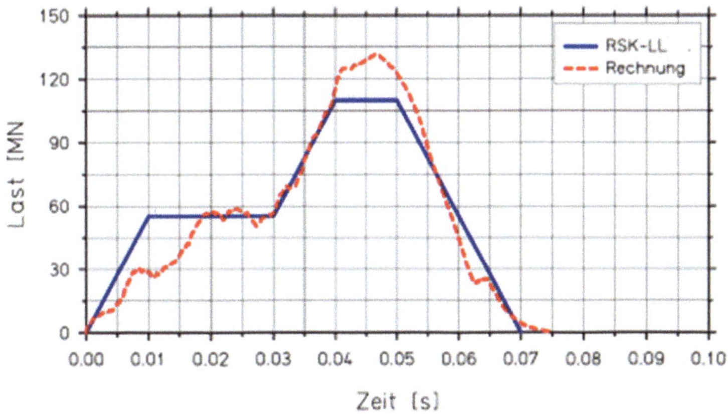

Fig. 14.12 Load-time function of the Phantom, mass 20 t, v = 215 m/s. Comparison calculation and idealized function according to RSK-LL

14.4.1 Load Approach for a Long-Range Aircraft of the Type Boeing 747

Based on the explanations of the previous chapter the analysis of a passenger aircraft of the type Boeing 747 is shown as an example. To create a realistic mathematical model a considerable amount of information about the layout of the particular aircraft is needed. This information is partially available in subject-specific publications. Additional specifications were made available in extracts by the manufacturers and airline companies due to growing demands after the attacks on September 11, 2011. In order to make use of the mathematical models that are based on the approach presented by Riera the compilation of the masses for individual structural and machine components, such as the landing gear and the engines, for permitted payload in the cargo compartment and the cabin and for the aviation fuel distribution over the individual tanks as well as information on the bursting load of the load-bearing fuselage structure are important.

A further important source of information to aid the development of a mathematical model was discovered by the author by chance. Lufthansa donated a discharged but still functioning Boeing 747 to the technical museum in Speyer (town in Germany) for the purposes of exhibition. This aircraft was flown to the regional airport Baden-Airpark close to Karlsruhe and then partially disassembled so that it could be transported on the Rhine to Speyer. As a result it was possible to follow the disassembly over the course of several weeks and to collect additional information on the design of the aircraft with respect to the necessary data on the mass and rigidity distributions. Extracts of the collected information are presented below.

Figure 14.13 shows the fuselage of the Boeing 747 before and after the disassembly of the wings. This illustrates how large this type of aircraft truly is.

Fig. 14.13 The Boeing 747 during disassembly (with and without wings)

The important elements of the fuselage of the aircraft are the outer shell, the longitudinal beams, the pilasters and the cross bracings, so-called structural ribs. The cabin floor consists of a lightweight girder grill construction with panelling. In the mathematical models the theoretical maximum of the mass of the cabin and the cargo compartment are reduced and assumed to be distributed over the length of the aircraft. In this regard it is important to account for the fact that the Boeing 747 uses a double deck configuration at the front of the airplane. The front section up to about the middle of the wings must therefore be more strongly weighted. Regarding the fuselage the additional cabin floor and the slightly larger cross-section have to be considered, with respect to the mass of the cabin the increased live load due to the second deck. The mass of fuselage increases slightly towards the rear part due to the horizontal and vertical tail and the additional rear turbine.

The masses of the two cargo compartments can be conservatively assumed to be approx. 80 % of their maximum live load. Directly under the cargo compartment the center tank is situated which has a capacity of up to 52 t of aviation fuel. The terrorists responsible for the attacks on September 11, 2001 specifically chose long routes airplanes to maximize the damage caused by the largest possible amount of aviation fuel. The center and stabilizing tanks are only filled when necessary, i.e. for long distances. The reason for this is the significantly better relationship between the buoyancy and the dead weight of the wings. Due to this the center tank is usually the first to be emptied. This means that for a long-distance flight it might be assumed that the center tank is initially full when considering the mass distribution. Take-off and the approach to the target can be assumed to require approx. 20 t of aviation fuel. The "Jumbo-Jet" uses about 8 t of fuel per hour and about 10 t are required for take-off and climbing flight. Since only some airplanes have a tank in the rear only a partially filled rear tank can be assumed with respect to the mass distribution.

The wings are connected to the fuselage in the central part of the aircraft laterally to the center tank. Due to the necessity of transporting the above mentioned airplane to Speyer via the Rhine the wings had to be separated from the fuselage. As a result a relatively detailed inspection of the design of the wing including the internal structure was possible (cf. Fig. 14.14). The main load carrying action of the wing occurs in the central section where the tanks are also located. This section is reinforced with numerous longitudinal and transverse girders.

The connection between the wing and the fuselage, called the wing root, consists of a two-celled box girder cross section and is only connected to the fuselage as a load-bearing cross section around the tank. The rear edges of the wings show a bent close to the fuselage at which point the wing widens towards the fuselage. A solid girder is installed here to decrease the force on the outer landing gear and to stiffen the wing. In addition to the outer shell longitudinal girders are arranged on the upper and lower surfaces of the wing. When considering the distribution of the wing mass it is important to keep in mind that the structural and fuel distributions need to be known not only in the direction of the wing, but also the longitudinal direction of the airplane. The jet engines connected to the wings also contribute significantly to the distribution of the mass of the wings.

The landing gears represent fairly large individual masses. The nose landing gear (cf. Figs. 14.15 and 14.16) has a width of approx. 1.20 m when measuring the wheels and is connected to a hydraulic cylinder of approx. 2 m in length which points in the direction of flight when retracted. When viewing the landing gear the important additional function of damper for the landing process also becomes clear. The vertical hydraulic cylinder has a diameter of around 30 cm and a length of about 2 m. Since the landing gear has to be assumed to be retracted in the case of a deliberate forced crash, its mass has to be distributed over a length of just over 2 m. In addition to the wheels and the axle the mass of the suspension assembly in the aircraft fuselage must be taken into account.

The main landing gear consists of four landing gear groups, one pair of which is arranged slightly behind the other directly in the fuselage. These inner landing gears are retracted forward such that the approx. 2.5 m long hydraulic cylinder faces in the direction of flight.

Fig. 14.14 View of the contact section of the wing to the fuselage; only the central third is designed as a load-bearing cross section and is simultaneously designed as a tank

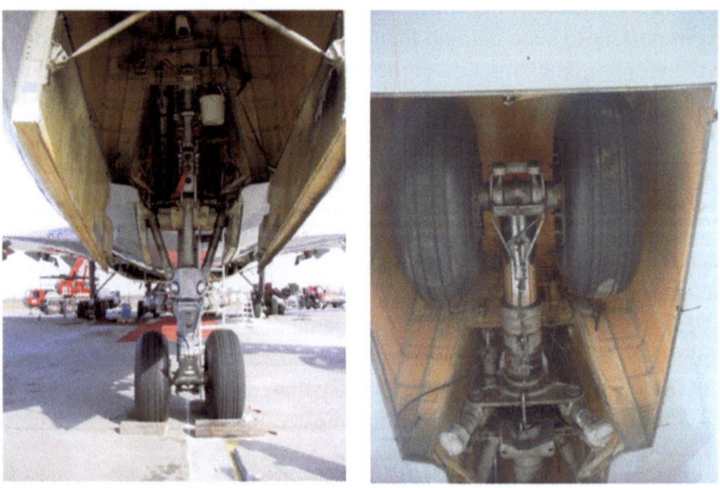

Fig. 14.15 Nose landing gear; on the left in the extended position, on the *right* in the retracted position; the hydraulic cylinder points forward in the direction of flight

Fig. 14.16 Massive structure of the nose tire including steering mechanism

Fig. 14.17 Inner and outer landing gear groups with cylinder and struts

The outer landing gears which are located slightly in front are outwardly suspended from the wing. They retract inward such that the wheels, axle and longitudinal axle lay in the fuselage directly in front of the inner landing gears. The approx. 4.5 m hydraulic cylinder comes to rest perpendicular to the longitudinal axis of the plane in the rear section of the wings. About half of the mass can be assigned to the fuselage and the inner wing section each (Fig. 14.17).

Figure 14.18 shows a simplified plot of a possible total mass distribution along the aircraft axis. There is an abrupt rise in the mass distribution curve at around 22 m from the nose of the aircraft due to the beginning of the wings and tanks.

To determine the impact load-time function the distribution of the bursting loads for the individual aircraft sections is needed in addition to the mass distribution. As mentioned above this determination is relatively difficult and brings with it a degree of uncertainty. The integral bursting load distribution over the longitudinal axis must be derived from the respective structural dimensions and rough estimates concerning the rigidity of individual cross sections. As a reminder it should be noted that the bursting load distribution only has a secondary influence on the results of the calculations at higher speeds which are of interest, such that the uncertainty of the assumptions seems acceptable.

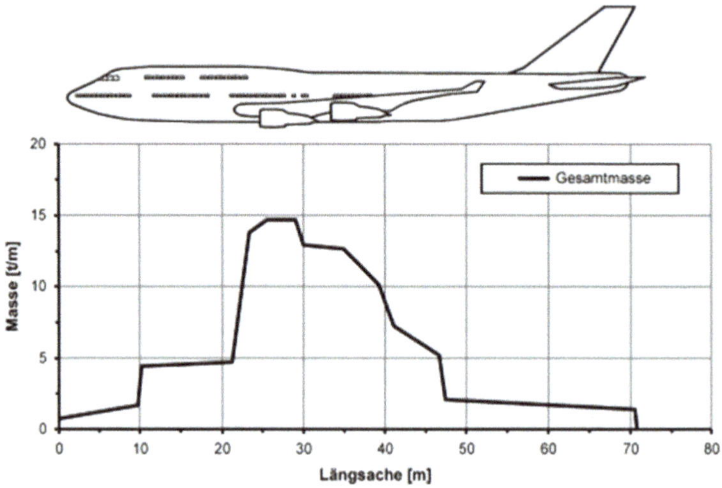

Fig. 14.18 Schematic representation of the total mass distribution along the longitudinal aircraft axis for a Boeing 747

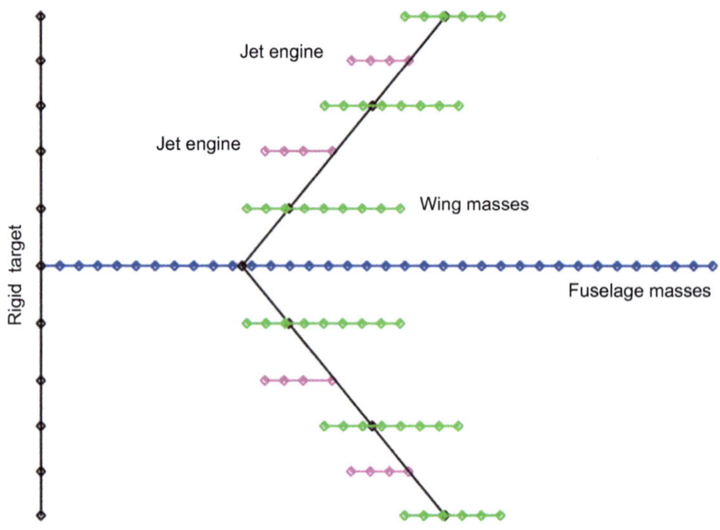

Fig. 14.19 Computational model for a Boeing 747 impacting on a rigid straight target

To determine an impact load-time function the two-dimensional mathematical model presented in Sect. 14.2 can be used among others in conjunction with the gathered data for a Boeing 747. Figure 14.19 shows a top view of the spring-mass system. A straight rigid target is located on the left side. Upon impact of the respective components of the aircraft structure they are deaccelerated due to the contact springs. A speed of 175 m/s is assigned to all masses as an initial condition.

During the impact initially only the fuselage hits the target and is compressed. Subsequently the wing sections are included in the process and make contact with

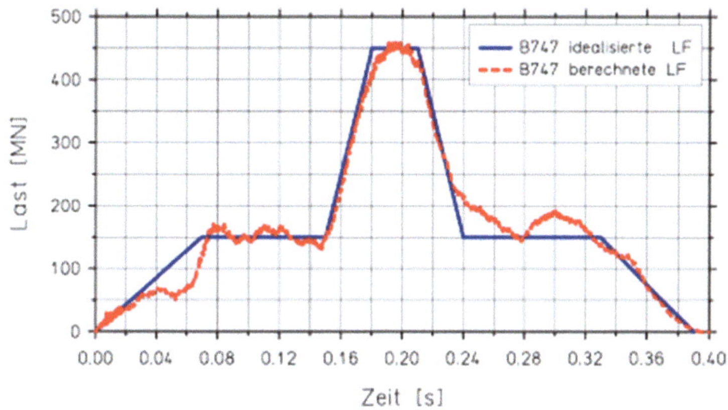

Fig. 14.20 Example of a load function for a Boeing 747, mass 380 t, v = 175 m/s, comparison between calculation results and idealized function

the target. After t = 400 ms all masses of the fuselage are compressed except those in the rear section. The aircraft has come to rest. An examination of the contact forces to the target over time shows that initially only the spring extending along the fuselage axis transmits a force. Only after approx. 130 ms the wings hit the obstacle, at which point the springs located there take part of the load. The sum of all contact spring forces is the integral impact force that acts upon the target or structure.

For the Boeing 747 with an initial speed of 175 m/s the impact duration is approx. 400 ms. Within the framework of the investigation the mass and bursting load distributions were varied with respect to magnitude and distribution in different computational runs. A variation of the bursting load showed only insignificant differences in the end result: the impact load-time function. Within the scope of the overall uncertainties of the model and the assumptions these can be ignored. The mass distribution therefore plays the deciding role at speeds of about 175 m/s.

Figure 14.20 presents the impact load-time function as the result of a calculation for a given set of conditions (mass and bursting load distributions). It shall be noted here that the function shown is only to be considered as an example. With different boundary conditions and other mathematical models different load function curves can be determined. An impact load-time function that is calculated in this way is badly suited for use in further calculations of the structural behaviour. It would have to be specified as a polygon with many nodes. It also includes many high frequency oscillation which are completely irrelevant for the loading of a massive structure. Therefore the loading function can be well described by an idealized function with only a few nodes. The peak value of the load, the duration and the momentum as an integral over the time of the impact load-time function should largely agree. With these considerations in mind, Fig. 14.20 also shows an idealized function. The momentum of the idealized function shown is equal to I = 66.75 MNs.

Within the scope of parameter studies taking into account the site-specific factors different impact situations can be simulated. For example, the straight target can be replaced by one with a cylindrical geometry to correspond to the

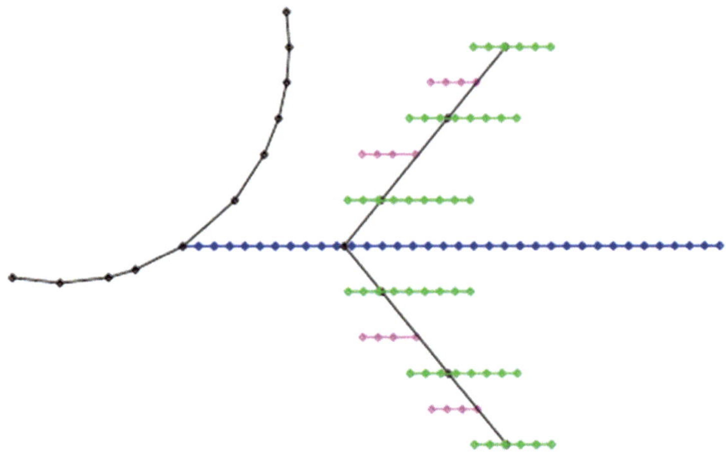

Fig. 14.21 Impact with an offset cylindrical target

impact on a convoy reactor building. In such a simulation the aircraft has been almost entirely compressed to the contour of the cylindrical target after approx. 350 ms. The load function only changes slightly in comparison to the straight target, however. The wing sections hit the obstacle a little later in the case of a cylindrical obstacle than in the case of a straight one. The mass density of the outer wing sections is relatively small compared to the fuselage and the middle wing sections with the integrated tanks, however such that the impact load-time function only shows small differences in comparison to the rigid straight target and is covered by this case.

In additional studies the rigid cylindrical target was offset to the flight direction. Figure 14.21 shows the initial state. The load on the target is decreased compared to the central impact. This raises the question of whether the aircraft could be diverted and deflected by the impact surface that is tilted with respect to the fuselage axis.

14.4.2 Impact Areas Boeing 747

To calculate the structural behaviour it is necessary to analyze and define the possible impact areas in addition to the temporal progression of the impact force. Using the geometric data of the B747 and the knowledge gained through the calculations of the load function the following areas can be used for example:

$A_R = 39\ m^2$: Substitute area for the fuselage only: circle with a diameter of 7 m.
$A_G = 64\ m^2$: Substitute area for the entire aircraft including wing components: ellipse with the semi-diameters 17.00 m and 5.00 m or alternatively a circle with a diameter of 9 m.

Fig. 14.22 Effective impact areas for a B747—complete aircraft

For illustration the effective impact areas above are shown in Fig. 14.22.

14.4.3 Load Approach for the Medium-Range Aircraft of the Type Airbus A320

In the same manner as for the long-range aircraft Boeing 747 the impact load-time function for an Airbus A320 as a typical medium-range aircraft will be shown here for comparison. Exemplary results for the impact with a straight target are shown in Fig. 14.23 in the form of an integral load function. The impact duration is 165 ms, the peak value of the impact force reaches approx. 165 MN after 115 ms. As a substitute for a calculated function with many polygon nodes an idealized impact load-time function is shown. The effective impact surface for the entire aircraft can be described by a circle with a diameter 6.50 m. The fuselage diameter of an A320 of 4.50 m is smaller than that of a B747 of 7 m.

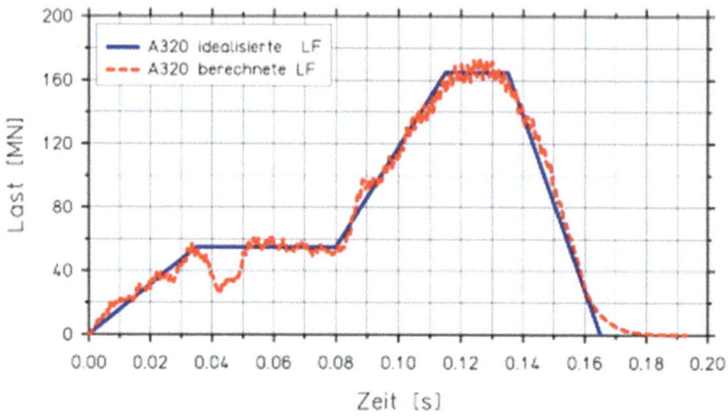

Fig. 14.23 Example of a load function for a medium-range aircraft A320; mass 74.5 t, v = 175 m/s; comparison between calculation results and idealized functional

14.5 Compilation of the Load Approaches

In the following figures the impact load-time functions of the different types of aircraft mentioned above are compiled for comparison. The load functions are plotted once as absolute force values, once taken over the particular effective impact area and once over the perimeter of the impact area. It shall be noted that the values presented here for the Boeing 747 and the Airbus A320 are only to be treated as examples. Using different boundary conditions and other mathematical models deviating functions can be determined (Figs. 14.24, 14.25, and 14.26).

This compilation clearly shows that the loads are very different with respect to the absolute force values. In comparison to the design standards of the newer nuclear power plants that use the Phantom, the B747 has an impact force that is four times higher and a momentum that is 15 times higher. These values can be relevant with respect to the global structural stability and the induced vibrations.

For the local load concentration on the impact area the scaled quantities are more meaningful. A comparison of the loads divided by the impact area shows that the Phantom is the most unfavorable case. The quantities scaled to the perimeter describe the sectional load of a hit structure in a first approximation. The integrated load of the impact area has to be distributed to the perimeter and transferred to the remaining structure. This results in values that are only 30 % higher for the B747 than for the Phantom.

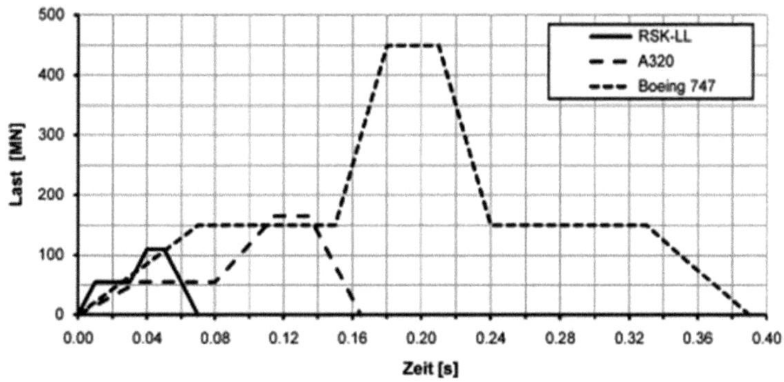

Fig. 14.24 Comparison of load functions according to RSK-LL, Boeing B747 and Airbus A320—resulting total force

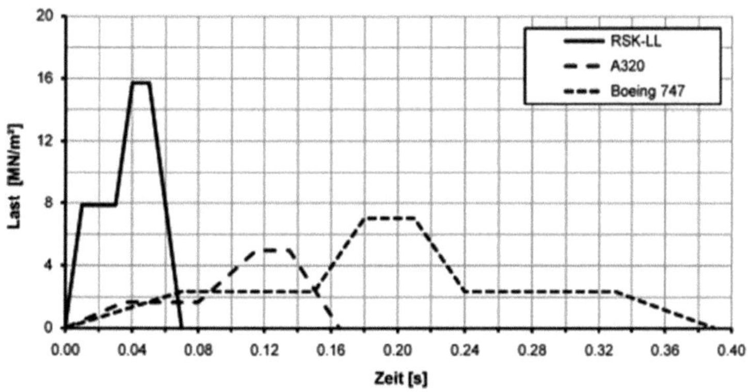

Fig. 14.25 Comparison of load functions according to RSK-LL, Boeing B747 and Airbus A320—resulting total force divided by the particular impact area

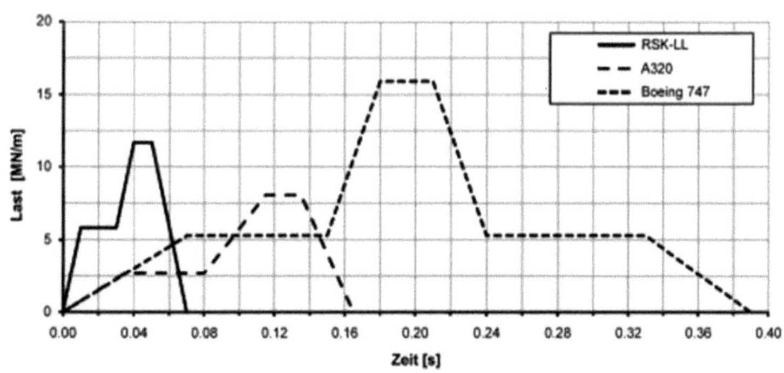

Fig. 14.26 Comparison of load functions according to RSK-LL, Boeing B747 and Airbus A320 – resulting total force divided by the particular perimeter

References

1. Riera JD (1968) On the stress analysis of structures subjected to aircraft impact forces. Nucl Eng Des 8:415–426
2. Drittler K, Gruner P, Sütterlin L (1973) Zur Auslegung kerntechnischer Anlagen gegen Einwirkungen von außen. Teilaspekt Flugzeugabsturz – Zwischenbericht. Bericht des Instituts für Reaktorsicherheit der Technischen Überwachungsvereine e.V, Cologne
3. Drittler K, Gruner P, Sütterlin L (1975) Zur Auslegung kerntechnischer Anlagen gegen Einwirkungen von außen. Teilaspekt Flugzeugwrackteile – Zwischenbericht. Bericht des Instituts für Reaktorsicherheit der Technischen Überwachungsvereine e.V, Cologne
4. Drittler K, Gruner P (1975) Zur Auslegung kerntechnischer Anlagen gegen Einwirkungen von außen. Teilaspekt Berechnung von Kraft-Zeit-Verläufen beim Aufprall deformierbarer Flugkörper auf eine starre Wand. Bericht des Instituts für Reaktorsicherheit der Technischen Überwachungsvereine e.V, Cologne
5. Drittler K, Gruner P (1976) Krivy: Berechnung des Stoßes eines deformierbaren Flugkörpermodells gegen ein deformierbares Hindernis. Bericht des Instituts für Reaktorsicherheit der Technischen. Überwachungsvereine e.V, Cologne
6. Jonas W, Rüdiger E (1982) Kinetische Grenztragfähigkeit von Stahlbetonplatten. Berichte zum Forschungsvorhaben RS165. Hochtief AG, Abteilung Kerntechnischer Ingenieurbau, Frankfurt
7. Stangenberg F, Schwarzkopp D (1984) Stahlbetonkonstruktionen von kerntechnischen Anlagen unter extremen Anprallbelastungen. Jahrestagung Kerntechnik 1984, Frankfurt, 22.-24.05.1984. Deutsches Atomforum e.V./Kerntechnische Gesellschaft
8. Eibl J (1985) Design of concrete structures to resist accidental impact. Struct Eng 65A(1)
9. Eibl J, Henseleit O, Schlüter F-H (1988) Baudynamik. In: Betonkalender 1988, Teil II. Verlag Ernst & Sohn, Berlin
10. Eibl J, Schlüter F-H (1985) The design of impact endangered concrete structures. In Transactions of the 8th international conference on structural mechanics in reactor technology, Brussels, 19–23 August 1985. North-Holland P.P., Amsterdam
11. Eibl J, Schlüter F-H (1987) Numerical analysis of thick reinforced concrete slabs under impact load. In Proceeding of the IABSE-Colloquium Delft 1987 'Computational Mechanics of Concrete Structures'. IABSE Report, vol 54, pp 591–598
12. Eibl J, Schlüter F-H (1987) Local behaviour of thick reinforced concrete slabs under impact loading. In Transactions of the 9th international conference on structural mechanics in reactor technology, Lausanne, 17–21 August 1987. A.A. Balkema, Rotterdam
13. Schlüter F-H (1987) Dicke Stahlbetonplatten unter stoßartiger Belastung – Flugzeugabsturz. Dissertation Universität Karlsruhe
14. Von Riesemann et al. (1989) Full-scale aircraft impact test for evaluation of impact forces – Parts 1 and 2. In 9th International conference on structural mechanics in reactor technology (SMiRT 9), Anaheim, USA, August 1989
15. Eibl J, Schlüter F-H (1989) Vergleichsuntersuchung zum Lastfall Flugzeugabsturz auf Kernkraftwerke –Last-Zeit-Funktion nach RSK-Leitlinie und SANDIA-Test. Institut für Massivbau und Baustofftechnologie, Universität Karlsruhe, Bericht
16. Eibl J, Schlüter F-H, Hoffmann D (1989) Unterschiede in ausgewählten Etagen-Antwort-Spektren eines typischen Reaktorgebäudes beim Lastfall Flugzeugabsturz mit Last-Zeit-Funktion nach RSK-Leitlinie und SANDIA-Test. Institut für Massivbau und Baustofftechnologie, Universität Karlsruhe, Bericht
17. Prof. Eibl und Partner GbR (1998) Beurteilung des lokalen Verhaltens von Stahlbetonstrukturen von KKW-Anlagen bei Einwirkung der aus den Versuchen abgeleiteten Belastungsfunktionen bei Flugzeug- bzw. Triebwerksaufprall. Karlsruhe
18. Bundesanstalt für Materialforschung und -prüfung (BAM) (1998) Verifikation des strukturellen Responses von Konvoi-Anlagen unter Annahme einer gemessenen Flugzeugabsturzbelastung.

Einsatz eines nichtlinearen 3D-Modells unter Berücksichtigung wirklichkeitsnaher Boden-Bauwerk-Kopplung zur Überprüfung des 0.5–g-Konzeptes. BAM, Berlin

19. Stangenberg und Partner, Ingenieur-GmbH (1998) Bewertung der Ergebnisse der bei SANDIA durchgeführten japanischen Stoßversuche hinsichtlich der Sicherheit von Kernkraftwerken bei Flugzeugabsturz. Teilaufgabe: Folgerungen für das globale Verhalten (induzierte Erschütterungen) von KKW-Gebäuden mit DWR bei Einwirkung der aus dem Versuch abgeleiteten Belastungsfunktion bei Flugzeugabsturz. Abschlußbericht zum BMU-Vorhaben SR 2216, G-F ~ 73.1/G1
20. GRS (1998) Bewertung der Ergebnisse der bei SANDIA durchgeführten japanischen Stoßversuche hinsichtlich der Sicherheit von Kernkraftwerken bei Flugzeugabsturz – Teilaufgabe 1: Bewertung von Versuchs- und Analyseergebnissen zur Ermittlung von Belastungsfunktionen infolge von Flugkörperaufprall. Abschlußbericht SR 2070, GRS-A-2148
21. Siefert A, Henkel FO (2011) Nonlinear analysis of commercial aircraft impact on a reactor building. Transactions, SMiRT 21, 6–11 November 2011, New Delhi, India. Div. III, Paper ID#144
22. Kostov M, Henkel FO, Andonov A (2011) Safety assessment of A92 reactor building for large commercial aircraft crash. Transactions, SMiRT 21, 6-11 November 2011, New Delhi, India. Div. VII, Paper ID#222
23. Ruch D (2010) Bestimmung der Last-Zeit-Funktion beim Aufprall flüssigkeitsgefüllter Stoßkörper. Dissertation Karlsruher Instituts für Technologie (KIT)

Chapter 15
Verification of the Structural Behaviour in the Event of an Airplane Impact

Abstract The structural behaviour in the event of an airplane impact has to be analysed. The local resistance as well as the global stability has to be verified. The integrity and functional reliability of all safety related structures and components has to be guaranteed, including the effects of induced vibrations.

15.1 General

The mechanical actions of an airplane crash on a building can be divided into three categories:

(a) local loads (resistance to penetration),
(b) global loads (structural and position stability) and
(c) loads of assemblies and components due to induced vibrations (integrity and functional reliability).

In practice different computational models are used for all three categories in order to be able to include all the relevant influences and boundary conditions for the specific verification. It is not possible to discuss all verification procedures as part of this contribution. Rather only a broad overview will be given here.

The first regulations in Germany for the design of building structures against an airplane crash were published by the Institut für Bautechnik in 1974 (cf. [1]). Further rules were to be established by the Nuclear Safety Standards Commission (KTA). Drafts were proposed (cf. [2, 3]) but never adopted as final KTA rules. Because there have been no plans for new construction projects since the 80s work was not continued. There are several references to airplane crashes in the current version of DIN 25449 (cf. [4]) such as specifications for the resistance to punching for impact loads.

G. Kessler et al., *The Risks of Nuclear Energy Technology*, Science Policy Reports,
DOI 10.1007/978-3-642-55116-1_15,

The structural behaviour and the resistance of safety enclosures of a nuclear power plant in the case of an aircraft impact shall be examined here using the example of the containment of a pressurized water reactor of the convoy type. The outer form of the containment consists of a cylindrical shell with a superimposed dome with an outer diameter of 66.8 m. This rotationally symmetric form is only disrupted by the materials lock, its front building and the fresh steam and feed water armature chambers. The wall thickness of the reinforced concrete construction is 1.80 m. The internal structures such as the inner steel containment etc. are only connected to the outer structure via the foundation.

Figure 15.1 schematically shows the design of a reactor building of the convoy type. In the sectional cut the 1.80 m thick reinforced concrete shell is clearly visible. The structural design of the outer shell is not uniform for all locations. The spectrum of the flexural reinforcement of the high-strength reinforcing steel BSt 1100 ranges from approx. 26 cm^2/m to 60 cm^2/m per side and flexural direction, depending on the impact location (cylinder or sphere) and the location of the nuclear power plant. The shear reinforcement of the ductile stell BSt 420/500 varies between 44 and 66 cm^2/m^2. For illustration of the geometric dimensions of an aircraft impact see Fig. 14.2.

15.2 Local Structural Behaviour: Resistance to Penetration

The impact load of an aircraft is a dynamic effect with a very short duration. For military aircraft this effect lasts approx. 70 ms, for commercial aircraft up to approx. 400 ms. Due to inertia effects only the directly impacted areas are initially affected, the rest of a large building does not "experience" anything as a result of the impact in this phase. The entire building only responds after a certain time lag. This applies especially to the internal structures, which are separated from the impacted outer shell and only coupled to it by the foundation. It is therefore possible to analyze the local behaviour independent of the overall structure.

Figure 15.2 shows how a steel reinforced concrete structure behaves in principle in the case of an impact. In the case of a soft impact, which can be assumed for an aircraft impact, local shear failure (punching) or flexural failure is possible. In the case of a hard impact on the other hand, such as in the case of a massive engine component, penetration or perforation can occur depending on the ratio of the plate thickness to the diameter of the impacting body in addition to spalling on the front and rear side of the target. The term "inherent protection" is used when no spalling occurs on the reverse side of the impact area.

To determine the resistance of the safety enclosures against the penetration of an aircraft different models are available ranging from simple equivalent static approaches against punching to complex non-linear finite element analyses.

In the design praxis for the impact of a military aircraft a simplified static method was predominately used to examine the local resistance. Figure 15.3 shows the assumptions of the model used to examine punching according to DIN 25449 [4].

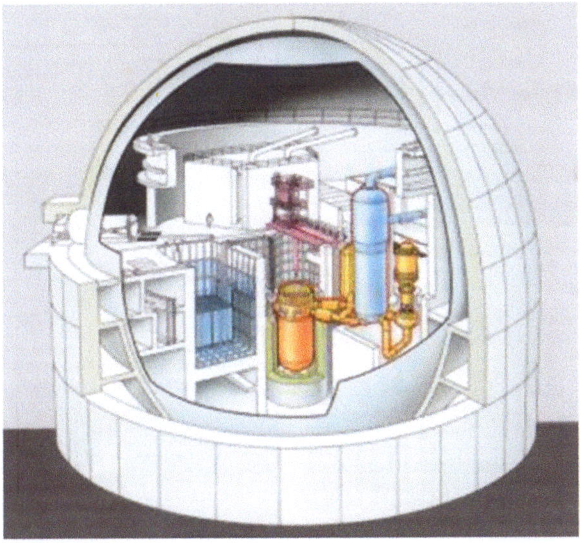

Fig. 15.1 Schematic illustration of the design of a reactor building of the convoy type: outer reinforced concrete containment with a wall thickness of 1.80 m, steel safety structure (*sphere*); internal structures are only coupled with the outer containment through the foundation

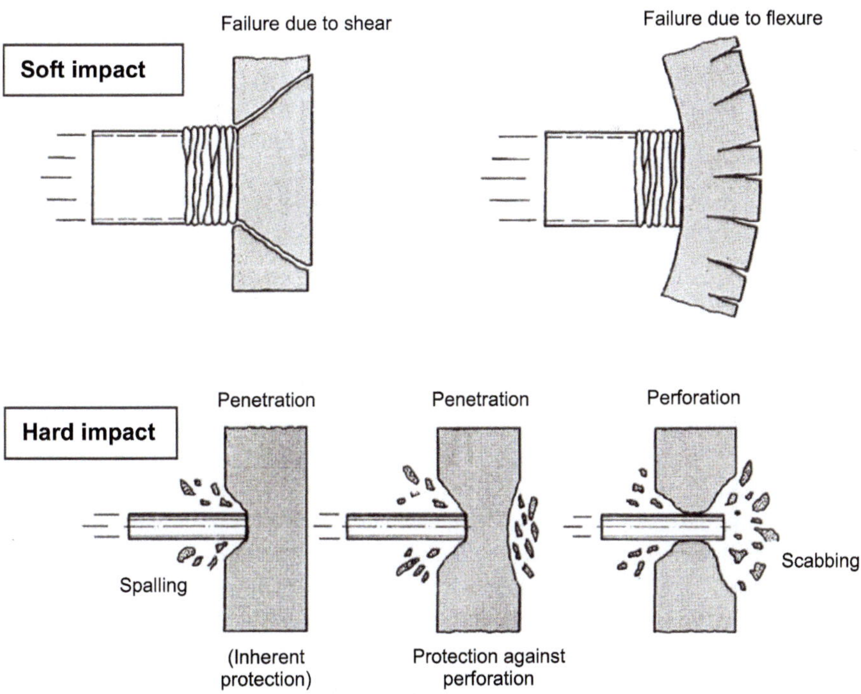

Fig. 15.2 Soft and hard impact

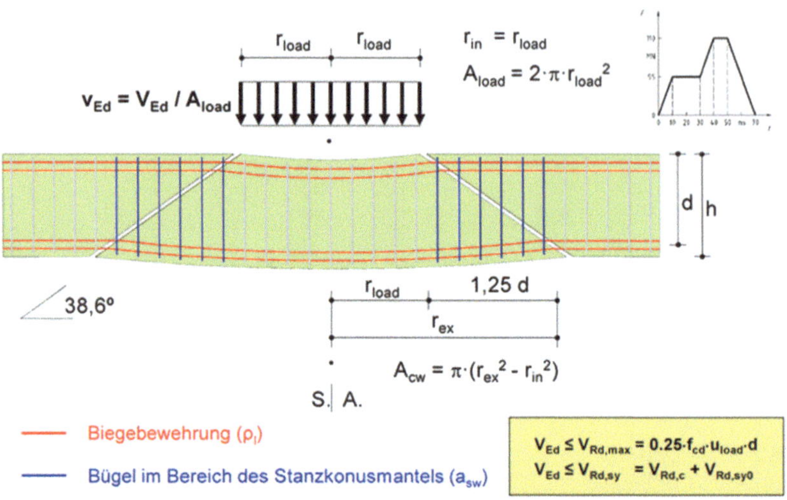

Fig. 15.3 Local transverse force measurements in the impact area according to DIN 25449 [4]

A punching cone is assumed which starts at the edge of the impacted area and has a determined punching angle which was derived from dynamic punching tests. The stirrup reinforcement within the area of influence of the cone must be able to take the maximum impact load of the impact load-time function. Requirements for the plastic straining are not explicitly stated.

More complex models take into account the masses, stiffnesses and plastic deformation characteristics of the structures which are directly affected. Figure 15.4 shows in principal the functionality of such a model. To determine the loads on the concrete and rebars in the impact zone the adequate vibration characteristics and the beneficial effects of inertia are taken into consideration.

The author has derived a simplified design model in [5]. The characteristic curves of the resistance in the punching zone are shown in Fig. 15.5. In this model of a double mass oscillator the inertia, the resistance of the concrete and the contribution of the shear reinforcement as well as the flexural reinforcement including energy dissipation due to elastoplastic material behaviour is taken into account. The model was published amongst others in [6]. This model was used by one of the teams participating in the IRIS project (cf. [7, 8]) in Finland as a pre-calculation method for ballistic tests and correlated fairly well with the results of the experiments, even in comparison to other complex numerical models.

In addition to the simplified models the use of complex FE models is possible. To determine the local loads it is often sufficient to use a rotationally symmetric model with appropriate boundary conditions. For a convoy reactor building a section from a sphere or rather from a hemisphere can be used as a model as shown in Fig. 15.6. An average radius of 32.50 m and a thickness of the steel reinforced concrete of 1.80 m were chosen for the sectional model shown. The reinforcements are also considered in detail in the model. When realistic material

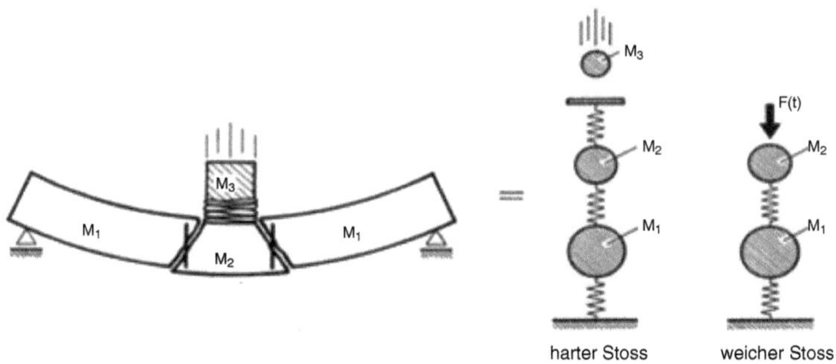

Fig. 15.4 Simplified model for a directly hit structure during impact

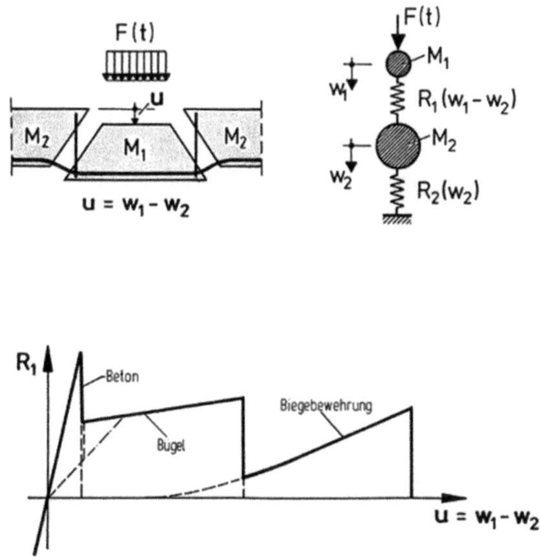

Fig. 15.5 Replacement model to determine the local load on the impact area as proof of resistance to punching (from [5])

laws for steel and concrete with non-linear stress-strain behaviour in the compression regime and cracking above the tensile strength are taken into consideration, the local structural behaviour can be very accurately predicted with such models.

In the past numerous studies on the local behaviour of the relevant structures of German nuclear power plants were carried out using the computational models described above. A detailed description of the calculations and results would go beyond the scope of this contribution. Only a general summary with respect to the necessary wall thickness is presented here.

Depending on the strength of concrete and level of reinforcement a wall thickness of between 1.50 and 1.80 m is necessary to prevent punching, i.e. the entry of

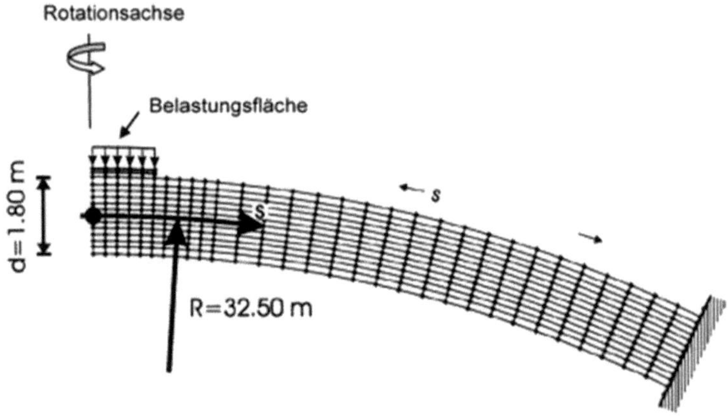

Fig. 15.6 Basic representation of the discretization of a rotationally symmetric model as a section of a spherical shell

the aircraft into the interior of the building when considering the impact of a fast flying military aircraft according to the approach detailed in the RSK-LL (Phantom). A small amount of spalling on the reverse side cannot be ruled out, however. All enclosing walls of the safety-relevant constructions of the convoy complex are designed for this eventuality and have walls that are either sufficiently thick or are equipped with a redundant design with spatial partitioning. Thinner walls cannot absorb the load according to the RSK-LL without significant local destruction or can only absorb an impact with a lower impact speed than 215 m/s. The latter can be assumed if there are structures in front of the wall that would decelerate the aircraft.

Assuming an increased amount of reinforcement, a wall thickness of at least 60 cm is necessary for the impact of a Starfighter, in the case of normal reinforcement this increases to close to 80 cm. All facilities of a nuclear power plant with a lower wall thickness only offer marginal or no constructive protection in the case of an aircraft impact. Due to this the boiling water reactors of the series 69 that were built in the 1970s are to be viewed critically, even after they were shut down in 2011 because radioactive inventory remains in these plants.

After the attacks on September 11, 2001 the computational models and load approaches described in the previous chapter were used to determine whether a large commercial aircraft, such as a Boeing 747 or an Airbus A320, could penetrate the enclosure walls of the reactor building. These analyses were originally limited to the reactor building of convoy facilities with a wall thickness of 1.80 m. Using refined mathematical models it was shown that penetration could be prevent with a high degree of confidence. In the impact zone the deformations of the wall can be up to 50 cm, including intensive cracking but without rupture of the reinforcement. These examinations were not carried out for every individual facility, but rather generalized models were considered with reinforcement grades towards the lower end of the scattering range presented at the beginning of this chapter. With respect to the load-bearing capacity of the concrete it was taken into consideration that the real concrete used in the construction has a higher strength than originally planned. This can be explained by the fact that, with the goal of achieving low stripping times, often a higher strength concrete is installed and additionally over the course of years a further increase of the concrete strength can be expected. Extensive experience with concrete samples from existing structures proves this phenomenon.

In the case of thin walls and low reinforcement grades the probability that a large commercial aircraft will cause heavy damages on the structures increases. When appropriate verifications with lower impact speeds, i.e. smaller than 175 m/s, can be carried out. In some publicly available studies from pertinent conference papers, e.g. SMiRT, speeds of 150 m/s or even 100 m/s are assumed. At these speeds even walls with a thickness of 1.50 m may still be sufficiently sound.

15.3 Global Structural Behaviour: Structural Stability

It must be demonstrated that the global structural stability of a building is assured in the case of a postulated aircraft impact. To be on the safe side, dynamic analyses are carried out with a global model without taking into account the local effects but including the mounting in the foundation. The global structural stability of a large structure such as a reactor building with a very large mass in comparison to the impacting aircraft does not pose a large problem from a structural engineering standpoint. For the disadvantageous impact directions and locations the appropriate verifications of the load-bearing capacity and position stability can be carried out on a convoy reactor building for a fast flying military aircraft as well as for a large commercial aircraft. Due to the fact that the outer enclosing walls are designed to withstand the local effects of a random impact position they are usually sufficiently reinforced to ensure the global structural stability. Figure 15.7 shows an example of such a global mathematical model.

Small buildings whose mass is small in comparison to the impacting aircraft constitute an exception. In these cases it can become necessary to secure the building against tipping and possibly to reinforce the connection to the foundation. This could become difficult to achieve with respect to the load due to the impact of a large commercial aircraft.

15.4 Induced Vibrations

In addition to the evaluation of the resistance of the outer structure the effects of vibrations due to a crash onto the structure must be considered in the design of the interior assemblies and components. Although the regulations do not explicitly cover the case of an airplane crash the analyses and verifications used for the design case earthquakes can be adapted. In combination with appropriate engineer-like adaptations, KTA 2201, Parts 3 and 4 ([9, 10]) can be used as a regulatory reference.

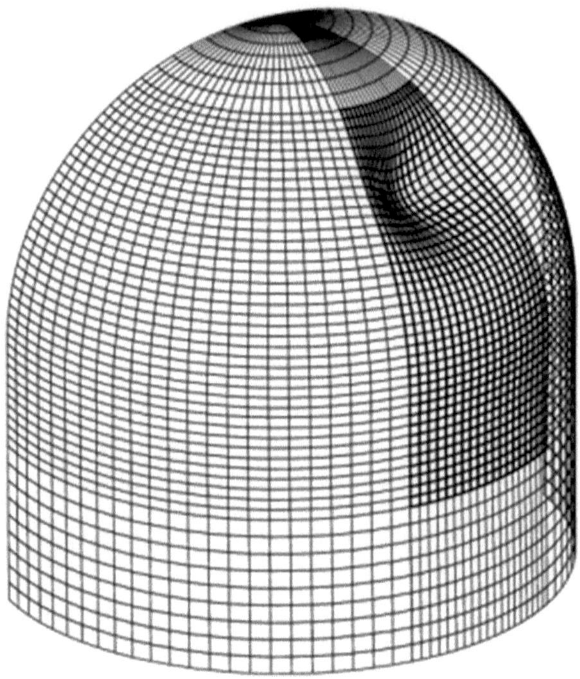

Fig. 15.7 Global model of a reactor building—deformed structure due to an aircraft impact

For some convoy facilities the 0.5 g concept was used for the verification of the integrity of the primary circuit and other components after an aircraft impact according to RSK-LL due to induced vibrations. The loads of the components were replaced by static loads consisting of masses and accelerations of 0.5 g in all directions. This represents a simplified approach.

Furthermore floor response spectra can be determined to account for the dynamic processes for the load case airplane crash in the design. Using complex models which calculate the vibration characteristics of the entire construction as realistically as possible, the system responses from all possible impact directions and impact locations are calculated. The foundation and the masses and stiffnesses of the actual construction play an important role in this. To account for the inevitable variations in e.g. the foundation parameters the relevant parameters of the model are varied according to engineering best estimate and covering response spectra are developed. Comparable to the design case earthquakes a close cooperation between civil and mechanical engineering is necessary. The impact action effects are transferred to the components via the structure including the ground where they induce vibrations. The resulting forces from the reaction of the components must be reintroduced into the structure and dissipated via the ground.

For those facilities that were explicitly designed in accordance with the RSK-LL against an aircraft impact the induced vibrations of the components were either calculated and designed accordingly or the 0.5 g concept was applied. The necessary

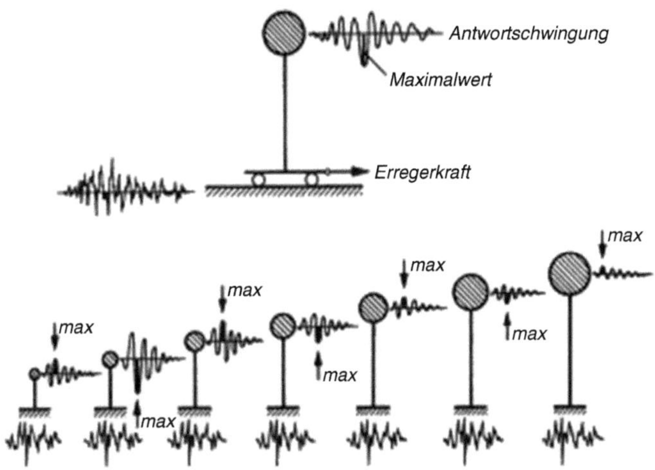

Fig. 15.8 Schematic diagram for the determination of response spectra from dynamic systems with different eigenfrequencies

integrity and functionality verifications for all convoy facilities in the case of an impact of a military aircraft have been performed.

This raises the question of whether these facilities could also dissipate the induced vibrations caused by the impact of a heavy commercial aircraft. To the knowledge of the author, only preliminary examinations have been carried out for certain examples.

To identify the principle influences and tendencies the following simplified examinations are carried out. First of all a linear response spectrum of the load function for a commercial aircraft is determined and compared to the response spectrum of the load function according to the RSK-LL. The time-dependent load is transformed into the frequency domain. This shows which systems are excited in particular and to which loads they are exposed.

Figure 15.8 illustrates this principle. To determine the response spectra a time variant load is applied to a multitude of single mass systems, characterized by their respective eigenfrequency, and the maximum dynamic system response is registered and plotted over the eigenfrequency.

The results of the comparison between the load function according to Fig. 14.20 a Boeing 747 and the function according to the RSK-LL as given in Fig. 14.5 are shown in the diagram in Fig. 15.9. For the function according to the RSK-LL the resulting effective load as an equivalent static load is approx. 200 MN at a frequency of 32 Hz. The load function of the Boeing therefore particularly excites structures in the low frequency domain. For frequencies between 5 and 10 Hz this load yields values approx. 5 times higher. This can be a deciding factor with regard to induced vibrations and the consequences on the integrity and functionality of components and assemblies.

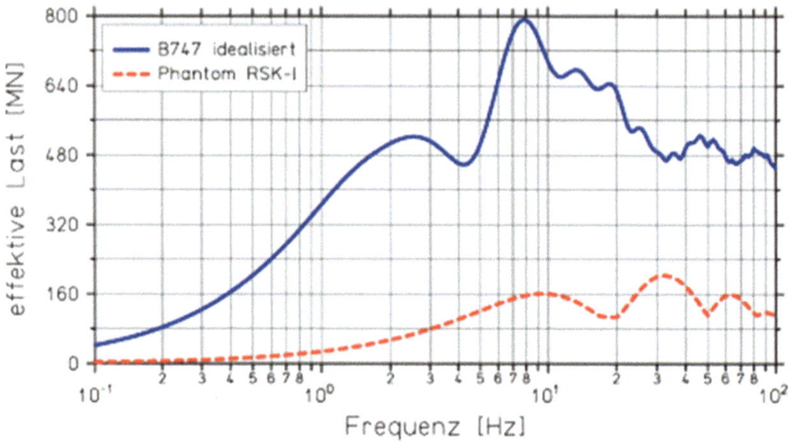

Fig. 15.9 Comparison of the undamped response spectra of the load function: Idealized load functions for B747 versus Phantom RSK-LL

The statements made above are universally valid and not specific to one specific system. It remains to be clarified whether the comparison of the results are as clear for real structures. To answer this question, a simplified 3D computational model of a convoy containment in accordance with Fig. 15.10 is loaded with the load function for the Boeing 747 and also with the function according to RSK-LL. Then the time response of the accelerations at different locations within the secondary shielding is calculated. Subsequently floor response spectra are established for different damping ratios using these time responses.

As an example Fig. 15.11 compares a chosen spectrum for a typical damping ratio of 4 %, which describes the situation for the interior at a height of approx. 20 m. Depending on the frequency range the design accelerations are up to four times higher for an excitation due to a commercial aircraft than for the Phantom according to the RSK-LL, in extreme cases up to eight times higher. In order to make a reliable statement about the protection status of German nuclear power plants, especially of convoy facilities, continued research is necessary on the effects on component integrity, an important element in the chain of evidence. It should be noted however that the first approximations made here lies on the safe side. The calculations are based on a simplified model and assume a linear system. Due to the local destruction (formation of cracks) in the impact area, however, a significant portion of the energy is consumed and not transferred into the building, such that with respect to the induced vibrations in a realistic non-linear examination of the impact area a reduction of the floor response spectra is to be expected.

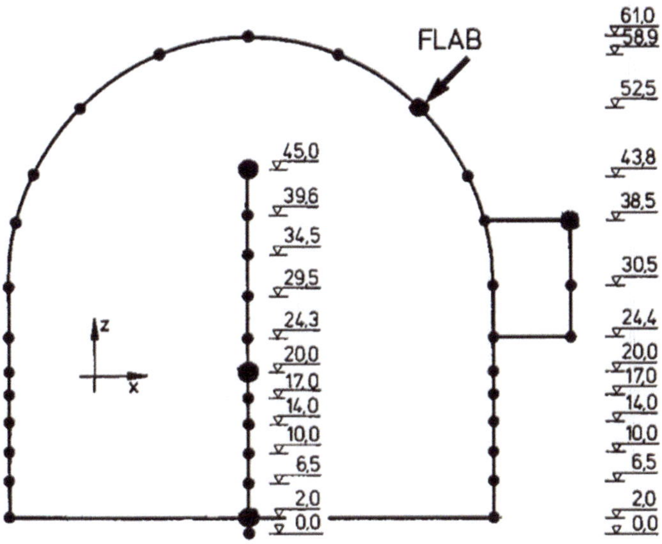

Fig. 15.10 Symbolic representation of a simple computational model for a typical convoy containment for the calculation of induced vibrations caused by an airplane crash

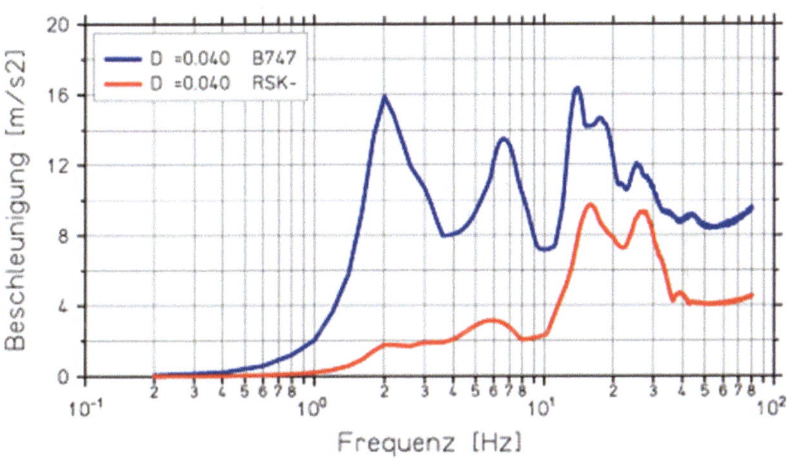

Fig. 15.11 Floor response spectrum with 4 % damping, interior height 20 m

References

1. Institut für Bautechnik (1974) Richtlinien für die Bemessung von Stahlbetonbauteilen von Kernkraftwerken für außergewöhnliche äußere Belastungen (Erdbeben, äußere Explosion, Flugzeugabsturz). Berlin
2. KTA 2202.1 (1980) Schutz von Kernkraftwerken gegen Flugzeugabsturz. Grundsätze und Lastannahmen. Regelentwurfsvorschlag, Version March 1980

3. KTA 2203 (1983) Schutz von Kernkraftwerken gegen Flugzeugabsturz. Auslegung der baulichen Anlagen (bei vorgegebenen Lastannahmen). Regelentwurfsvorschlag
4. DIN 25449 (2007) 2007-11 Bauteile aus Stahl- und Spannbeton in kerntechnischen Anlagen – Sicherheitskonzept, Einwirkungen, Bemessung und Konstruktion. Beuth Verlag, Berlin
5. Schlüter F-H (1987) Dicke Stahlbetonplatten unter stoßartiger Belastung – Flugzeugabsturz. Dissertation Universität Karlsruhe
6. CEB (1988) Concrete structures under impact and impulsive loading. Comite Euro-International du Beton, Synthesis Report. Bulletin d'Information No. 187, Dubrovnik
7. Rambach J-M, Orbovic N, Tarallo F (2011) IRIS_2010 – Part I: General overview of the benchmark. SMiRT-21 Transactions, New Dehli, India, SMiRT-21
8. Vepsa A, Saarenheimo A, Tarallo F, Rambach J-M, Orbovic N (2011) IRIS_2010 – Part II: Experimental data", SMiRT-21 Transactions, New Delhi, India, SMiRT-21
9. KTA 2201.3 (1990) Auslegung von Kernkraftwerken gegen seismische Einwirkungen. Teil 3: Auslegung der baulichen Anlagen. Regelentwurf
10. KTA 2201.4 (1990) Auslegung von Kernkraftwerken gegen seismische Einwirkungen. Teil 4: Anforderungen an Verfahren zum Nachweis der Erdbebensicherheit von maschinen- und elektronischen Anlagenteilen. Regel.

Chapter 16
Special Cases

Abstract Special cases like engine impacts, the effects of flying wreckage or small aircraft and debris have to be considered in the design. Jet fuel fire may also cause damage and must be regarded.

16.1 Engine Impact

The impact of an individual jet engine as a side effect of an airplane crash represents a special case and must be considered separately. However, the effect of the impacting engine is already an integral part of the respective impact load-time function of a specified aircraft and is therefore included in the calculations of the local and structural behaviour.

In addition to the local and global structural behaviour due to an aircraft impact the question must be asked whether an engine can penetrate the reinforced concrete shell of the reactor building locally. Aspects such as vibrations do not play a role in connection to this. The topic of the possibility of perforation has already been extensively examined by a variety of sources in the past (cf. e.g. [1–5]), in large part using information gained from military expertise.

A turbojet engine of the type GE-J79 built by General Electric (cf. Fig. 16.1) is installed in the Phantom and many other military aircraft. Within the framework of the experiments carried out at Sandia the effects of this engine were analyzed in detail. For example an impact load-time function as shown in Fig. 16.2 was developed. In addition appropriate perforation formulas were tested and evaluated. Figure 16.3 for example shows the results of the necessary penetration thickness that the author has compiled in conjunction with other projects. The results of the examinations show that a 1.80 m thick steel reinforced concrete wall cannot be penetrated by an engine.

G. Kessler et al., *The Risks of Nuclear Energy Technology*, Science Policy Reports,
DOI 10.1007/978-3-642-55116-1_16,

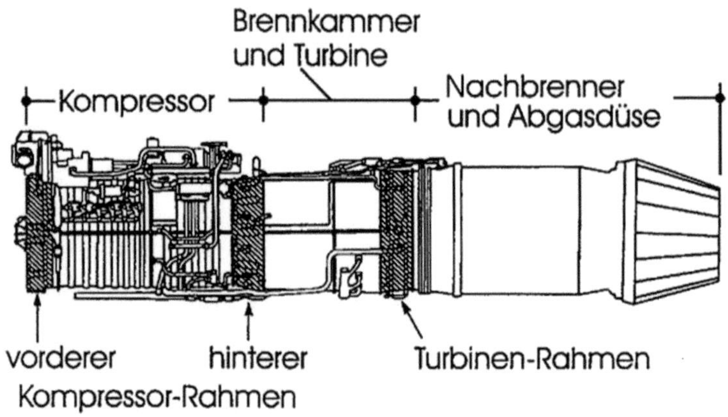

Fig. 16.1 Jet engine GE-J79 of a Phantom (Sandia Test)

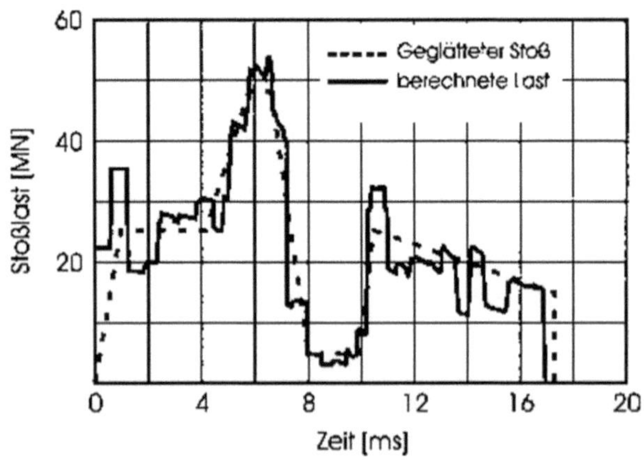

Fig. 16.2 Impact load-time function of a GE-J79 engine; function measured and calculated within the framework of the examinations at Sandia (cf. [14] of Chap. 14)

Commercial aircraft use a diverse variety of jet engines. However a comparison shows that jet engines currently in use have a much lower mass-to-volume and mass-to-sectional-area ratio than the GE-J79, the jet engine used in the Phantom. As an example Fig. 16.4 shows the jet engine of a Boeing 747. Using the established sources it was determined that these engines are also not able to penetrate a 1.80 m thick steel reinforced concrete such as those used in convoy containment. This case is not seen as a particular threat potential for the containment but there is danger of penetration in the case of thinner walls.

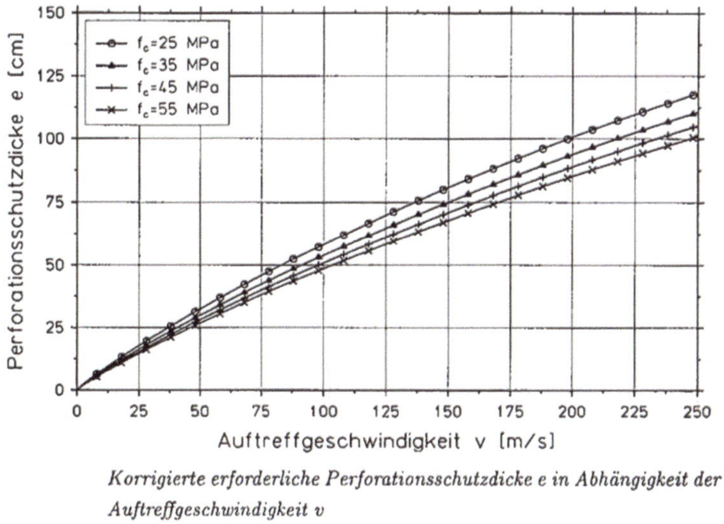

Fig. 16.3 Perforation protection thickness for a deformable projectile (e.g. engine impact)

Fig. 16.4 Jet engine of a Boeing 747

16.2 Wreckage, Small Aircraft and Debris

In principle the same assertions are relevant for wreckage and small aircraft as for the impact of a jet engine. Convoy facilities, built with relatively thick enclosure walls, are able to cope with the effects of impacting wreckage. Penetration is prevented by the massive enclosure. Thinner walls, i.e. smaller than 1.50 m, result in higher penetration probabilities.

The effects of debris loads due to the collapse of structures that do not withstand an aircraft crash and topple over (cf. e.g. Fig. 16.5) are considered in the design of the convoy facilities and therefore do not present a relevant threat potential.

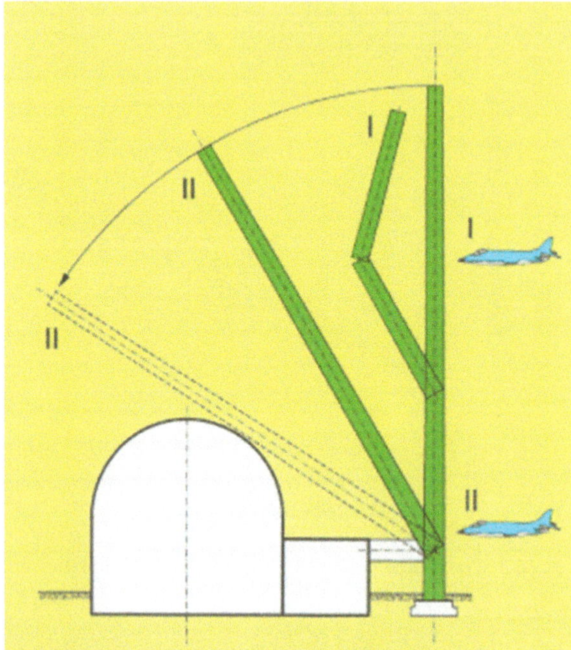

Fig. 16.5 Example of debris loads due to the toppling of a chimney (Sketch by Hochtief IKS)

This statement also holds true for the crash scenarios of heavy commercial aircraft, because it is irrelevant if the debris is created by the crash of a military aircraft or a passenger aircraft.

16.3 Jet Fuel Fire

Military aircraft only carry a comparatively small amount of kerosene that can ignite and combust in a fireball in the case of an impact. The consequences in this case are relatively minor, however. The concrete structures, so long as they are not penetrated, can withstand such a fire without any problem. Precautionary measures are planned for secondary effects. Standards are set forth in the rules and regulations for these measures.

The situation is different in the case of a deliberate forced crash of a large commercial aircraft. The amount of carried jet fuel is considerable. On impact, a large part is atomized and directly combusted. A further part is spread over the site. There is a danger that kerosene pools with significant depths are formed. Depending on the rate of combustion the effects of the fire can last for a significant amount of time up to half an hour. So long as no kerosene enters the interior of a building that is to be protected, however, the concrete structures are able to withstand the effects of the fire. Even if spalling of parts of the concrete structure occurs due to the effects of the fire, the structural stability is not endangered.

The topic of jet fuel fires and their consequences were examined in detail after September 11, 2001. Details cannot be reproduced here. To the knowledge of the author, however, the statement can be made that the physical structures of a convoy system are sufficiently robust and that appropriate measures have been taken for secondary effects such as smoke, etc.

References

1. Sliter GE (1980) Assessment of empirical concrete impact formulas. ASCE J Struct Div 106 (ST5):1023–1045
2. Berriaud C (1981) Survey of current formulas used in nuclear power plant design for impact on concrete. In 6th SMiRT – Post seminar on extreme load design of nuclear plant facilities, Paris
3. Chang WS (1981) Impact of solid missiles on concrete barriers. ASCE J Struct Div 107 (ST2):257–271
4. Degen PP (1980) Perforation of reinforced concrete slabs by rigid missiles. ASCE J Struct Div 106(ST7):1623–1642
5. Sugano T et al (1993) Local damage to reinforced concrete structures caused by impact of aircraft engine missiles. J Nucl Eng Des 140:387–423

Chapter 17
Evaluation of the Security Status of German and Foreign Facilities

Abstract The security status of German and foreign nuclear facilities is discussed. For the actual operating German reactors of type convoy it is not expected that an impacting aircraft—even under the assumption of a large commercial aircraft—would penetrate the shell of the building or that a significant amount of kerosene would enter the interior of the building.

17.1 Security Status of German Reactors

Those facilities that are currently still in service in Germany are designed to withstand the crash of a fast flying military aircraft according to the specifications set forth in the RSK-LL. This results in large wall thicknesses of the enclosure walls of ≥ 1.80 m and very high levels of shear and bending reinforcement in order to ensure local protection against penetration. For example Fig. 17.1 shows the roof of the recently completed vitrification facility in Karlsruhe, which is also designed according to the RSK-LL with regards to aircraft crashes. This example illustrates the high reinforcement density, especially of the shear reinforcement to prevent local punching.

Due to this design of the convoy facilities (cf. Table 12.1) there is a large potential for resistance to the impact of a large commercial aircraft without major damage. Even though the structural design of convoy reactor buildings in Germany is not homogeneous—there are differences in the amount of reinforcement according to the location—the examinations carried out to date show that the reinforced concrete shell, consisting of a dome built on a cylindrical shell, would most likely withstand the impact of a large commercial aircraft in the scenarios presented in this contribution. This also holds true when considering the impact of a jet engine and the effects of debris. Despite cracking and spalling on the front and back as well as plastic deformation of the bending and shear reinforcements the

G. Kessler et al., *The Risks of Nuclear Energy Technology*, Science Policy Reports, DOI 10.1007/978-3-642-55116-1_17,

Fig. 17.1 Reinforcement of the roof of the Vitrification Facility Karlsruhe (Ger.: VEK)—the load case airplane crash is the design basis for the massive reinforcement

integrity is guaranteed. It is not expected that an impacting aircraft would penetrate the shell of the building or that a significant amount of kerosene would enter the interior of the building. The global structural stability of the reactor building, which is designed to withstand earthquakes and the pressure waves of explosions, also seems to be assured according to the hitherto existing examinations.

In addition to the local and global mechanical robustness the induced vibrations must also be examined. The above given examples of the calculations of the response spectra show that their effects can be considerable. The accelerations for the loading of components are expected to be up to four times as high for the impact of a commercial aircraft than for a Phantom. Because the safety of a facility depends on the weakest link in the chain of evidence, a differentiated examination and possibly additional investigations are necessary to address this problem. For the facilities that use the so-called 0.5 g concept, the applicability of the process for a postulated impact of a heavy passenger aircraft should be verified.

The situation is judged to be more critical for the older nuclear power facilities (cf. Table 12.2). The physical structures of the enclosure were not designed for the crash of a Phantom. The thicknesses of the reinforced concrete walls of these facilities are relatively small. Even without detailed investigations it can be said from experience in the design and robustness of such structures that these facilities are not able to withstand the impact of an aircraft without additional measures. Until the end of the decay phase and the transfer of the radioactive inventory into Castor containers there is a threat from an aircraft impact. By installing a supplementary protective superstructure it would however be possible from a technical viewpoint to realize an effective safeguard against an airplane crash in older facilities as well.

17.2 Design of Foreign Reactors

Few nuclear facilities in Europe and worldwide are designed for the impact of a fast flying military aircraft. The European pressurized water reactor (EPR) Olkiluoto 3 currently under construction in Finland is designed for the crash of a large commercial aircraft. Official specifications and details are not common available, however. Switzerland and Belgium adopted the German design standards in the 1980s. In France, where a large amount of nuclear power plants are operating, the design only incorporates small civilian aircraft such as the Lear Jet 23 or the Cesna 210. However now in the design of the EPR-Reactor currently under construction in Flamanville, Normandy, the crash of a military aircraft and a large commercial aircraft are included in the design specifications. The effects of an aircraft impact have only recently been regarded in the construction of new projects around the world. It is not known to the author whether or not induced vibrations are also considered in the design of these facilities. For these reasons, even the nuclear power plants that were shut down last year in Germany have a standard of safety that is comparable or superior to the majority of facilities operating around the world.

Chapter 18
Summary

The presented contribution deals with the design of nuclear power plants in Germany in the case of a postulated aircraft impact. Both the accidental crash of a fast flying military aircraft and the deliberate forced crash of a large commercial aircraft are considered. Requirements, effect types and crash scenarios are discussed. Using examples the development of load approaches are explained. Subsequently the basic procedure for the verification of the structural integrity of the building as well as the determination of induced vibrations is shown. Also briefly discussed are special considerations such as the effect of debris and jet fuel fires. An opinion is delivered to what extent the containments of German nuclear power plants are able to withstand a terrorist attack using a commercial aircraft.

G. Kessler et al., *The Risks of Nuclear Energy Technology*, Science Policy Reports, DOI 10.1007/978-3-642-55116-1_18,

Part III
The RODOS System as an Instance of a European Computer-Based Decision Support System for Emergency Management after Nuclear Accidents

C. Landman, J. Päsler-Sauer, W. Raskob

Chapter 19
Introduction

Abstract The introduction outlines the field of application and the historical evolution of decision support systems for use in nuclear or radiological emergencies.

When in the course of some nuclear or radiological emergency or threat radioactive material is being released, or a release can be expected or has already occurred, radiation protection measures can become necessary or at least seem indicated. These measures aim at preventing deterministic radiation effects and minimizing stochastic effects to a reasonably achievable extent. In this context, decision support systems shall offer a sound knowledge and facts base on which rational, reasonable and verifiable decisions can be taken in a given situation. A further field of application lies in assisting in the preparedness for such situations by allowing the identification and elaboration of appropriate countermeasure strategies in advance, and by providing research and training scenarios.

Assessing the radiological situation in a real emergency requires firstly a diagnosis of the prevailing radiological conditions, and secondly, a forecast of their potential future development, both as fast and reliable as possible under the given circumstances. Formerly, only manual methods were available in terms of blank forms, data tables, diagrams, and computation rules, allowing relevant radiological quantities to be estimated by manual calculations. The necessary material was compiled in textbooks, in Germany for example in a "Compendium for Radiation Protection Consultants" [1] and a "Catalog of Measures" [2]. With the emerging era of computing, the manual calculations were supplemented or replaced by computer codes; however, the latter represented mainly a computer-suited transformation of the manual methods and not a new methodology.

G. Kessler et al., *The Risks of Nuclear Energy Technology*, Science Policy Reports, DOI 10.1007/978-3-642-55116-1_19, © Springer-Verlag Berlin Heidelberg 2014

Nowadays there exist advanced model- and data-based computer systems for assisting decision makers. They differ from those basing on the manual methodology not only in the use of more complex models for assessing all relevant phenomena and quantities, but in particular in their possibility to couple up to emission and pollution measurements from monitoring networks or stationary or mobile local stations. Some systems are devoted only to specific topics and tasks. In Germany, for example, the fields of emergency management and radiation protection are covered by distinct systems, the remote surveillance systems for nuclear reactors operated by the German federal states (KFÜs) [3], and the integrated measurement and information system (IMIS) [4] of the federal government, respectively. There are also comprehensive systems that cover all radiological protection and recovery aspects following a nuclear accident, for example the European systems RODOS [5] and ARGOS [6], or the American emergency response consequence assessment tool RASCAL of the U.S. Nuclear Regulatory Commission [7].

Our article presents the Real-time On-line Decision Support System RODOS which was developed as consequence of the Chernobyl disaster in 1986 for off-site emergency management after nuclear accidents for use in national emergency centers in Europe. In this context the article also gives insight into the importance and application of computer-based decision support systems in general and into concepts and recent developments in German and European radiological emergency management.

Knowledge of basic radiological concepts such as radioactivity, radiation and dose, is assumed. To facilitate the understanding for readers not familiar with the subject, the following chapter summarizes relevant radiological phenomena, the fundamentals of radiological emergency management, and the modeling of the radiological situation in computer programs. Topics as the data requirements of the models and the actual availability of data in the different phases of an accident and the respective uncertainties are also addressed.

All descriptions in the article refer to the assumption of some hypothetical accidental release of radioactive material. Statements about the probability of releases from a given facility or specific types of facilities are not made and not intended.

References

1. Berichte der Strahlenschutzkommission (SSK) (2004) des Bundesministeriums für Umwelt, Naturschutz und Reaktorsicherheit, Heft 37, Leitfaden für den Fachberater Strahlenschutz der Katastrophenschutzleitung bei kerntechnischen Notfällen. Elsevier Urban & Fischer
2. Berichte der Strahlenschutzkommission (SSK) (2010) des Bundesministeriums für Umwelt, Naturschutz und Reaktorsicherheit, Heft 60, auch als CD, Übersicht über Maßnahmen zur Verringerung der Strahlenexposition nach Ereignissen mit nicht unerheblichen radiologischen Auswirkungen (Maßnahmenkatalog). H. Hoffmann GmbH, Berlin
3. Eberbach F (1993) Possible contributions of the KFÜ systems to decisions for off-site emergency management. Radiat Prot Dosimetry 50:107–112

4. Weiss W, Leeb H (1993) IMIS – the German integrated radioactivity information and Decision support system. Radiat Prot Dosimetry 50:163–170
5. RODOS Homepage: http://www.rodos.fzk.de
6. ARGOS Homepage: http://www.pdc.dk/Argos/decision.asp
7. RASCAL 4.2 Summary: https://rsicc.ornl.gov/codes/ccc/ccc7/ccc-783.html

Chapter 20
Relevant Radiological Phenomena, Fundamentals of Radiological Emergency Management, Modeling of Radiological Situation

Abstract The chapter summarizes relevant radiological phenomena, the fundamentals of radiological emergency management, and the modeling of the radiological situation in computer programs. Topics as the data requirements of the models and the actual availability of data in the different phases of an accident and the respective uncertainties are also addressed. The chapter is mainly intended for readers without deeper familiarity with the respective scientific field.

For the emergency management in case of a nuclear or radiological accident, radioactive releases into the atmosphere are the primary point of interest as they can jeopardize the population quickly by external exposure and by inhalation, and also may lead to a rapid contamination of food products. Subject of the present chapter are the "terrestrial exposure pathways", namely the exposure of people by airborne radioactive substances or those deposited on the ground or on other surfaces. The "aquatic pathways" resulting from direct or indirect contamination of rivers, lakes, and coastal waters are described in the chapter on "The Hydrological Model Chain."

The descriptions of the relevant radiological phenomena and of the fundamentals of radiological emergency management are oriented on the German "Radiological Principles of Decisions about Measures Protecting the Population against Accidental Releases of Radionuclides" [1] that contain also many further details and references to other literature.

On the internet, searching e.g. for "Dictionary of Radiation Terms" yields several sites that provide reliable and understandable information and links to other sites related to the topics described in the current chapter, for example from the U.S. Department of Health and Human Services.[1]

---

[1] http://www.remm.nlm.gov/dictionary.htm

G. Kessler et al., *The Risks of Nuclear Energy Technology*, Science Policy Reports,
DOI 10.1007/978-3-642-55116-1_20, © Springer-Verlag Berlin Heidelberg 2014

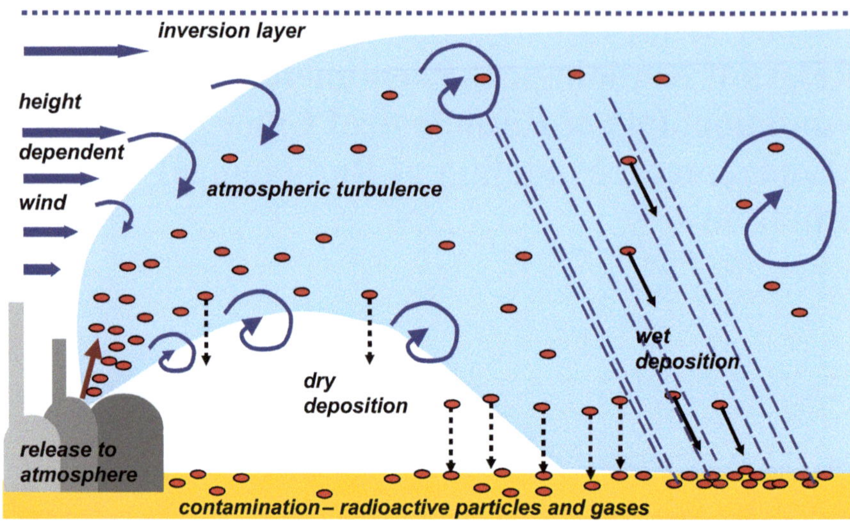

Fig. 20.1 Release, atmospheric dispersion and deposition, and resulting contamination

20.1 From Atmospheric Radioactivity Releases to Human Radiation Exposure

Atmospheric dispersion begins with the release of gaseous and aerosol-type radioactive fission products from a reactor. Figure 20.1 illustrates the processes leading to a radioactive contamination of the air and of surfaces, and Fig. 20.2 the resulting radiation exposure from terrestrial exposure pathways. The term "radiological situation" comprises the contamination of air and surfaces as well as the resulting radiation fields and doses.

Depending on the type of accident, there may be controlled releases through monitored openings that are designed for such purpose and typically equipped with filters, or uncontrolled releases from other parts of a building. In any case, the released volumes come under the influence of external atmospheric processes: they follow the mean wind flow (advection) and get mixed with the outside air by turbulent diffusion. The final outcome is an increasing expansion of the cloud during the transport, and a corresponding dilution of the nuclide concentration in the cloud. The vertical spreading and mixing of the cloud is limited by the inversion layer that sets a limit to the mixing layer. If the released volumes have high temperature, the cloud will rise until equilibrium with the surrounding air has been achieved.

Unlike the concentration of chemical tracers in air, which is quantified as "microgram per cubic meter", the concentration of radionuclides in air is specified as "decays per second per cubic meter," Bq/m^3, because the biological effect of radiation is caused by the nuclear disintegrations. The total number of

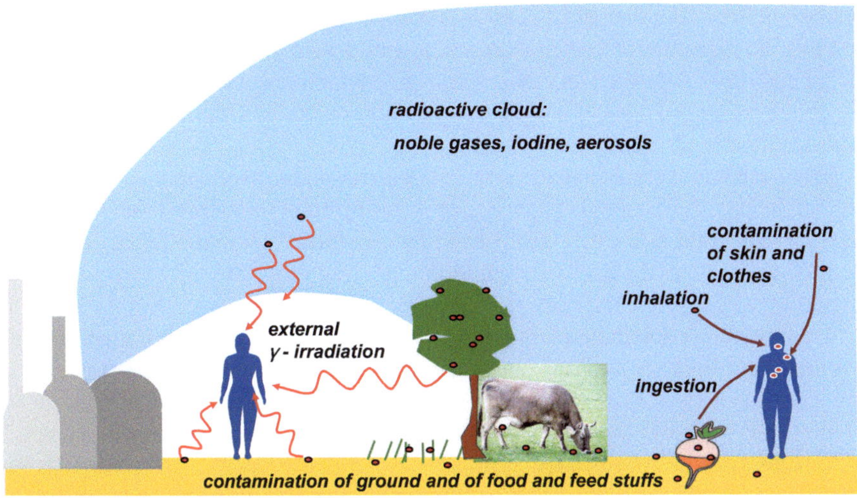

Fig. 20.2 Radiation exposure of humans, animals, and vegetation, by terrestrial exposure pathways

disintegrations as the radioactive cloud passes through a specific volume element for a given time period is referred to as the time-integrated nuclide concentration, Bq s/m^3.

The time-integrated concentration in air close to the ground determines the contamination of surfaces by dry deposition and the inhalation of radioactive material from the cloud by humans and animals; cf. Figs. 20.1 and 20.2. However, it is in particular the wet deposition resulting from the cloud being washed out by precipitation that can considerably increase the contamination of surfaces and thus lead to a much higher longer-term external γ-radiation exposure from the ground and contamination of food and feed stuffs; cf. Figs. 20.1 and 20.2. Deposition processes are specific to nuclide groups: noble gases are not deposited at all while aerosols and iodine gas (elemental or bound organically) are subject both to dry and wet deposition.

External γ-radiation from the cloud and from contaminated surfaces is partly absorbed by the human body, and results in organ doses.

Skin contamination of humans arises from radionuclide deposition onto the uncovered (bare) skin. Especially wet deposition of beta emitters on the skin can cause significant doses to the skin itself. Contamination of both skin and clothes causes γ-exposure of the skin and the whole body. α-particles, however, have such a short depth of penetration that they will not reach the radiation-sensitive layer of the skin, thus not giving rise to a relevant skin or organ dose.

Inhalation of iodine and aerosols from the air leads to an uptake of radionuclides into the body and subsequently to organ doses by internal α-, β-, and γ-radiation. A well-known example is the internal β-exposure of the thyroid by inhaled radioactive iodine nuclides. Internal α-, β-, and γ-exposures and subsequent organ doses also follows the ingestion of contaminated food and drinking water.

Inhalation of deposited and then resuspended nuclides plays a secondary role in moderate climate zones, such as Central Europe, unless the released material consists mainly of α-emitters.

Once radioactive substances have entered the body, they become excreted or stay in the blood stream or organs or tissues for different periods of time. As long as they remain in the body they give rise to the so called "committed dose" which is defined as the dose for adults and children resulting from internal exposure over a period of 50 and 70 years, respectively.

Doses evaluated without taking into account any shielding effects or actions for avoiding or mitigating exposure are referred to as a "potential doses." For instance, the dose calculated from external γ-radiation, inhalation, and contamination of clothes under the assumption of permanent unprotected residence in the open air is a "potential dose" (also called "open air dose)". Actual doses may differ from the potential ones; this depends on the surroundings and thus on the lifestyle habits of the considered individuals. Doses derived by assuming some representative average lifestyle are called "normal living doses".

The total dose rate that results from γ-exposure from the cloud and from the ground or other contaminated surfaces is referred to as the "local (γ-)dose rate"; it is mostly expressed in terms of micro Sievert or nano Sievert per hour (μSv/h $= 10^{-6}$ Sv/h and nSv/h $= 10^{-9}$ Sv/h, respectively). The German Federal Office for Radiation Protection operates an automatic local γ-dose rate measuring network as part of the German IMIS system with approximately 2,000 measuring points. According to this measuring network, the average natural local γ-dose rate in Germany is about 0.1 μSv/h, roughly corresponding to an annual dose of 1 mSv.

Significant deviation from the average natural background level may be the first indicator that a nuclear accident is going on. In case of a real accident, a comparison of readings from stationary networks or mobile monitors with computed local γ-dose rate values allows conclusions about the source term and the real atmospheric transport situation.

20.2 Effects on Health from Radiation Exposure

This chapter briefly recapitulates the phenomena and terms in connection with the possible effects on health from radiation exposure. Information about radiation damage also for non-experts can be found for example on web sites of national health offices.[2]

[2] In German see e.g. the web site of the German Federal Office for Radiation Protection, http://www.bfs.de/. In English see e.g. the web site of the American CDC Centers for Disease Control and Prevention, http://www.bt.cdc.gov/radiation/glossary.asp

The energy of ionizing radiation impacting on living creatures can be absorbed partly or completely inside the cells. It is primarily the absorption by ionization processes in the cell's molecules, especially in the deoxyribonucleic acid (DNA) that causes biological radiation effects. Many DNA defects are fully repaired by endogenous repair mechanisms. If the repair turns out to be faulty, however, there may be permanent changes in the DNA (mutations) which can lead to health detriments after a latency period of years or even decades. If cell regeneration is impossible at all or takes too long because the damage is too severe or too extended, there may be cell death either right after exposure or somewhat later.

DNA mutations in cells of the body can cause somatic (physical) effects, such as leukemia, tumors or cancer; they affect the individual exposed. Mutations in germ cells cause genetic (hereditary) effects; they affect potential progeny of an exposed individual. With respect to all radiation-induced detriment caused by DNA mutations it is assumed that there is no threshold dose[3] for the manifestation and that the severity does not depend on the dose—an increase in radiation dose will add to the probability of occurrence but not change the type of effect. This type is thus referred to as a "stochastic radiation effect". The risk of damage is expressed as the probability of occurrence per unit dose, with the risk numbers taken from epidemiological studies of larger populations exposed to radiation and dependent on a multitude of parameters, such as gender and age of the individuals.

Acute radiation detriments occur either immediately or within days or weeks[4] after the exposure to high radiation doses; the manifestation can be temporary or permanent. Unlike stochastic radiation effects, such damage can be related to the radiation exposure of the affected individual, and is therefore referred to as a "deterministic radiation effect". As a rule, these effects are the consequences of a massive killing of cells in an organ or tissue; if this exceeds a certain level, the affected organ or tissue will lose its functionality. Once a threshold dose has been exceeded, the severity of the injury increases with the dose, and the injury occurs earlier as the dose increases. According to ICRP-103 [2], (Sect. 3.1), no clinically relevant loss of function as a result of radiation exposure is expected below an absorbed dose of approx. 100 mGy.[5] A survey of deterministic effects and their threshold levels can be found, e.g., in ICRP-103 in Annex A.

[3] Very low threshold levels are under discussion, too. Not finally clarified, this discussion plays no fundamental role in emergency management.

[4] Radiation-induced cataracts manifest only after some years.

[5] The absorbed dose [Gy] is to be used for assessing deterministic effects, and the dose equivalent [Sv] for determining stochastic effects. For loosely ionizing radiation, the numerical values are the same.

20.3 Emergency Management and Emergency Measures

20.3.1 Basics of Emergency Management

Emergency management in case of nuclear or radiological events associated with radionuclide releases comprises acute disaster management as well as preventive and long-term rehabilitation management. Depending on national regulations, both areas can be organized strictly separated or closely connected; however, they are united by two fundamental safety goals:

1. Prevention of severe deterministic health effects in the population, emergency crews and auxiliary teams
2. Minimization of collective and individual risks of stochastic health effects on the basis of the principle of proportionality

Because scientifically supported "hard" threshold levels for stochastic effects are lacking, especially the second objective has been the subject of various considerations and interpretations. For this reason, there are different concepts for planning and implementing emergency management measures in the international literature that have developed historically and are being in use in different countries.

The principles of radiation protection currently being valid require compliance with the following regulations, pertaining to each individual event and to the planned measures, as individual measures and in totality:

- Justification—All positive and negative aspects must be taken into account, i.e. the reduction of the radiation-induced health risk aimed at on the one hand, and on the other hand further health and societal aspects, such as health risks potentially caused by the measures, direct and indirect costs of a measure, public acceptance of a measure, etc.
- Optimization—Taking account of the safety goals as well as the prevailing constraints, such as the availability of resources
- Proportionality—The radiation risk after the measures should be kept "as low as reasonably achievable" (ALARA), accounting also for the impact of the intervention on personal life

20.3.2 Distinction of Accident Phases from the Emergency Management Point of View

Precondition for decisions about emergency measures is the diagnosis of the existing and the prognosis of the future development of the radiological situation. In case of accidental releases of radioactivity into the air, both are essentially determined by the type and quantity of the released radionuclides and the atmospheric dispersion and deposition conditions. From the emergency management point of view, three situations can be distinguished.

- An accident is taking place, but no radioactive substances have been released yet (pre-release phase)
- There are releases going on that influence the acute emergency management in the affected area (acute release phase)
- A release takes or took place, but the atmospheric dispersion and deposition processes are over or no longer significant or relevant (later phase)

Occasionally, the term "early phase" is used for designating the pre-release phase and the acute release phase together.

The pre-release phase begins when the possibility of a relevant release is recognized, and terminates either when the event is brought under control or with the onset of a major release. By definition there have not yet been any releases, thus, estimates of a potential future radiological situation can only base on forecasts of the source term and the meteorological conditions. With respect to the latter, the meteorological data measured at the facility can be used for the near future and the nearby surroundings; otherwise, numerical weather prognosis data are to be employed. In the pre-release phase, preventive measures against an expected release can be carried out in order to exclude any subsequent radiation exposure.

In the acute release phase, the available information are meteorological data measured on site and in the surroundings, prognoses from numerical weather forecast models, and maybe measured source term data from the plant instrumentation (mostly, however, only source term estimates). Moreover, increasingly more measurements of the local γ-dose rate and of soil and food contamination levels will become available as time goes on. In the acute phase, preventive measures should be carried out preferably in areas not situated in the current or predicted transport directions or those not yet reached by the radioactive cloud. Unforeseen changes in the release pattern or in the meteorological conditions can necessitate changes in, or additions to, the measures already initiated. Also, special attention must be given to the radiation protection of emergency crews and auxiliary teams.

The later phase, when the external radiation and inhalation of radioactive material from the cloud and also the deposition processes are over or no longer relevant in the area under consideration, can be divided into a transitional phase and a long-term phase.

At the beginning of the transitional phase, there will be still the necessity for continuing the analyses of the radiological situation that were already started during the release phase but likely not finished because this requires a sufficient number and quality of measurements of the contamination levels in food, drinking water, surfaces, soils, plants, and water bodies. At the end, the necessary data, resources and time will be available for deciding upon the event-related justification and optimization of measures for controlling the exposure of the general population, of subgroups of the population with higher risks, and of the executing radiation protection and rehabilitation workers. It is now the time to decide about changes with regard to measures already undertaken, or about measures that become required in addition (e.g. resettlement), and also about the stepwise cancellation of existing measures. The transitional phase can last from days to several weeks.

The long-term post-accident phase can last up to several years or even decades. It is characterized by a long-term contamination and an associated low but long-lasting radiation exposure of people that can be regarded as an existing exposure situation in the sense of ICRP 103 and ICRP 111 [3]. The question is then how individual, social and economic life can be shaped in the affected areas. This requires a culture of treating radiation protection issues that needs to be communicated with the affected population and economic agents, and the further optimization of measures has to be realized in a social consensus. In most cases, medical surveillance of the population in the contaminated areas is also necessary.

20.3.3 Off-Site Radiation Protection Measures and Their Initiation

Measures for avoiding or reducing radiation doses can take effect only on anticipated doses; once a dose has been received it cannot be undone. This means that the doses underlying the decisions about future measures can only come from prognostic assessments, including extrapolations from existing contamination patterns and exposure histories; they cannot be measured in advance.

At present, a given measure is initiated when the estimated value of a dose quantity, calculated with rules specific for the measure, exceeds a given limit (the "intervention level") or is within a given limiting interval, respectively, also specific for the measure. There are significant international differences in the rules for determining the dose quantity, i.e. in the exposure pathways and time periods to be taken into account, and whether the "projected dose" or the "avertable dose" shall be used. But even if the same concept is in operation, the dose limits may differ between different countries.

Since a couple of years, the International Commission on Radiological Protection has developed a new concept and published it in ICRP-103 which implies that the individual dose in the population, expected from all exposure pathways—food ingestion included and typically integrated over 1 year—should not exceed a given dose level, the so called reference level. Strategies of countermeasures have to be defined and optimized in the planning stage to assure that in an acute event this reference level will not be exceeded. The advantages and disadvantages of this concept and the practical implementation are currently being debated both on national and international levels.

The next two chapters outline the measures which can be taken for avoiding or mitigating off-site radiation exposure of the population in the early and late phases of an accident.

20.3.3.1 Measures in the Early Phase

In the early phase, sheltering and evacuation are measures for protecting the population partially or completely against external irradiation from the passing cloud, against external irradiation from activity deposited onto the ground, and against internal irradiation following inhalation of radioactive material from the cloud. The intake of stable iodine tablets is a measure for blocking the thyroid against inhaled radioactive iodine in cases where the inhalation itself can or could not be prevented e.g. by evacuation before cloud arrival. Various measures addressing the ingestion pathways should also be taken early, for instance, public recommendations not to consume locally produced milk or vegetables in order to prevent the uptake of radioactive substances by ingestion of these food stuffs.

If possible, evacuation, or the distribution and intake of iodine tablets, should be ordered and carried out as a preemptive measure in the pre-release phase. While a release is going on, it may become impossible to carry them out at all, or makes little sense due to uncertainties about the future development of the accident and the weather. The request for staying indoors (with windows closed and ventilation systems turned off) is meaningful as a preventive measure even when a release has already begun.

In Germany, the intervention level for evacuation is 100 mSv, to be applied to the projected effective dose resulting from external cloud gamma exposure during the cloud passage, ground gamma exposure over the first seven days plus the projected effective committed dose from inhalation during cloud passage. The intervention level for sheltering is the projected effective dose of 10 mSv, considering the same exposure pathways and integration times as for evacuation.

Timely evacuation can entirely avoid any radiation exposure of the population. On the other hand, evacuation constitutes a major disruption of normal life, and the risks associated with this measure must not be disregarded in decision making.

For indoor locations, the achievable shielding against external exposure depends on the type of building and the construction materials. As an example, Table 20.1 shows the mean shielding factors derived for open air without surrounding structures and for various types of buildings that are used in the RODOS decision support system; a value of 1 means no shielding effect. Contaminated surfaces in- and outside buildings and of vegetation—in particular trees—in the vicinity of a given location can increase the exposure in comparison with the open-air irradiation, thus giving rise to shielding factors greater than 1.

In Germany, the intervention level for the distribution of iodine tablets is 50 mSv for children under the age of 18 and pregnant women, and 250 mSv for adults, to be compared with the committed dose to the thyroid resulting from the inhalation of radioactive iodine nuclides. Preemptive intake before the inhalation takes place offers 100 % protection; a slightly delayed intake, e.g. 3 h later, still reduces the thyroid dose commitment by 50 %. In iodine deficiency areas, however, this measure can lead to health problems in particular for the elderly, therefore, in

Table 20.1 Mean shielding factors in use in RODOS

Shielding factor against	Open air (without trees or buildings in the vicinity)	Lightweight flat module building	Semi-detached or terraced houses	Multi-storey block building
Ground γ-irradiation	1.0	0.5	0.1	0.01
Cloud γ-irradiation	1.0	0.8	0.5	0.2
Inhalation from the cloud	1.0	0.5	0.5	0.5

many countries there are age limits for the application of this measure; e.g. in Germany where tablets are not distributed if the age exceeds 45 years.

Measures associated with the ingestion pathways can be issued either as non-binding precautionary warnings and recommendations irrespective of any threshold levels, for example, warnings against the intake of locally produced freshly harvested food, or the requests to put grazing cattle off pasture and on uncontaminated feed if possible. Such measures can be taken already in the early phase.

Binding interventions into the supply of food and feed stuffs, however, must be issued on the basis of maximum permitted activity levels, and play a major role only in the later phase after an accident.

20.3.3.2 Late-Phase Measures in Urban and Rural Areas

The primary measure for protecting the general public against external irradiation from activity deposited onto the ground and other surfaces in the post-release phase is relocating the people from the area, either temporarily for weeks or months, or permanently for an indefinite period of time. In general, relocation will be initiated only when measured data for the entire area are available.

Aside from relocation, in the post-release phase there are on the one hand agricultural measures with the aim of mitigating the uptake of radioactivity via contaminated food and drinking water, and on the other hand decontamination and other measures in urban environments with the aim of reducing daily life exposure of the general public.

After the Chernobyl accident in 1986, Europe undertook the effort of systematically identifying and cataloguing all decontamination and rehabilitation measures which could reasonably be applied in Europe following a nuclear or a radiological event [4]. The survey yielded in roughly 100 measures, and a range of typical criteria that allow selecting the most appropriate ones from the list:

- Extension of the contamination in space and time
- Effectiveness of a measure
- Amount and treatment of generated waste

- Radiation exposure of the involved workers
- Damage to the environment
- Costs
- Social and ethical aspects
- Communication with and expectations from the public

For many of the late phase measures other than relocation identified in [4], there is only a limited time window in the order of days or weeks after the accident where they stay efficient, so that the effectiveness of a given measure in general will depend on the delay until carrying-out becomes possible by radiological, logistical or other reasons. Moreover, agricultural measures also depend very much on the time of year when the accident takes place, the types of radionuclides released, the local soil types, and the given food item.

The aim of agricultural measures is to reduce the activity concentration in a given foodstuff below a given limit so that the product can be marketed. Typical limits in the European Union are in the range of a few hundred to several thousand Bq/kg fresh weight; they were introduced after the Chernobyl accident [5, 6] and are under discussion again after the Fukushima accident in 2011.

Measures against the contamination of milk by iodine or cesium, which are typical fission products present in accidental releases from reactors, have to be initiated at a very early stage. When milk cows are grazing outside, the peak contamination of iodine in the milk, for example, occurs already after a few days after the deposition on pasture grass. Apart from banning the consumption, measures can also be taken for reducing the transfer of the contamination from the cows into the milk (by feeding sorbents), or for reducing the contamination of the end-product by processing the milk to cheese or milk powder.

Another important fact with respect to the timing is that it is the effect of direct deposition of activity on existing vegetation that causes high activities in food stuffs and feed stuffs, so consumption bans are mostly needed only in the 1st year. In the following years, the activity is transmitted to the plants only by means of root uptake, which reduces the effectiveness of uptake by at least one or two orders of magnitude [7].

Decontamination measures in urban areas can be subdivided into two categories:

- Measures shielding the population from the contamination
- Measures removing contamination from the area as far as possible

Measures applied in Chernobyl, now under discussion also for Fukushima, include in particular the removal of top soil and the cleaning of paved or solid surfaces, such as roads, sidewalks, walls, and roofs. However, decontamination in urban environments clearly requires always aggregates of measures for the variety of surfaces present in a given area. Depending on the level of urbanization, specific surfaces will contribute most to the dose. In low-density areas and especially in suburbs, for example, the doses of the local population can be dominated by contributions from plants in the surroundings, such as trees, shrubs, and lawns [8].

Decontamination measures in urban areas almost always involve waste problems or a carry-over of contamination. The actually very effective measure of top layer removal, be it lawn or road pavement or roof, for instance, always gives rise to large volumes of waste that must be stored under controlled conditions (i.e. protected from the weather). These measures are very costly, too. Hosing down paved surfaces, on the other hand, is less effective but relatively inexpensive and applicable for large areas. However, it often causes activity to be flushed into the sewer system, thus shifting removed contamination over to sewage treatment plants.

Last but not least, all these actions will not eliminate entirely the contamination from an affected area, as their effectiveness is limited, if not the entire surface together with all buildings is removed. Hence, there is always the necessity to weigh measures which can still be carried out without affecting the quality of life too much towards the extent to which the goals of protecting the population by reducing the dose can be attained.

20.4 Modeling the Radiological Situation (Terrestrial Pathways)

20.4.1 Atmospheric Dispersion Models

The mathematically simplest atmospheric dispersion model is the Gaussian plume model [9]. It can be applied in plain topographies within a range of approximately 20 km under steady state conditions, i.e. a uniform release with a constant rate, geometry, and altitude, and constant atmospheric conditions.

In the Gaussian plume model, the horizontal and vertical concentration profiles of the dispersing plume are modeled by Gaussian distributions, cf. Fig. 20.3. The widths of these distributions are sensitive quantities in the dispersion calculation, because they describe the dilution of radioactive material during transport in the air, with direct consequence for the resulting doses. The widths are described by diffusion parameters that depend on the distance from the origin and on the turbulent state of the atmosphere. There are numerous field experiments in which diffusion parameters were determined [10, 11]. The Gaussian plume model is for example used in the manual methodology for assessing the radiological situation.

Dispersion calculations for variable conditions require models able to take into account the variability in space and time of both release and atmosphere. Mathematical realizations are the so-called "puff models" which decompose the release into time steps, i.e. a continuous release is replaced by a sequence of puff emissions, and describe the nuclide concentrations within the puffs by three-dimensional Gaussian distributions. Each puff is passed on a trajectory step by step through a

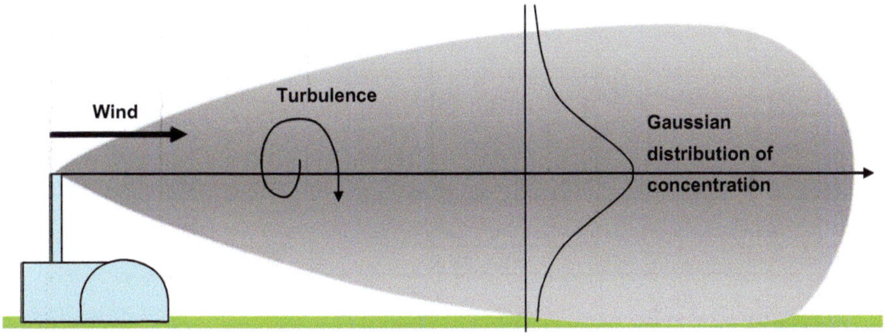

Fig. 20.3 Schematic representation of a Gaussian plume model

spatial 3D grid of time-dependent meteorological data (wind vectors, turbulence). All puffs will be superimposed, thus forming an aggregate plume.

In a Gaussian puff model, the horizontal and vertical concentration profiles of the puffs are modeled by Gaussian distributions. As in the Gaussian plume model, the widths of these distributions are described by distance and turbulence dependent diffusion parameters, cf. Fig. 20.4.

For both Gaussian plume and Gaussian puff models, the thermal rise of a plume or puff, respectively, in releases with thermal energy can be modeled by plume rise formulae, see [12].

In case of rather complex meteorological conditions, such as wind directions highly variable in space, spatially extended puffs can no longer be considered because they will lose spatial resolution. Here, so-called particle models offer a solution: They describe the releases of gases and aerosols by a "mathematical" particle cloud. Each particle is propagated in time steps in accordance with the local wind vector. Additionally a stochastic motion component is added that corresponds to the turbulence (random walk model), cf. Fig. 20.5. In principle, a particle model can deliver highly detailed simulations of dispersion processes in the atmosphere. However, high resolutions in space and time will be achieved only when the calculations are carried out with large numbers of particles and small time steps. This, in turn, requires long computation times. In addition, the quality of the dispersion calculation cannot be any better than that of the input data required, here, the computed three-dimensional time-dependent wind vector and the turbulence and precipitation fields from meteorological prediction models.

The German Weather Service for example operates a Lagrange particle dispersion model, LPDM [13], for dispersion calculations of radioactive substances over long distances.

A fundamentally different approach to calculating atmospheric transport and dispersion is used in Eulerian grid models. They are based on a general equation for the transport of matter in turbulent fluids, the advection—diffusion differential equation. The wind vector fields contained in meteorological data correspond to advection while the turbulence fields correspond to diffusion. Solutions of these

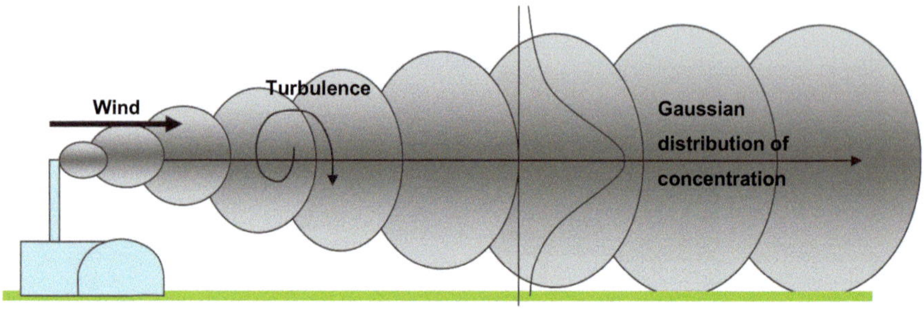

Fig. 20.4 Schematic representation of a Gaussian puff model

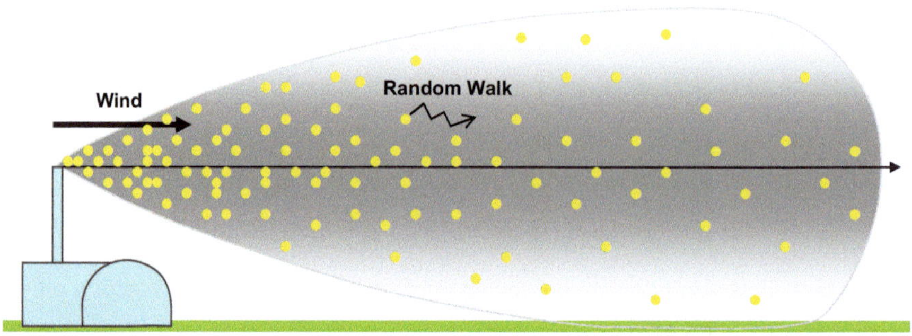

Fig. 20.5 Schematic representation of a particle model

differential equations are time-dependent distributions of concentration in the atmospheric boundary layer. They are computed by codes in time steps on a space grid; the resolution corresponds to the size of the grid cells. Figure 20.6 illustrates the outcome of a computation of this type. Eulerian grid models are frequently used for long distance dispersion calculations (scale 100–2000 km).

20.4.2 Modeling Radionuclide Deposition onto Surfaces

20.4.2.1 Dry Deposition

Dry deposition of airborne aerosols and gases on surfaces is characterized by a deposition velocity v_d. The deposition velocity is the increase of surface contamination over a given time interval divided by the concentration in the air above the surface element. The value of v_d depends on the turbulent transport capacity of the air layers above the surface and on the adhesion properties of the involved materials. For instance, v_d is low for stable atmospheric stratification, smooth surfaces, and aerosols with low binding capacity. On the other hand, v_d is high for pronounced vertical mixing, vegetation surfaces with high roughness lengths, and

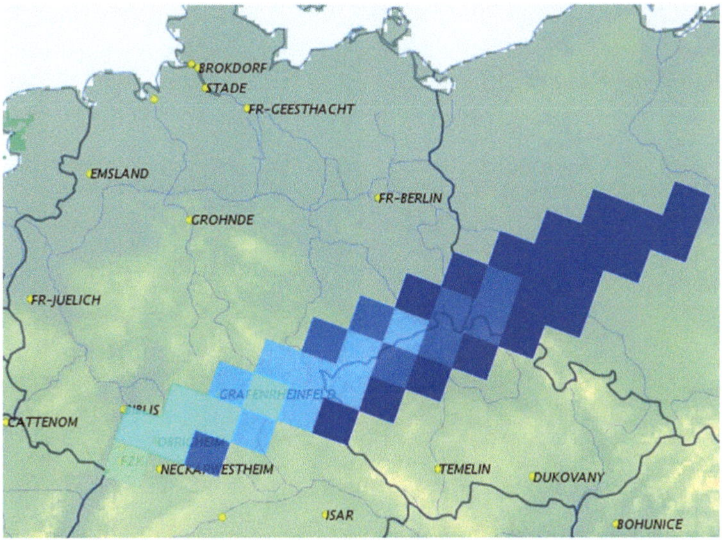

Fig. 20.6 Result of an Eulerian grid model calculation (JRodos screen shot)

gaseous elemental iodine with high binding capacity. For modeling v_d, the prevailing meteorological and flow conditions and data about the land use (urban, agricultural, forestry areas, vegetation, types of plants, leaf sizes, etc.) is taken into account.

Dry deposition on a given plant type is calculated with the time-integrated air concentration close to the plant surface and a deposition rate that depends on plant type and the seasonal development of the plant, making the resulting deposition velocity clearly dependent on the season. The seasonal plant status is described by the current leaf surface index that is defined as the aggregate leaf area per unit area of soil.

The radionuclides are grouped with respect to their deposition properties: noble gases (not depositing), aerosols (including iodine in aerosol form), and gaseous iodine (elemental, organically bound).

20.4.2.2 Wet Deposition

Wet deposition is described by a wash-out model, which assumes that the entire vertical volume of the radioactive cloud is washed out in the area covered by the rain, whereby the radionuclide content gets deposited partly on the ground and partly on vegetation or on other surfaces. The intensity of wash-out is determined by the rain duration and intensity and by the wet deposition characteristics of the nuclides.

When modeling wet deposition on leaf surfaces, account must be taken of the fact that incipient rain initially increases the leaf contamination, but parts of that contamination gets washed off with continuing rain. This process depends on the rain intensity and is modeled by using data describing the retention capacity of leaves for different precipitation classes.

20.4.2.3 Cloud Depletion

Due to mass conservation, deposition of radionuclides from the cloud leads to a depletion of the concentration in the cloud—the material deposited on surfaces is no longer present in the cloud.

20.4.3 Processes and Models for the Transport of Activity Through the Human Food Chain

Airborne activity is deposited onto the ground and onto vegetation parts above ground by dry and wet deposition processes; the latter causing the external contamination of plants, to begin with. A fraction of the deposited activity gets carried into the interior of the plant and transported to other parts of the plant (translocation), also to those underground, while the activity remaining on the above-ground plant surfaces gets removed by weathering processes as time goes on, and will end up finally on the topmost soil layer. The direct deposition onto plants and the processes following afterwards depend significantly on the seasonal development of the plant. This is the reason why the date of an accident plays a crucial role for the ingestion pathways in regions with pronounced seasonal variations of the climate.

Because direct deposition can occur only during the passage of the radioactive cloud, all respective processes act only in the affected plants of the current vegetation and harvesting cycles, respectively, that is, typically within the first year after the deposition event. On the other hand, there are the processes of root uptake (the transport of activity from the soil into the plant via the roots) and resuspension (the repeated external pollution by raised-up radioactive soil particles) that lead to a longer term contamination of plants. As a rule of thumb one may say that the contamination by direct deposition will exceed the contamination by root uptake by one or more orders of magnitude, provided there are any plant parts above ground at the time of the accident.

For the human consumption of plant products, only the contamination of the edible parts plays a role—leaves (salad, vegetables, herbs); fruit and fruit vegetables; roots (e.g. potatoes, carrots); and cereals.

An uptake of activity by animals followed by biological transport and excretion processes leads to a contamination of animal food stuffs—milk and dairy products, meat, and eggs. Activity uptake via feed stuffs is the most important pathway for animal contamination. However, also inhalation from the cloud may play a role for a short period of time, in particular in winter, or on the long-term the inadvertent ingestion of contaminated soil during the grazing period. The processes leading to a contamination of plants in general have already been described, but for forage plants, other plants are relevant (such as grass) or other parts of plants (such as entire corn plants, not only the cobs). Again, the time of accident is very important, on the one hand with respect to the forage plants, and on the other hand, because feeding practices may vary with the season (such as grazing on pasture in summer and keeping the animals indoors on stored feed in winter).

All processes described so far do not only depend on the time of an accident, but also on the food item itself and the type of radionuclides causing the contamination, and on regional climatic and soil conditions and agricultural practices.

Before culinary use, vegetable products for human consumption are washed or peeled (salad, vegetables, herbs, potatoes) and, possibly, converted into a secondary form of product (such as flour from grain) or stored (potatoes, flour, frozen vegetables, herbs). Animal food stuffs also get stored, usually for at least a couple of days prior to consumption, or are converted into secondary products with longer shelf life (e.g. cheese from milk). During storage there is radioactive decay, and the food processing can also influence the contamination levels; therefore, the contamination of a product ready for consumption in general differs from that of the raw initial animal or plant product.

The modeling of terrestrial ingestion pathways may pursue two aims:

1. Estimating the aggregate activity intake of humans resulting from the consumption of specific food items or a representative food basket
2. Estimating the time-dependent contamination in food and feed stuffs as a function of the time of accident

On the whole, models for estimating the aggregate activity supply are less sophisticated. They mainly provide factors for manual calculations that are applicable for stationary or quasi-stationary conditions (e.g. in normal operation of nuclear power plants), or for rough estimates of the activity intake following accidental releases; here, usually two sets of factors are provided, one each for assumed accidents during and outside the vegetation period. Ingestion models of this type are included for example in the RESRAD Family of Codes (http://web.ead.anl.gov/resrad/home2/resrad.cfm) and in the American WASH-1400 Risk Study [14].

Knowledge of the contamination of food and feed stuffs as a function of the time after an accident is required for detailed planning and preparation of agricultural measures as well as for the analysis of the possible effectiveness of the measures. Models developed for such purposes must take into account the dynamic processes described in this chapter. Dynamic approaches decompose the food chain into transport-related components, or "compartments," and describe the seasonal development stage of plants and animals at the time of the accident analytically or by datasets. The transport between the compartments, and their development over time, is calculated on the basis of transfer factors derived from measurements, and biological half-lives. The German model ECOSYS [7] is an example of a dynamic food chain transport model. For obtaining realistic results for a given locale, dynamic models need a region-specific data base containing information about soil types, vegetation periods, agricultural production habits and production rates, and food consumption rates.

20.5 Calculation of Doses for the Terrestrial Exposure Pathways

For deriving pathway-specific doses on the basis of the concentration and contamination fields determined with the atmospheric dispersion models, further steps are necessary for covering the different terrestrial exposure pathways:

- Calculation of the cloud gamma irradiation field and of the resulting dose
- Calculation of the gamma irradiation from the radionuclides deposited by dry and wet deposition processes on urban and agricultural areas (summarily referred to as ground irradiation) and of the resulting dose
- Determining the internal exposure from inhalation of contaminated air from the cloud and of activity resuspended from the ground, and the resulting dose
- Determining the internal exposure from ingestion of contaminated food, and the resulting dose

The different exposure pathways require different model approaches which are described below, separately for "Doses from the Cloud and from Contaminated Surfaces" and "Doses from the Food Chains".

20.5.1 Doses from the Cloud and from Contaminated Surfaces

Figure 20.7 illustrates the different steps for calculating gamma dose rates and doses from the cloud and from the contaminated ground.

An analytical calculation of the gamma-radiation field from the cloud requires sophisticated mathematical solutions of complicated integral equations and is therefore often replaced by approximations or interpolations between data tables.

Calculating gamma-irradiation from the ground is easier; most models assume "(ideal) lawn" as a reference surface and the geometry of a plane and infinite contaminated surface. Any contaminated structures (trees, buildings) deviating from the plane surface are taken into account by local factors.

The transit from radiation fields to absorbed doses (unit: Gray = Gy = Joule/kg) and to biological equivalent doses (unit: Sievert = Sv) requires the modeling of the physical absorption processes and the biological effects in the organs of the human body for the nuclides involved. The results of such model calculations are then

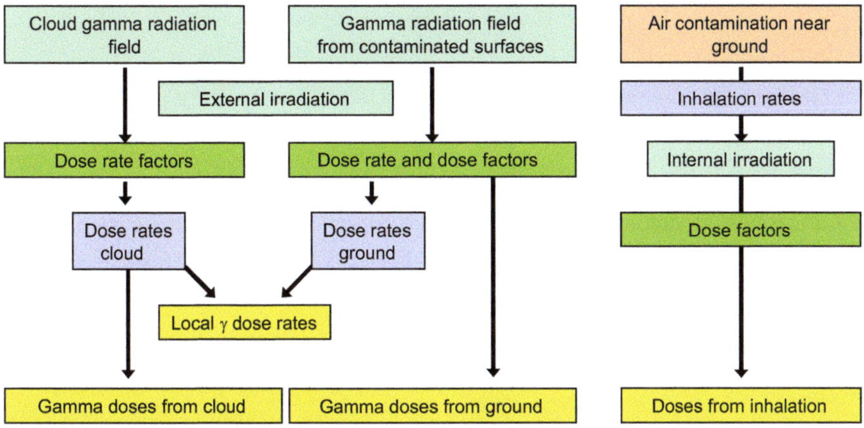

Fig. 20.7 Scheme for calculating dose rates and doses (as function of space and time, specific for organs and nuclides)

summarized as so-called "dose factors" and are available from various radiation protection institutions. With such dose factors, the ground gamma dose rate for example can be obtained with the following simple equation:

$$\text{Dose rate}\ \ (\text{Sv/h}) = \text{Dose rate factor}\ \ (\text{Sv/h})/\left(\text{Bq/m}^2\right) \\ \times \text{soil contamination}\ \ \left(\text{Bq/m}^2\right).$$

If the soil contamination is known, the dose rate factor allows direct calculation of the dose rate. The dose rate factor for cesium-137, e.g., is approximately 2.5×10^{-12} (Sv/h)/(Bq/m^2). A soil contamination of 10^6 Bq/m^2 of Cs-137 then results in an exposure rate of 2.5 μSv/h.

The local dose rate is calculated as the sum of the dose rate from the soil and the dose rate from the cloud near to the ground.

Besides the dose rate factors, there are also dose factors for ground irradiation. They are used for calculating doses over longer exposure times, i.e. the time spent on the contaminated ground surface. Ground dose rate and dose factors are nuclide-specific and include radioactive decay, ground dose factors also the absorption resulting from radioactive nuclides penetrating into the soil.

The uptake of radionuclides by inhalation of air from the cloud near ground or from resuspension depends on the breathing rate and the inhalation time period. Once in the lungs, the complex transport of the radionuclides within the body, the effects of the internal irradiation in the body organs, and the resulting dose commitment is calculated with complicated models and summarized as nuclide- and organ dependent dose factors for various age groups. The organ doses are then obtained by multiplying the activity intake (Bq) by inhalation with the organ-specific dose inhalation dose factor (Sv/Bq).

20.5.2 Doses from the Food Chain

The activity that enters the human body as a result of consuming a given food stuff over a given period of time depends on the food consumption rate (e.g. in the unit of gram/day), the specific activity in the food stuff (e.g. Bq/g) as a function of time, and the ingestion time period (e.g. days). As for inhalation, the resulting dose commitment is calculated with complicated models and summarized as nuclide- and organ dependent dose factors for various age groups. The organ doses are then obtained by multiplying the activity intake (Bq) by ingestion with the organ-specific ingestion dose factor (Sv/Bq).

The basic problem in determining a realistic ingestion dose for a specific individual in industrialized societies lies in the ever-changing distribution pattern between the food production origin and the consumer—consuming food produced locally is the exception rather than the rule. This implies that a realistic calculation of the aggregate individual ingestion dose on the basis of a representative food basket makes sense only for locally autonomous self-suppliers.

For these reasons, individual ingestion doses are usually calculated only for selected food stuffs on the assumption of local production and consumption, and doses for an individual representing the average population are usually not quantified.

The aggregate collective dose from ingestion is determined usually on the assumption that all food produced somewhere is also consumed somewhere else. Then, the collective dose for a given food stuff can be calculated by multiplying the local production rate for the foodstuff taken from a database and the contamination level in the food stuff calculated by a food chain transport model, and summing up over all locations and time periods considered.

References

1. Berichte der Strahlenschutzkommission (SSK) des Bundesministeriums für Umwelt, Naturschutz und Reaktorsicherheit, Heft 24 (2000) Radiologische Grundlagen für Entscheidungen über Maßnahmen zum Schutz der Bevölkerung bei unfallbedingten Freisetzungen von Radionukliden. München, Urban & Fischer, München · Jena
2. Annals of the ICRP, Publication 103 (2007) The 2007 recommendations of the International commission on radiological protection. Elsevier
3. Annals of the ICRP, Publication 111 (2008) Application of the commission's recommendations to the protection of people living in long-term contaminated areas after a nuclear accident or a radiation emergency. Elsevier
4. Nisbet AF et al (2010) Decision aiding handbooks for managing contaminated food production systems, drinking water and inhabited areas in Europe. In Raskob W, Hugon M (eds) Enhancing nuclear and radiological emergency management and rehabilitation: Key results of the EURANOS European project. Radioprotection 45(5 Supplément): S23–S38
5. Council Regulation (Euratom) No 3954/87 of 22/12/1987 laying down maximum permitted levels of radioactive contamination of foodststuffs and of feedingstuffs following a nuclear

accident or any other case of radiological emergency. Official Journal of the European Communities L146 of 30/12/1987, Luxembourg
6. Council Regulation (Euratom) No 2218/89 of 18/7/1989 amending Regulation (Euratom) No 3954/87 laying down maximum permitted levels of radioactive contamination of foodstuffs and of feedingstuffs following a nuclear accident or any other case of radiological emergency, Official Journal of the European Communities L211 of 22/7/1989
7. Müller H, Pröhl G (1993) ECOSYS-87: a dynamic model for assessing radiological consequences of nuclear accidents. Health Phys 64(3):232–252
8. Charnock T W (2010) The European model for inhabited areas (ERMIN) – developing a description of the urban environment. In Raskob W, Hugon M (eds) Enhancing nuclear and radiological emergency management and rehabilitation: Key results of the EURANOS European project. Radioprotection 45(5 Supplément):S55–S62
9. Veröffentlichungen der Strahlenschutzkommission des Bundesministeriums für Umwelt, Naturschutz und Reaktorsicherheit, Band 17 (1992) Modelle, Annahmen und Daten mit Erläuterungen zur Berechnung der Strahlenexposition bei der Ableitung radioaktiver Stoffe mit Luft oder Wasser zum Nachweis der Einhaltung der Dosisgrenzwerte nach § 45 StrlSchV. Fischer Verl, Stuttgart
10. Thomas P, Dilger H, Hübschmann W, Schüttelkopf H, Vogt S (1981) Experimental determination of the atmospheric dispersion parameters at the Karlsruhe Nuclear Research Centre for 60 m and 100 m emission heights, Part 1: Measured Data. Kernforschungszentrum Karlsruhe, KfK-Bericht 3090
11. Thomas P, Nester K (1981) Experimental determination of the atmospheric dispersion parameters at the Karlsruhe Nuclear Research Centre for 60 m and 100 m emission heights, Part 2: Evaluation of Measurements. Kernforschungszentrum Karlsruhe, KfK-Bericht 3091
12. Briggs GA (1975) Plume rise predictions. In: Haugen DA (ed) Lectures on air pollution and environmental impact analysis. Workshop proceedings. AMS, Boston, MA
13. Glaab H, Fay B, Jacobsen I (1998) Evaluation of the emergency dispersion model at the Deutscher Wetterdienst using ETEX data. Atmos Environ 32(24): 4359–4366
14. USNRC (U.S. Nuclear Regulatory Commission) (1975) Reactor safety study, Appendix VI. WASH-1400 (NUREG-75/014)

Chapter 21
The Decision Support System RODOS

Abstract The chapter begins with an outline of the historical development from the first UNIX-based RODOS system until the most recent Java-based version JRodos. This is followed by an overview of the models contained in RODOS, and a description of the RODOS Center in Germany, where RODOS operates since 2005 at a central location for use by the federal government and the federal states.

21.1 History

The Chernobyl reactor accident on April 26, 1986, showed how badly European countries were prepared for an emergency like this. Assessments of the radiological situation and decision-making about actions for protecting the public were partly determined by actionism or inadequate, often simply due to a lack of standardized and reliable information. Furthermore, there was no cross-border coordination of emergency measures. These reasons finally led to the development of the decision support system RODOS[1] that combines all relevant data, produces diagnoses and prognoses, and compares the efficiencies of various measures. In addition to the objective of assisting decision makers and consultants in case of an emergency with often tremendous time pressure and great psychological stress, the system also was to be used for training and education in radiological and emergency-related issues.

In 1988, the development started at the former Institute of Neutron Physics and Reactor Engineering of the Karlsruhe Research Center, which is now the Institute of Nuclear Technology and Energy Technology of the Karlsruhe Institute of Technology. Until late 1998, the German Federal Ministry for Environment, Nature Conservation and Nuclear Safety funded the RODOS/RESY subsystem, whose functionality was restricted to emergency measures in the vicinity of nuclear

[1] RODOS: Real-time On-line Decision Support.

G. Kessler et al., *The Risks of Nuclear Energy Technology*, Science Policy Reports,
DOI 10.1007/978-3-642-55116-1_21, © Springer-Verlag Berlin Heidelberg 2014

facilities. The first operational version, RODOS/RESY PV3.0, became available in 1997 [1].

Since 1990 and in parallel the development of the comprehensive RODOS real-time on-line decision support system was advancing, promoted by the European Commission. This system contains RODOS/RESY as an integral component and has additionally been designed for large-area consequences and later accident phases [2]. The first fully functional operational version, PV4.0F, was issued in late 2000 [3].

One focal point of the EURANOS[2] project as part of the European Commission's 6th Framework Programme in 2003–2008 [4] consisted in improving RODOS with respect to contents, user-friendliness and facilitated maintenance, including the adaptation to national conditions, and in creating the possibility to operate the system under modern information technology platforms. Suggestions and requests from users became increasingly more important. The new demands resulted first in a novel user interface[3] and then in a complete redesign of the entire operating software,[4] which took shape in 2009 with the first Java-based JRodos version. JRodos [5] was accepted very well in the RODOS community. Since late 2010 it is the basis for all further developments and the final version RODOS PV7.0 for HP-UX and Linux is only maintained. At present, JRodos can run under Microsoft Windows, Linux, and Mac OS.

In Germany, RODOS operates since 2005 at a central location for use by the federal government and the federal states; see chapter "The RODOS Center in Germany." Moreover, it is currently (date 2013) operational in several national emergency centers (Bulgaria, Croatia, Czech Republic, Finland, Poland, Portugal, Slovakia, Slovenia, Spain, Russia); it will soon become operational in the Netherlands, Switzerland, and Ukraine. In addition, institutions like universities and local municipal organizations are applying the system on their own initiative in Europe and overseas. The adaptation of the system to local conditions is described in the chapter on "Adapting to National Conditions".

21.2 Overview of the Models Contained in RODOS

Besides the core models referred to in the text, the system also contains further models for specific applications not described in this article. When in the text explicit reference is made to "JRodos," and not to "RODOS," this means that the mentioned models are contained exclusively in the developing JRodos version and no longer in the HP-UX and Linux versions.

---

[2] EURANOS: European Approach to Nuclear and Radiological Emergency Management and Rehabilitation Strategies.

[3] cf. [4], pp. 171–179.

[4] cf. [4], pp. 181–189.

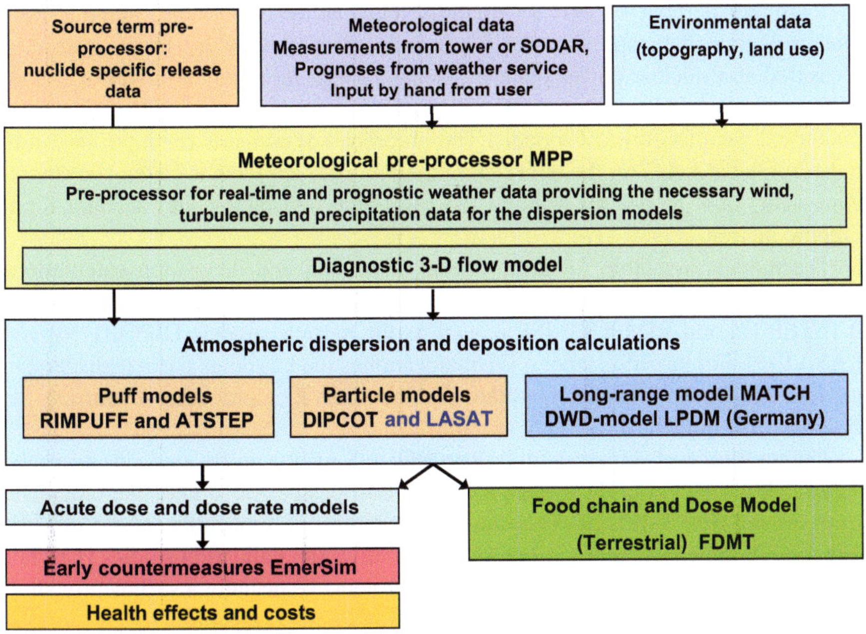

Fig. 21.1 The terrestrial model chain of JRodos

21.2.1 The Terrestrial Model Chain

Figure 21.1 illustrates the components of the terrestrial model chain of JRodos.

21.2.1.1 The Source Term Preprocessor

The source term preprocessor prepares in each calculation time step the amount of released radionuclides for the atmospheric dispersion models. Accepted as input are source term data put in by hand, or measurements from the German remote surveillance systems for nuclear reactors, or archived source terms from the system's data base. The activity release may be specified in terms of nuclide-specific data or aggregated activities of nuclide groups; if necessary this becomes converted into the release rates of individual nuclides required by the models, taking into account the type of reactor and the type of accident.

21.2.1.2 Meteorological Preprocessor and Diffusion Models

The meteorological preprocessor [6] prepares in each calculation time step a three-dimensional wind vector and turbulence field as well as a precipitation field in a format suitable for the atmospheric dispersion and deposition models.

Meteorological input of varying quantity and quality is possible, ranging from meteorological data applying for a given location entered by hand, over data measured at a nuclear power plant by means of a meteorological tower or sound/sonic detecting and ranging (SODAR) equipment, to extensive data fields from numerical weather forecast models. The preprocessor converts the input data into meteorological fields on the entire computation grid, applying a diagnostic (mass-consistent) flow model for refining the resolution and taking into account topographic factors.

The models available for atmospheric dispersion and deposition calculations over a range of up to about a few hundred km are the Gaussian puff models ATSTEP [7] and RIMPUFF [8] as well as the particle models DIPCOT [6] and LASAT [9]. For larger distances, Germany applies the Lagrange particle dispersion model, LPDM [10] of the German Weather Service, and other countries the Eulerian grid model, MATCH [11].

The multitude of diffusion models in RODOS results on the one side from the historic development of the system as a joint European effort, and on the other side from the different levels of performance and ranges of application of the models. The model properties were verified in comparative and validation studies [12]. For the future, it is considered to cover all requirements (especially a sufficiently short computation time) with one single short-range model, if possible.

21.2.1.3 The Model for Early Countermeasures, EmerSim

With the EmerSim model [13], the need and the extent of early countermeasures and the effects of such actions on doses are determined, aiming at the optimization of early countermeasure strategies. A strategy scenario in EmerSim consists of a combination of three specific actions: Staying indoors; evacuation; taking iodine tablets. In a first step, those areas are estimated where the potential doses—that is, the doses without consideration of actions for avoiding or reducing exposure—exceed the respective intervention levels. In a second step, the dose reduction by the actions is simulated in these areas by applying location- and time-dependent shielding factors for the respective actions. The calculations result in time series of modified organ doses for all cells of the calculation grid that can be compared with the potential doses. Comparing the numbers of people in given dose ranges without any action and with different countermeasure strategies provides a measure for the radiological effectiveness, thus helping to discern between the strategies.

EmerSim can be applied in many countries, as the database includes corresponding country-specific intervention levels and dose criteria for early countermeasures.

21.2.1.4 The Terrestrial Food Chain and Dose Model, FDMT

The Terrestrial Food Chain and Dose Model, FDMT, calculates the activity transport through the terrestrial food chains into the human body. It further estimates shorter term and lifetime doses for various age groups that result from food consumption and from all other exposure pathways.

The transport model in FDMT is based on the dynamic model ECOSYS, see chapter "Processes and Models of Activity Propagation in the Human Food Chain". The results are maps showing the specific activity concentrations in food and feedstuffs at the time of the peak concentration and the development of the concentrations with time. FDMT offers results for 17 vegetable and 16 animal-based human food stuffs as well as for 21 feed stuffs for animals.

An additional module, DepoM, is situated between the atmospheric dispersion and deposition models and FDMT (not shown in Fig. 21.1) and determines contamination of the soil and plant surfaces by dry and wet deposition processes as a function of the season.

21.2.2 The Models for Radiological Consequences in Contaminated Inhabited and Agricultural Areas, ERMIN and AGRICP

One task of the European EURANOS project was the development and implementation of a new kind of flexible and consistent methodology for a dynamic assessment of the radiological consequences in contaminated inhabited and agricultural areas and their mitigation by appropriate actions. This resulted in two models: ERMIN ("EuRopean Model for INhabited Areas") for inhabited areas and AGRICP ("AGRIcultural Countermeasure Program") for agricultural areas that became implemented in JRodos and in the ARGOS decision support system [14].

Both models contain dynamic activity transport models as an integral part, thereby differing significantly from previous methodologies that employ pre-calculated—hence static—datasets. Embedding dynamic transport models into the simulation codes enables dynamical calculations and thus achieves unmatched flexibility in the simulation of the effects of late-phase actions on contamination and dose levels and the associated waste and costs. Taking account of measurements and the coupling to data assimilation models is foreseen but not yet realized to a sufficient extent.

The two models and the EURANOS manuals mentioned below are covered in reference [4] in several contributions; therefore, no individual references are given here.

ERMIN contains a transport model describing weathering, retention, and re-suspension processes on inner and outer surfaces, taking into account soil and grassland, trees, other vegetation, horizontal paved surfaces, and outer and inner surfaces of buildings. Basing on the results of a preceding calculation with one of the atmospheric dispersion and deposition models, a start-up step determines time-dependent contamination levels of surfaces and the resulting doses due to external exposure and inhalation without assuming any actions. In a second step, practically any combination of measures described in the "EURANOS Inhabited Area Handbook" can be considered. The effect of the selected actions on the development of the contamination and the doses are then recalculated by the transport model.

AGRICP contains an adapted version of the food chain and dose model described in the chapter "The Terrestrial Food Chain and Dose Model, FDMT". The agricultural measures that can be considered base on the "EURANOS Handbook for Assisting the Management of Contaminated Food Production Systems". As ERMIN, also AGRICP calculates dynamically the development of contamination without and with actions and the resulting doses in the respective transport and dose model.

21.2.3 The Hydrological Model Chain

Simulation models for the aquatic pathways are described in [15]; however, the reference is limited to fresh water systems. Generally one can say that any such model needs two components. On the one hand, the hydrological component with the transport and diffusion of radionuclides must be simulated and on the other hand exchange processes among different phases must be taken into account: dissolved radionuclides, those bound to particles, and radionuclides deposited in the sediment (see, e.g. [16]).

The JRodos system contains a complete hydrological model chain [17], whose components are illustrated in Fig. 21.2.

Starting from surface runoff, the RETRACE box model simulates how radionuclides are washed out of the top soil and carried into rivers or lakes. Then, the one-dimensional model RIVTOX computes the distribution and transport in the rivers by means of pre-calculated information about hydrodynamic and sediment-specific characteristics of the rivers [16]. Radionuclide transport on flooded areas, in reservoirs, lakes and coastal waters is described with the two-dimensional model COASTOX, or with the three-dimensional model THREETOX for more complex geometries. For applications to oceans there is the compartment model Poseidon.

The resulting contamination of water and fish is used in the aquatic food chain and dose module, FDMA (Food chain and Dose Module Aquatic), for simulating the radiation exposure of people by intake of contaminated drinking water and fish. Hereby the use of contaminated water for animals as well as irrigating agricultural areas is taken into account.

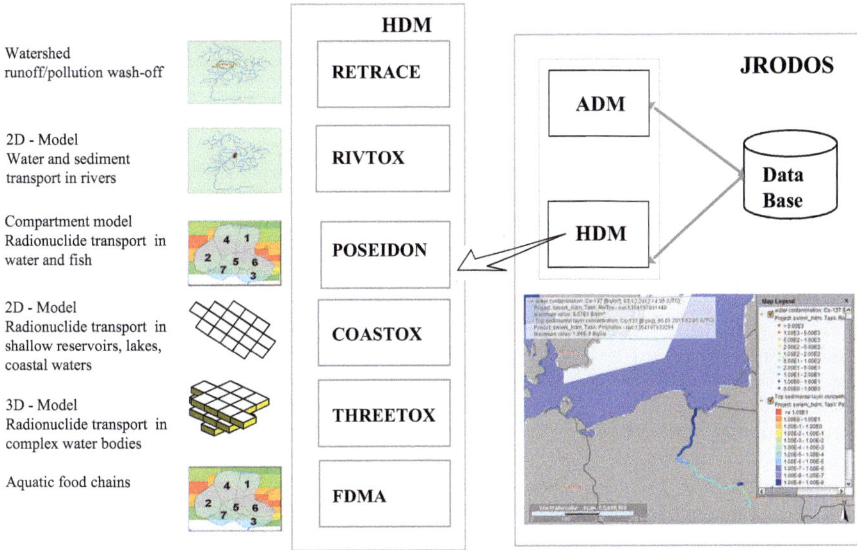

Fig. 21.2 Hydrological model chain in JRodos

21.3 Representation of Location-Dependent Results in RODOS

Location-dependent results in RODOS are calculated in the centre points of the cells of a square Cartesian grid; results that develop with time are available for each time step. The computational grid is centered at the point of the release; and around the latter lie the cells with the finest resolution, where users can choose between various sizes. The cell sizes then increase step by step as a function of the radial distance from the location of the source.

By default, results are represented as color-coded grid cells superimposed on maps. Map-type results can be displayed on the screen and also stored as image or data files.

A visible picture in JRodos is made up of a variety of layers consisting of results and maps that can be combined arbitrarily. On delivery the system contains a default set of maps; in addition, user-customized maps or map layers from the Internet[5] can be used.

Figure 21.3a shows a screen shot of such a map-type result in JRodos with the system's default layers used for the background. The computation grid with the coarser cells as the distance from the source increases can be discerned. The result

[5] At present, "ESRI shapefile" and "geotiff" geographic data formats is supported. Alternatively, one may use geodata from a PostGIS database or from the Web Map Service (WMS) server. The use of Google Maps (hybrid view) or OpenStreetMaps layers is also possible.

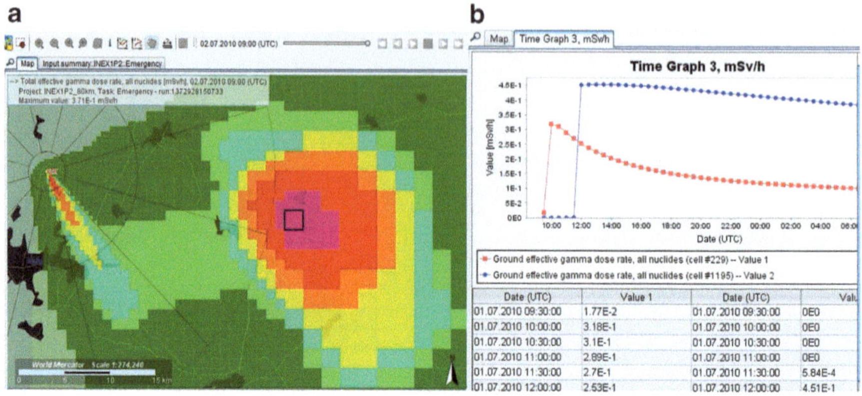

Fig. 21.3 Result representation in JRodos (**a**) map (**b**) time functions at selectable grid points

visible in Fig. 21.3a is the total gamma dose rate at a given time point; the upper menu band contains the time slider for displaying this result for each time step. For time-dependent results, the time function at user-selectable locations can be displayed in graphical form, as shown in Fig. 21.3b, or saved as Microsoft Excel or plain text files.

21.4 The RODOS Center in Germany

21.4.1 Data and User Concept

In 1997 the German Federal Ministry for the Environment took a fundamental decision about setting up a decision support system for radiation protection in Germany with the federal government and the federal states governments at a central location and using RODOS as system—the so-called "RODOS Center." The necessary activities were coordinated and carried out by the Accident Consequence Group at the Institute for Nuclear Technology and Energy Technology of the then Karlsruhe Research Center. After the concept had been worked out [18], the RODOS Center initially was established at the Federal Office for Radiation Protection in Bonn in 2001. In 2003 it was moved to Neuherberg near Munich where it is fully operational since 2005 using the Linux-based RODOS system. After a successful test phase with JRodos under operating conditions transition to the Java-based system is foreseen for the end of 2013.

Figure 21.4 illustrates the conceptual structure and the external data flow of the RODOS Center.

The RODOS Center couples to the German integrated measurement and information system of the federal government (IMIS) and to the remote surveillance systems for nuclear reactors operated by the German federal states (KFÜs), and

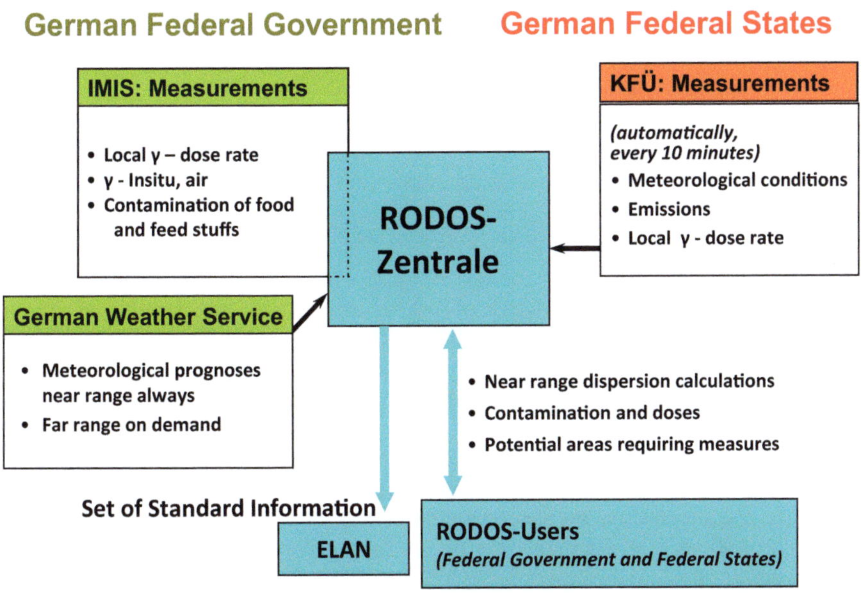

Fig. 21.4 RODOS Center—structure and data flow (source: BfS)

additionally to the German Weather Service (DWD). The KFÜ data come in every 10 min, also in routine operation. In case of a real emergency, measurements from the IMIS system are transmitted automatically in real time. The German Weather Service automatically supplies the latest meteorological forecasts twice a day, and far-range dispersion calculations on demand.

The incoming data together with information from the internal databases of RODOS are used for running diagnoses and prognoses of the radiological situation which is the basis for estimating the potential for protective actions. In addition, health consequences and other quantities are assessed without and with actions being taken into account. All this information assists the decision making team in evaluating potential measures in a way that the most suitable ones can be implemented.

In case of a real emergency, according to German legislation, only a so-called "lead user" has the right to initiate the emergency operation mode of the RODOS Center and to enter a set of standard information into the electronic situation description system (ELAN). In case of a reactor accident in Germany, the lead user is the government of the federal state in which the respective plant is situated. For foreign reactors the lead user is the German Federal Ministry for the Environment, except for some facilities close to the border where special arrangements are in effect. All other users are then entitled either to run interactive calculations with RODOS on their own or let run by the Center (A- and B-users), or to look at results of RODOS calculations released by the lead user for access for this purpose (C-users); different user interfaces are available for this purpose. The RODOS

Center is on alert around the clock which is put to trial once a week. A-users exercise with the system roughly once or twice per week. National exercises are conducted approximately once or twice a year, and there is also participation in international exercises.

21.4.2 Modes of Operation in the RODOS Center

In the pre-release phase, interactive prognosis calculations are carried out on the basis of meteorological forecast data provided by the German Weather Service and assumed source terms.

In the release phase, RODOS performs the following functions:

- Progressing diagnosis calculations are carried out automatically in ten-minutes cycles, based on measurement data from the German KFÜ surveillance system
- Prognoses calculations are issued automatically every 60 min, based on current meteorological forecast data from the Germany Weather Service and assumptions about the source term
- Supplementary interactive calculations are carried out on demand

In the case that new information about the source term is received during an ongoing accident, RODOS repeats the diagnostic calculations from the beginning of the accident until the current "now time", using the new source term together with existing weather data. When this is finished all calculations proceed in real time in the usual way.

The post-release phase is the domain of repeated interactive calculations with RODOS using the results calculated for the release phase, measurements from the IMIS system and from other sources as they become available.

21.5 Adaptation to National Conditions

Comprehensive computer-based decision support systems can provide realistic results only if the data required for the calculation sufficiently reflect the situation in the area under consideration. The default data outfit provided with the standard release of JRodos consists of a geo-referenced database covering the whole world[6] in a relatively coarse resolution, a comprehensive data base about nuclear power plants and other facilities of the nuclear fuel cycle, country-specific intervention levels for emergency measures, and the possibility to switch between several languages.[7]

[6] New feature in 2013; until then the default data base covered Europe only.

[7] Available in 2011 in English, Russian, Ukrainian.

Furthermore, JRodos includes possibilities and tools for facilitating the adaptation to regional or national conditions. The extent of adaptation can be very different, depending on the demands on the system. For instance, when automatic diagnostic calculations belong to the task list, a connection to local measurement systems is necessary. Prognostic calculations require the connection to national weather forecasting data for the near-range, and, if necessary, also for the far range.

In general, local maps can be added in JRodos without major problems. Whether or not the geo-referenced computation data base (elevation, land use, population number, soil type and agricultural production) needs refinement depends on the degree in which local data differ from the data contained in the default database, and also on the availability of region-specific data. It can turn out to be quite difficult to obtain data especially about soil types, agricultural production and consumption habits, as became apparent, for instance, during the customization for Russia.

Incorporating a new language into JRodos is straightforward. However, integrating a new language can be demanding when not only the user interface but also all result trees and manuals are to be translated.

References

1. Benz G, Ehrhardt J, Fischer F, Päsler-Sauer J, Rafat M, Schichtel T, Schüle O, Steinhauer C (1994) Inhalte und Funktionen der Pilotversion I von RODOS/RESY. Bericht KfK 5259
2. Ehrhardt J, Päsler-Sauer J, Schüle O, Rafat M, Richter J (1993) Development of RODOS, a comprehensive decision support system for nuclear emergencies in Europe - an Overview. Radiat Prot Dosimetry 50:195–203
3. Ehrhardt J, Weis A (eds) (2000) RODOS: Decision support system for off-site nuclear emergency management in Europe, Final report of the RODOS project, European Commission, Brussels. Report EUR 19144, ISBN No. 92-828-9773-7, inclusive. 2 CDs with all technical documents of the RODOS system
4. Raskob W, Hugon M (eds) (2010) Enhancing nuclear and radiological emergency management and rehabilitation: Key Results of the EURANOS Project. Radioprotection 45(5 Supplément) (© EDP Sciences)
5. Ievdin I, Trybushnyi D, Zheleznyak M, Raskob W (2010) RODOS re-engineering: aims and implementation details. In Raskob W, Hugon M (eds) Enhancing nuclear and radiological emergency management and rehabilitation: key results of the EURANOS European project. Radioprotection 45(5 Supplément): S181–S189
6. Andronopoulos S, Davakis E, Bartzis J G, Kovalets I (2010) RODOS meteorological preprocessor and atmospheric dispersion model DIPCOT: a model suite for radionuclide dispersion in complex terrain. In Raskob W, Hugon M (eds) Enhancing nuclear and radiological emergency management and rehabilitation: key results of the EURANOS European project. Radioprotection 45(5 Supplément): S77–S84
7. Päsler-Sauer J (2007) Description of the atmospheric dispersion model ATSTEP, Forschungszentrum Karlsruhe GmbH, on Wikipedia, including a link to the model description report: http://en.wikipedia.org/wiki/ATSTEP, or directly at: http://www.rodos.fzk.de/Documents/Public/HandbookV6f/Volume3/ATSTEPfinal20.pdf
8. Mikkelsen T, Larsen S E und Thykier-Nielsen S (1984) Description of the Risø puff diffusion model "RIMPUFF". Nucl Technol 67: 56–65

9. (2011) LASAT - Ein Programmsystem zur Berechnung von Schadstoffausbreitung in der Atmosphäre. http://www.janicke.de/de/lasat.html
10. Glaab H, Fay B, Jacobsen I (1998) Evaluation of the emergency dispersion model at the Deutscher Wetterdienst using ETEX data. Atmos Environ 32(24):4359–4366
11. Robertson L, Langner J, Engardt M (1999) An Eulerian limited-area atmospheric model. J Appl Meteorol 38:190–210
12. Päsler-Sauer J (2007) Validation studies with RODOS/Atstep. In Proceedings of the 11th international conference on harmonization within atmospheric dispersion modelling for regulatory purposes, Cambridge
13. Päsler-Sauer J (1993) Evaluation of early countermeasures and consequences in RODOS/RESY. Radiat Prot Dosimetry 50:219–226
14. Hoe S, Müller H, Gering F, Thykier-Nielsen S, Havskov Sorensen J (2002) ARGOS 2001 A decision support system for nuclear emergencies. Am Nucl Soc Trans 87:574–579 (Winter Meeting)
15. Hofman D et al (2011) Computerised decision support systems for the management of freshwater radioecological emergencies: assessment of the state-of-the-art with respect to the experiences and needs of end-users. J Environ Radioact 102:119–127
16. Onishi Y, Voitsekhovich O, Zheleznyak M (eds) (2007) Chernobyl – what have we learned?: the successes and failures to mitigate water contamination over 20 years. Springer, Heidelberg. ISBN 978-1-4020-5348-1 (Environ Pollut 12)
17. Zheleznyak M et al (2010) Hydrological dispersion module of JRODOS: development and pilot implementation – the vistula river basin. In Raskob W, Hugon M (eds) Enhancing nuclear and radiological emergency management and rehabilitation: key results of the EURANOS European project. Radioprotection 45(5 Supplément): S113–S122
18. Ehrhardt J, Rafat M, Raskob W (2002) Errichtung und Betrieb des RODOS Systems an zentraler Stelle (RODOS Zentrale). Forschungszentrum Karlsruhe GmbH, Karlsruhe, Bericht FZKA 6765

Chapter 22
RODOS and the Fukushima Accident

Abstract The chapter describes the contribution of the JRodos Accident Consequence Group of the Institute of Nuclear Technology and Energy Technology (IKET) as part of the activities of the Karlsruhe Institute of Technology (KIT), Germany, in frame of the Fukushima reactor accident.

As soon as the Fukushima reactor accident became known, the Karlsruhe Institute of Technology (KIT) started collecting information and making it available to public authorities, the media, and the interested public. The Accident Consequence Group of the Institute of Nuclear Technology and Energy Technology (IKET)[1] participated in these activities by means of the JRodos system.

To begin with, the general purpose of the work was the generation and supply of information about the situation in Japan and in particular around Fukushima on the basis of the very sparse data initially available. As the IKET Accident Consequence Group coordinates all RODOS specific activities and supports the operational emergency centers using the system by means of maintenance contracts, the most pressing task was upgrading the JAVA-based JRodos—a system originally developed for Europe—for usage as a forecast instrument for detecting potentially threatening situations in Fukushima. The successful setting-up of JRodos for Fukushima within only a few days provided the basis that enabled all other users to conduct own investigations with JRodos about the situation there.

In a first step, necessary basic data were collected for northern Japan and processed for JRodos, such as topography, land use, type of soil, and population distribution. In a second step, the KIT Institute of Meteorology and Climatology in cooperation with the Weather Center[2] established the meteorological forecast

[1] IKET Homepage: http://www.iket.kit.edu/english/index.php (in English).

[2] Homepage: http://ww.wetterzentrale.de (in German).

G. Kessler et al., *The Risks of Nuclear Energy Technology*, Science Policy Reports, DOI 10.1007/978-3-642-55116-1_22, © Springer-Verlag Berlin Heidelberg 2014

model, WRF,[3] in a way that it generated JRodos-usable local meteorological fields for Japan on the basis of global prognostic and re-analysis meteorological prediction data publically available from the American NOMAD server.[4] Since March 17, 2011, the meteorological data field generation became operational and was used for carrying out prognostic calculations with JRodos for different points in time. The third step consisted in obtaining radiological source terms from the German Gesellschaft für Reaktorsicherheit[5] (GRS) and integrating them into JRodos. In all, GRS provided two source terms, one for a release from a few destroyed fuel elements ("gap release") and one for a core melt scenario, respectively.

The IKET Accident Consequence Group used the upgraded JRodos system for performing own dose assessments for potential releases from the Fukushima-Daiichi nuclear power plant. Only the first 300 km around the site were taken into account because emergency measures are limited to the vicinity of a site. The first assessments were devoted mainly to find out whether Tokyo was endangered and up to which distance from the facility a potential for emergency measures existed. Both questions were covered by calculations using the above-mentioned GRS source terms or in-house source term assumptions and a number of weather variations. The only reference values available consisted of dose rate measurements close to the facility that were provided by GRS and placed on publically accessible web pages.[6]

By middle/end of the first week local dose rates measurements (in micro-Sievert/h) for more locations became published. This allowed for the first time at least a rough estimate of the releases from the reactors and from the spent fuel storage pool. The source term estimated in this way was clearly higher than the GRS fuel element source term, but still considerably below the GRS core melt source term, indicating a release of a few parts per million of the initial core inventory of Fukushima-Daiichi Unit 2.

The measurements had shown that the source term was dominated by iodine and cesium,[7] and further estimates concentrated on these two radionuclides. Moreover, the releases decreased markedly after March 17, thus putting the emphasis on recalculations of the prior release period. Assuming a release of roughly 1.0×10^{16} Bq over four days (March 12, 2011–March 16, 2011) the re-analysis data of the meteorological WRF model were used to estimate the soil contamination by cesium 137 (Fig. 22.1).

The values calculated are comparable with data measured by U.S. and Japanese institutions published under http://energy.gov/articles/us-department-energy-releases-radiation-monitoring-data-fukushima-area, cf. Fig. 22.2.

[3] WRF stands for Weather Research and Forecasting, cf. http://www.wrf-model.org/index.php

[4] Data from the Global Forecast Systems (GFS), cf. http://www.emc.ncep.noaa.gov/FGS/

[5] http://www.grs.de/

[6] http://fukushima.grs.de/

[7] http://fukushima.grs.de/ and http://mext.go.jp/english/incident/

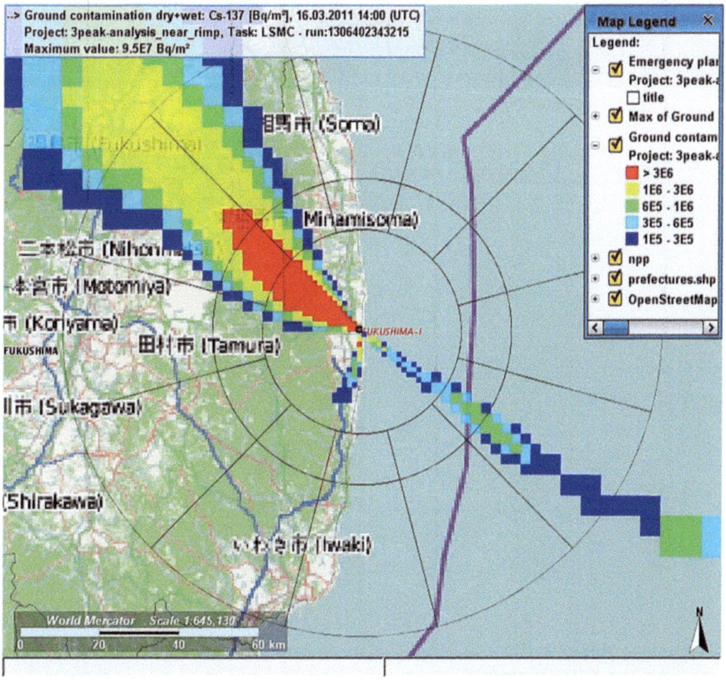

Fig. 22.1 Contamination of cesium 137 in Bq/m^2 (JRodos screen shot), realistic estimate encompassing 4 days of release beginning on March 12, 2011, 15:00 UTC (radii of circles: 20 km, 30 km, 60 km)

The maximum estimated soil contamination by cesium 137 was reflected relatively well in the prognostic calculations with JRodos. It also became obvious that the soil contamination by cesium amounts to some million Becquerel per m^2, an order of magnitude that is also measured in the highly contaminated areas of Russia, Belorussia and Ukraine after the Chernobyl accident. The model calculations overestimated the extension of the contaminated area in the northwesterly direction. This is due to the impossibility to determine precisely the release times and release quantities, but also to the meteorological data that were used.

Finally, estimates were made of the contamination of food, in particular of leafy vegetables (in Japan, spinach) and milk. Especially for a region northwest of the Fukushima-Daiichi nuclear power plant, levels clearly above Japanese and also European limits were calculated. It should be noted that milk is likely to be contaminated over a long period of time, as evident from Fig. 22.3. In particular with the Japanese limits for Cs-137 of 50 Bq/l in milk and baby food, the milk can hardly marketed in the highly contaminated areas for decades.

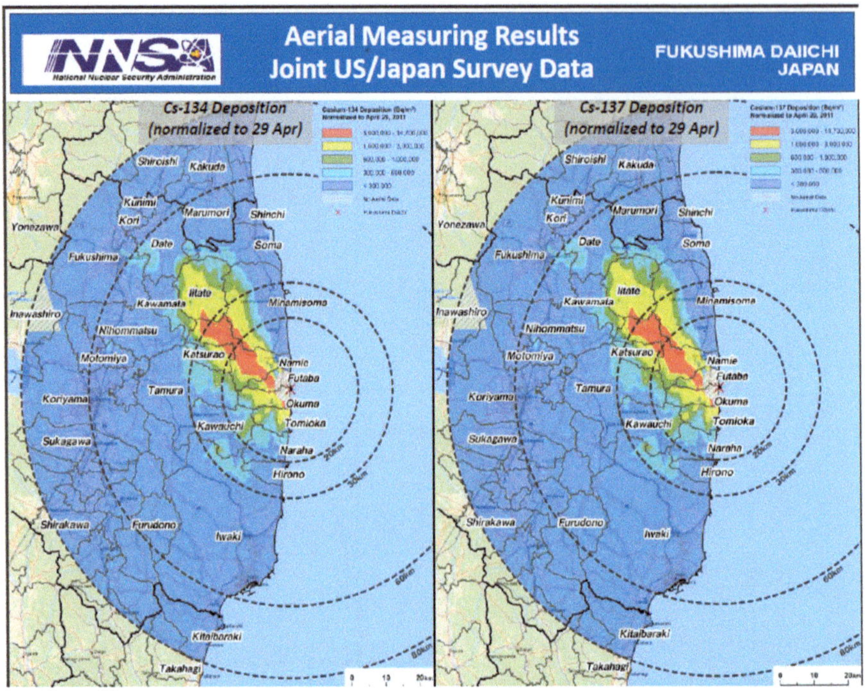

Fig. 22.2 Contamination of Cesium 134 and 137 in Bq/m^2 (U.S. and Japanese estimates) normalized to the end of April (*red* indicates a value of more than three million Becquerel per square meter) Source: http://energy.gov/articles/us-department-energy-releases-radiation-monitor ing-data-fukushima-area

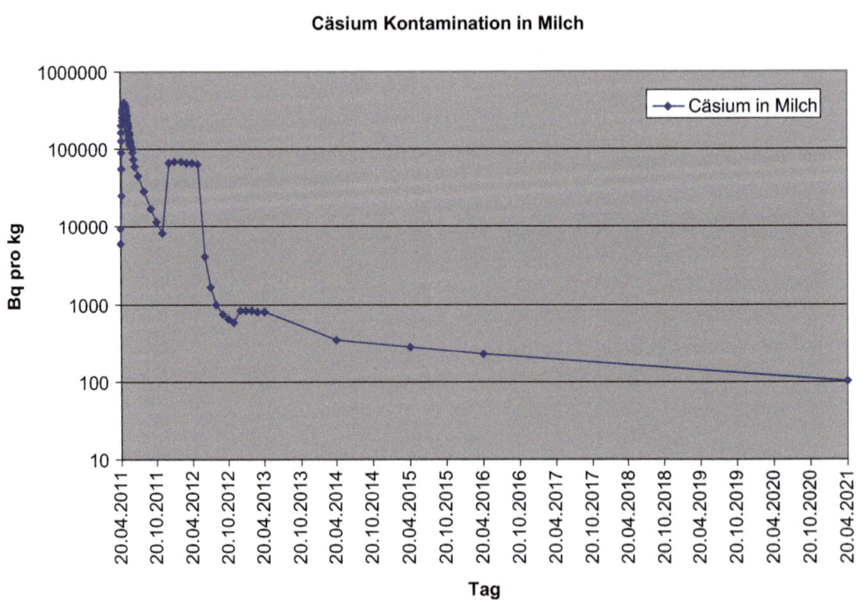

Fig. 22.3 Cesium 137 contamination in milk, based on a soil contamination of one million Bq/m^2

Chapter 23
Recent Developments in Nuclear and Radiological Emergency Management in Europe

Abstract The chapter outlines the Developments in Nuclear and Radiological Emergency Management from the 4th to the present 7th European Framework Program of the European Union.

Between 1994 and 2002, major progress was made with respect to the development of new methods and information technology (IT) tools for nuclear and radiological emergency management within the 4th and 5th Research Framework Programs of the European Union. However, by the end of the research period, the tools were not fully operational or to some extent difficult to distribute within the community.

For this reason, the European EURANOS project that started 2004 under the 6th Research Framework Program aimed at improving the coherence and effectiveness of European emergency management; furthermore, the long-term rehabilitation of contaminated areas was to be given increased consideration. A total of 17 national emergency organizations and 33 research institutions were involved in elaborating an improved common basis of European emergency management and testing the usefulness in demonstration projects. Particular emphasis was laid on the operational aspects of the methods and IT systems, and wishes and feedback of potential end users were taken into account more strongly than before. Key results of the project [1] were—among others—the development of European manuals for protective measures and countermeasures in urban and agricultural environments, the development of the two innovative simulation models ERMIN and AGRICP for coping with persistent problems in radioactively contaminated areas, and the development of the JAVA-based system JRodos that nowadays widely replaces the previous Unix/Linux-based RODOS system.

G. Kessler et al., *The Risks of Nuclear Energy Technology*, Science Policy Reports,
DOI 10.1007/978-3-642-55116-1_23,

The present 7th European Framework Program pursues further developments in this field within the NERIS-TP project[1]; in particular by the research and development efforts described below.

- Planning and establishment of a model for simulating the new concept of the International Commission on Radiological Protection, ICRP-103, as part of the European JRodos and ARGOS systems.
  A so-called screening model will serve for the detection of contaminated areas that require attention, and for proposing some suitable sequence of possible actions. Optimization of long-term measures will be subject for the existing ERMIN and AGRICP models that will be upgraded accordingly. Scenario preparation tools shall support the planning of possible radiological action scenarios.
- Improving radiological emergency management on a global scale
  The key element here is the development of a tool for processing publically available global meteorological forecast data so that they can be used by atmospheric dispersion and deposition models in e.g. JRodos and ARGOS, for calculations at any point on the globe.
- Coupling of the European decision support systems JRodos and ARGOS to early warning systems, such as the ECURIE system of the IAEA.[2]
- Strengthen the preparedness in Europe at the local/national level.

The usage of the products developed under NERIS-TP will be main subject of the training course on "Preparedness for Nuclear and Radiological Emergency Response and Recovery", to be held in October 2013 in Trnava, Slovak Republic.

Finally, NERIS-TP aims at the establishment of a platform for exchanging information, tools, and experience, among users and developers in nuclear and radiological emergency protection and long-term rehabilitation. The platform shall also serve for identifying and defining future necessary research and development topics and conducting the respective work with internal or European resources. This will introduce a self-sustaining institution and organization which, in the long run, will become a home to representatives of users, scientists, and victims alike, sharing the common vision to further improve the preparedness for nuclear or radiological emergencies in Europe (http://www.eu-neris.net/).

Reference

1. Raskob W, Hugon M (eds) (2010) Enhancing nuclear and radiological emergency management and rehabilitation – Key Results of the EURANOS Project. Radioprotection 45(5 Supplément) (© EDP Sciences)

[1] The full designation of the project is "Towards a self-sustaining European Technology Platform (NERIS-TP) on Preparedness for Nuclear and Radiological Emergency Response and Recovery."

[2] Council Decision 87/600/EURATOM on urgent information exchange in case of a radiological emergency (ECURIE), see for example http://eur-lex.europa.eu/LexUriServ/LexUriServ.do?uri=CELEX:31987D0600:DE:HTML

Index

G. Kessler et al., *The Risks of Nuclear Energy Technology*, Science Policy Reports, DOI 10.1007/978-3-642-55116-1, © Springer-Verlag Berlin Heidelberg 2014

MIX
Papier aus verantwortungsvollen Quellen
Paper from responsible sources
FSC® C105338

If you have any concerns about our products,
you can contact us on
ProductSafety@springernature.com

In case Publisher is established outside the EU,
the EU authorized representative is:
Springer Nature Customer Service Center GmbH
Europaplatz 3, 69115 Heidelberg, Germany

Printed by Libri Plureos GmbH
in Hamburg, Germany